Moments and Moment Invariants
in Pattern Recognition

Moments and Moment Invariants in Pattern Recognition

Jan Flusser

Institute of Information Theory and Automation,
Academy of Sciences of the Czech Republic,
Prague, Czech Republic

Tomáš Suk

Institute of Information Theory and Automation,
Academy of Sciences of the Czech Republic,
Prague, Czech Republic

Barbara Zitová

Institute of Information Theory and Automation,
Academy of Sciences of the Czech Republic,
Prague, Czech Republic

A John Wiley and Sons, Ltd, Publication

MATLAB® MATLAB and any associated trademarks used in this book are the registered trademarks of
The MathWorks, Inc.

For MATLAB® product information, please contact:

The MathWorks, Inc.
3 Apple Hill Drive
Natick, MA, 01760-2098 USA
Tel: 508-647-7000
Fax: 508-647-7001
E-mail: info@mathworks.com
Web: www.mathworks.com

Library of Congress Cataloging-in-Publication Data

Flusser, Jan.
  Moments and moment invariants in pattern recognition / Jan Flusser, Tomas Suk, Barbara Zitova.
     p. cm.
  Includes bibliographical references and index.
  ISBN 978-0-470-69987-4 (cloth)
1. Optical pattern recognition–Mathematics. 2. Moment problems (Mathematics) 3. Invariants.
I. Suk, Tomas. II. Zitova, Barbara. III. Title.
  TA1650.F57 2009
  515'.42–dc22   2009031450

A catalogue record for this book is available from the British Library.

ISBN 978-0-470-69987-4 (H/B)

Set in 10/12pt Times by Sunrise Setting Ltd, Torquay, UK.
Printed in Great Britain by CPI Antony Rowe, Chippenham.

We dedicate this book to our families

To my wife Vlasta, my sons Michal and Martin, and my daughters Janička and Verunka
Jan Flusser

To my wife Lenka, my daughter Hana, and my son Ondřej
Tomáš Suk

To my husband Pavel and my sons Janek and Kubík
Barbara Zitová

Contents

Authors' biographies

Professor Jan Flusser, PhD, DSc, is a director of the Institute of Information Theory and Automation of the ASCR, Prague, Czech Republic, and a full professor of Computer Science at the Czech Technical University, Prague, and at the Charles University, Prague. Jan Flusser's research areas are moments and moment invariants, image registration, image fusion, multichannel blind deconvolution and super-resolution imaging. He has authored and coauthored more than 150 research publications in these areas, including tutorials (ICIP'05, ICIP'07, EUSIPCO'07, CVPR'08, FUSION'08, SPPRA'09, SCIA'09) and invited/keynote talks (ICCS'06, COMPSTAT'06, WIO'06, DICTA'07, CGIM'08) at major international conferences. He gives undergraduate and graduate courses on digital image processing, pattern recognition, and moment invariants and wavelets. Personal webpage **http://www.utia.cas.cz/people/flusser**.

Tomáš Suk, PhD, is a research fellow of the same Institute. His research interests include invariant features, moment and point-based invariants, color spaces and geometric transformations. He has authored and coauthored more than 50 research publications in these areas, some of which have elicited a considerable citation response. Tomáš Suk coauthored tutorials on moment invariants held at international conferences ICIP'07 and SPPRA'09. Personal webpage **http://zoi.utia.cas.cz/suk**.

Barbara Zitová, PhD, is Head of the Department of Image Processing at the same Institute. Her research interest is mainly in image registration, invariants, wavelets, and image processing applications in cultural heritage. She has authored and coauthored more than 30 research publications in these areas, including tutorials at international conferences (ICIP'05, ICIP'07, EUSIPCO'07, FUSION'08 and CVPR'08). Her paper "Image Registration Methods: A Survey," *Image and Vision Computing*, vol. 21, pp. 977–1000, 2003, has become a major reference work in image registration. She teaches a specialized graduate course on moment invariants and wavelets at the Czech Technical University. Personal webpage **http://zoi.utia.cas.cz/zitova**.

Preface

Automatic object recognition has become an established discipline in image analysis. Moments and moment invariants play a very important role as features in invariant recognition. They were introduced to the pattern recognition community almost 50 years ago and the mathematical background they utilized is even older, originating from the second half of the nineteenth century. Nowadays, we may find thousands of references to journal and conference papers when searching SCOPUS, Web of Science, or IEEE Xplore databases for the keyword "moment invariants" and this number grows every year. Despite this, any comprehensive book covering the current state-of-the-art and presenting the latest development in this field in a consistent form has not so far been published. Our main purpose in writing this book is to bridge this gap. In this book, the reader will find both a survey of all the important theoretical results as well as a description of how to use them in various image analysis tasks.

The book presents a unique overview of recent as well as traditional image analysis and pattern recognition methods based on image moments. Invariants to traditional transforms – translation, rotation, scaling, and affine transform – are studied in depth from a new point of view. Recent results on invariants to linear filtering of the image, and on implicit moment invariants to elastic deformations are presented clearly and are well explained. Various classes of orthogonal moment (Legendre, Zernike, Pseudo-Zernike, Chebyshev, Fourier-Mellin, Krawtchouk, and other moments) are reviewed and compared, and their application to image reconstruction from moments is demonstrated. We review efficient numerical algorithms that can be used for moment computation in a discrete domain. Finally, we demonstrate practical examples of using moment invariants in real applications from the areas of computer vision, remote sensing, and medical imaging. Most of the book is related to two-dimensional (2D) images but generalization to three-dimensional (3D) is also shown in most cases.

The book is based on our considerable experience with moments and moment invariants gained from 15 years of research in this area, from teaching graduate courses on moment invariants and related fields of image analysis at the Czech Technical University and at the Charles University, Prague, Czech Republic, and from presenting several tutorials on moments at major international conferences.

The target readership of the book is academic researchers and research and development engineers from all application areas who need to recognize 2D objects extracted from binary/graylevel/color images and who look for invariant and robust object descriptors, as well as specialists in moment-based image analysis interested in new developments in this field. Last but not least, the book is also intended for university lecturers and graduate students of image analysis and pattern recognition. It can be used as a textbook for advanced graduate courses on invariant pattern recognition. The first two chapters can even be utilized as supplementary reading to undergraduate courses on pattern recognition and image analysis.

For the readers' convenience, we maintain a permanently updated accompanying website at **http://zoi.utia.cas.cz/moment_invariants**. It contains selected MATLAB codes, complete lists of the invariants, slides for those who wish to use this book for educational purposes, and errata.

Acknowledgments

The writing of this book was enabled thanks to financial support from the Czech Science Foundation (project no. 102/08/1593) and from the Czech Ministry of Education, Youth, and Sports (Research Center *Data-Algorithms-Decision Making*, project no. 1M0572).

We would like to express thanks to our employer, the Institute of Information Theory and Automation (ÚTIA) of the ASCR, Prague, and to the former director of the Institute, Professor Milan Mareš, for creating an inspiring and friendly environment and for the ongoing support of our research. We also thank both universities where we teach, the Czech Technical University, Faculty of Nuclear Science and Physical Engineering and the Charles University, Faculty of Mathematics and Physics, for their support and for the administration of all our courses. We would also like to thank the many students on our courses and our tutorial attendees who have asked challenging and stimulating questions.

Special thanks go to the staff of John Wiley & Sons Ltd, namely to Simone Taylor, who encouraged us to start writing the book, to our project editor Nicky Skinner, and publishing assistant Alexandra King for their continuous help and patience during the writing period.

Many of our colleagues and friends have helped us with the preparation of the book in various ways. We are most grateful to all of them, particularly to Jaroslav Kautsky for his significant contribution to Chapter 4 and willingness to share with us his extensive experience with orthogonal polynomials; Filip Šroubek for his valuable contribution to Chapter 4; Jiří Boldyš for his valuable comments and suggestions on 3D invariants; Pavel Šroubek for careful language corrections of the manuscript; Irena Váňová for creating the accompanying website and for performing several experiments used throughout the book; Jarmila Pánková for the cover design; Michal Šorel for help with the optical flow experiment in Chapter 8; Babak Mahdian for his contribution to the image forensics experiment in Chapter 8; Michal Breznický for his comments on the Gaussian blur invariants; Alexander Kutka for his help with the focus measure experiment in Chapter 8, and to Jan Kybic for providing us with his code for elastic matching. Finally, we would like to thank our families. Their encouragement and patience was instrumental in the completion of this book.

Jan Flusser
Tomáš Suk
Barbara Zitová
Prague, Czech Republic, 2009

1

Introduction to moments

1.1 Motivation

In our everyday life, each of us almost constantly receives, processes and analyzes a huge amount of information of various kinds, significance and quality and has to make decisions based on this analysis. More than 95% of information we perceive is optical in character. Image is a very powerful information medium and communication tool capable of representing complex scenes and processes in a compact and efficient way. Thanks to this, images are not only primary sources of information, but are also used for communication among people and for interaction between humans and machines.

Common digital images contain an enormous amount of information. An image you can take and send in a few seconds to your friends by a cellphone contains as much information as several hundred pages of text. This is why there is an urgent need for automatic and powerful image analysis methods.

Analysis and interpretation of an image acquired by a real (i.e. nonideal) imaging system is the key problem in many application areas such as robot vision, remote sensing, astronomy and medicine, to name but a few. Since real imaging systems as well as imaging conditions are usually imperfect, the observed image represents only a degraded version of the original scene. Various kinds of degradation (geometric as well as graylevel/color) are introduced into the image during the acquisition process by such factors as imaging geometry, lens aberration, wrong focus, motion of the scene, systematic and random sensor errors, etc. (see Figures 1.1, 1.2 and 1.3).

In general, the relation between the ideal image $f(x, y)$ and the observed image $g(x, y)$ is described as $g = \mathcal{D}(f)$, where $\mathcal{D}$ is a degradation operator. Degradation operator $\mathcal{D}$ can usually be decomposed into radiometric (i.e. graylevel or color) degradation operator $\mathcal{R}$ and geometric (i.e. spatial) degradation operator $\mathcal{G}$. In real imaging systems $\mathcal{R}$ can usually be modeled by space-variant or space-invariant convolution plus noise while $\mathcal{G}$ is typically a transform of spatial coordinates (for instance, perspective projection). In practice, both operators are typically either unknown or are described by a parametric model with unknown parameters. Our goal is to analyze the unknown scene $f(x, y)$, an ideal image of which is not available, by means of the sensed image $g(x, y)$ and a-priori information about the degradations.

Moments and Moment Invariants in Pattern Recognition   Jan Flusser, Tomáš Suk and Barbara Zitová
© 2009 John Wiley & Sons, Ltd

Figure 1.1 Perspective distortion of the image caused by a nonperpendicular view.

Figure 1.2 Image blurring caused by wrong focus of the camera.

By the term *scene analysis* we usually understand a complex process consisting of three basic stages. First, the image is preprocessed, segmented and objects of potential interest are detected. Second, the extracted objects are "recognized", which means they are mathematically described and classified as elements of a certain class from the set of predefined object classes. Finally, spatial relations among the objects can be analyzed. The first stage contains traditional image-processing methods and is exhaustively covered in standard textbooks [1–3]. The classification stage is independent of the original data and is carried out in the space of descriptors. This part is comprehensively reviewed in the famous Duda–Hart–Stork book [4]. For the last stage we again refer to [3].

Figure 1.3 Image distortion caused by a nonlinear deformation of the scene.

1.2 What are invariants?

Recognition of objects and patterns that are deformed in various ways has been a goal of much recent research. There are basically three major approaches to this problem – brute force, image normalization and invariant features. In the brute-force approach we search the parametric space of all possible image degradations. That means the training set of each class should contain not only all class representatives but also all their rotated, scaled, blurred and deformed versions. Clearly, this approach would lead to extreme time complexity and is practically inapplicable. In the normalization approach, the objects are transformed into a certain standard position before they enter the classifier. This is very efficient in the classification stage but the object normalization itself usually requires the solving of difficult inverse problems that are often ill-conditioned or ill-posed. For instance, in the case of image blurring, "normalization" means in fact blind deconvolution [5] and in the case of spatial image deformation, "normalization" requires registration of the image to be performed to some reference frame [6].

The approach using invariant features appears to be the most promising and has been used extensively. Its basic idea is to describe the objects by a set of measurable quantities called *invariants* that are insensitive to particular deformations and that provide enough discrimination power to distinguish objects belonging to different classes. From a mathematical point of view, invariant I is a functional defined on the space of all admissible image functions that does not change its value under degradation operator $\mathcal{D}$, i.e. that satisfies the condition $I(f) = I(\mathcal{D}(f))$ for any image function f. This property is called *invariance*. In practice, in order to accommodate the influence of imperfect segmentation, intra-class variability and noise, we usually formulate this requirement as a weaker constraint: $I(f)$ should not be significantly different from $I(\mathcal{D}(f))$. Another desirable property of I, as important as invariance, is *discriminability*. For objects belonging to different classes, I must have significantly different values. Clearly, these two requirements are antagonistic – the broader the invariance, the less discrimination power and vice versa. Choosing a proper tradeoff between invariance and discrimination power is a very important task in feature-based object recognition (see Figure 1.4 for an example of a desired situation).

Usually, one invariant does not provide enough discrimination power and several invariants $I_1, \ldots, I_n$ must be used simultaneously. Then, we speak about an *invariant vector*. In this way, each object is represented by a point in an n-dimensional metric space called *feature space* or *invariant space*.

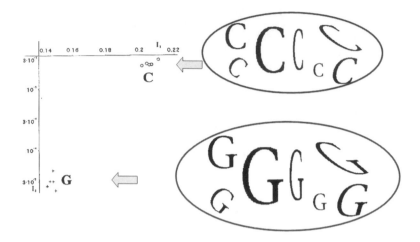

Figure 1.4 Two-dimensional feature space with two classes, almost an ideal example. Each class forms a compact cluster (the features are invariant) and the clusters are well separated (the features are discriminative).

1.2.1   Categories of invariant

The existing invariant features used for describing 2D objects can be categorized from various points of view. Most straightforward is the categorization according to the type of invariance. We recognize translation, rotation, scaling, affine, projective, and elastic geometric invariants. Radiometric invariants exist with respect to linear contrast stretching, nonlinear intensity transforms, and to convolution.

Categorization according to the mathematical tools used may be as follows:

- *simple shape descriptors* – compactness, convexity, elongation, etc. [3];

- *transform coefficient features* are calculated from a certain transform of the image – Fourier descriptors [7, 8], Hadamard descriptors, Radon transform coefficients, and wavelet-based features [9, 10];

- *point set invariants* use positions of dominant points [11–14];

- *differential invariants* employ derivatives of the object boundary [15–19];

- *moment invariants* are special functions of image moments.

Another viewpoint reflects what part of the object is needed to calculate the invariant.

- *Global* invariants are calculated from the whole image (including background if no segmentation was performed). Most of them include projections of the image onto certain basis functions and are calculated by integration. Compared to local invariants, global invariants are much more robust with respect to noise, inaccurate boundary detection and other similar factors. On the other hand, their serious drawback is the fact that a local change of image influences the values of all the invariants and is not "localized" in a few components only. This is why global invariants cannot be used when the object studied is partially occluded by another object and/or when a part of it is out of the field of vision. Moment invariants fall into this category.

- *Local* invariants are, in contrast, calculated from a certain neighborhood of dominant points only. Differential invariants are typical representatives of this category. The object boundary is detected first and then the invariants are calculated for each boundary point as functions of the boundary derivatives. As a result, the invariants at any given point depend only on the shape of the boundary in its immediate vicinity. If the rest of the object undergoes any change, the local invariants are not affected. This property makes them a seemingly perfect tool for recognition of partially occluded objects but due to their extreme vulnerability to discretization errors, segmentation inaccuracies, and noise, it is difficult to actually implement and use them in practice.

- *Semilocal* invariants attempt to retain the positive properties of the two groups above and to avoid the negative ones. They divide the object into stable parts (most often this division is based on inflection points or vertices of the boundary) and describe each part by some kind of global invariant. The whole object is then characterized by a string of vectors of invariants and recognition under occlusion is performed by maximum substring matching. This modern and practically applicable approach was used in various modifications in references [20–26].

Here, we focus on object description and recognition by means of moments and moment invariants. The history of moment invariants began many years before the appearance of the first computers, in the nineteenth century under the framework of group theory and the theory of algebraic invariants. The theory of algebraic invariants was thoroughly studied by the famous German mathematicians P. A. Gordan and D. Hilbert [27] and was further developed in the twentieth century in references [28] and [29], among others.

Moment invariants were first introduced to the pattern recognition and image processing community in 1962 [30], when Hu employed the results of the theory of algebraic invariants and derived his seven famous invariants to the rotation of 2D objects. Since that time, hundreds of papers have been devoted to various improvements, extensions and generalizations of moment invariants and also to their use in many areas of application. Moment invariants have become one of the most important and most frequently used shape descriptors. Even though they suffer from certain intrinsic limitations (the worst of which is their globalness, which prevents direct utilization for occluded object recognition), they frequently serve as "first-choice descriptors" and as a reference method for evaluating the performance of other shape descriptors. Despite a tremendous effort and a huge number of published papers, many problems remain to be resolved.

1.3   What are moments?

Moments are scalar quantities used to characterize a function and to capture its significant features. They have been widely used for hundreds of years in statistics for description of the shape of a probability density function and in classic rigid-body mechanics to measure the mass distribution of a body. From the mathematical point of view, moments are "projections" of a function onto a polynomial basis (similarly, Fourier transform is a projection onto a basis of harmonic functions). For the sake of clarity, we introduce some basic terms and propositions, which we will use throughout the book.

Definition 1.1 By an *image function* (or *image*) we understand any piece-wise continuous real function $f(x, y)$ of two variables defined on a compact support $D \subset \mathbb{R} \times \mathbb{R}$ and having a finite nonzero integral.

Definition 1.2[1] *General moment* $M_{pq}^{(f)}$ of an image $f(x, y)$, where p, q are non-negative integers and $r = p + q$ is called the *order* of the moment, defined as

$$M_{pq}^{(f)} = \iint_D p_{pq}(x, y) f(x, y) \, \mathrm{d}x \, \mathrm{d}y, \tag{1.1}$$

where $p_{00}(x, y)$, $p_{10}(x, y)$, ..., $p_{kj}(x, y)$, ... are polynomial basis functions defined on D. (We omit the superscript $^{(f)}$ if there is no danger of confusion.)

Depending on the polynomial basis used, we recognize various systems of moments.

1.3.1   Geometric and complex moments

The most common choice is a standard power basis $p_{kj}(x, y) = x^k y^j$ that leads to *geometric moments*

$$m_{pq} = \int_{-\infty}^{\infty} \int_{-\infty}^{\infty} x^p y^q f(x, y) \, \mathrm{d}x \, \mathrm{d}y. \tag{1.2}$$

Geometric moments of low orders have an intuitive meaning – m_{00} is a "mass" of the image (for binary images, m_{00} is an area of the object), m_{10}/m_{00} and m_{01}/m_{00} define the *center of gravity* or *centroid* of the image. Second-order moments m_{20} and m_{02} describe the "distribution of mass" of the image with respect to the coordinate axes. In mechanics they are called the *moments of inertia*. Another popular mechanical quantity, the *radius of gyration* with respect to an axis, can also be expressed in terms of moments as $\sqrt{m_{20}/m_{00}}$ and $\sqrt{m_{02}/m_{00}}$, respectively.

If the image is considered a probability density function (pdf) (i.e. its values are normalized such that $m_{00} = 1$), then m_{10} and m_{01} are the mean values. In case of zero means, m_{20} and m_{02} are *variances* of horizontal and vertical projections and m_{11} is a *covariance* between them. In this way, the second-order moments define the orientation of the image. As will be seen later, second-order geometric moments can be used to find the normalized position of an image. In statistics, two higher-order moment characteristics have been

---

[1]In some papers one can find extended versions of Definition 1.2 that include various scalar factors and/or weighting functions in the integrand. We introduce such extensions in Chapter 6.

commonly used – the *skewness* and the *kurtosis*. The skewness of the horizontal projection is defined as $m_{30}/\sqrt{m_{20}^3}$ and that of vertical projection as $m_{03}/\sqrt{m_{02}^3}$. The skewness measures the deviation of the respective projection from symmetry. If the projection is symmetric with respect to the mean (i.e. to the origin in this case), then the corresponding skewness equals zero. The kurtosis measures the "peakedness" of the pdf and is again defined separately for each projection – the horizontal kurtosis as m_{40}/m_{20}^2 and the vertical kurtosis as m_{04}/m_{02}^2.

Characterization of the image by means of geometric moments is complete in the following sense. For any image function, geometric moments of all orders do exist and are finite. The image function can be exactly reconstructed from the set of its moments (this assertion is known as *the uniqueness theorem*).

Another popular choice of the polynomial basis $p_{kj}(x, y) = (x + iy)^k(x - iy)^j$, where i is the imaginary unit, leads to *complex moments*

$$c_{pq} = \int\limits_{-\infty}^{\infty} \int\limits_{-\infty}^{\infty} (x + iy)^p (x - iy)^q f(x, y)\, dx\, dy. \tag{1.3}$$

Geometric moments and complex moments carry the same amount of information. Each complex moment can be expressed in terms of geometric moments of the same order as

$$c_{pq} = \sum_{k=0}^{p} \sum_{j=0}^{q} \binom{p}{k}\binom{q}{j}(-1)^{q-j} \cdot i^{p+q-k-j} \cdot m_{k+j,p+q-k-j} \tag{1.4}$$

and vice versa[2]

$$m_{pq} = \frac{1}{2^{p+q}i^q} \sum_{k=0}^{p} \sum_{j=0}^{q} \binom{p}{k}\binom{q}{j}(-1)^{q-j} \cdot c_{k+j,p+q-k-j}. \tag{1.5}$$

Complex moments are introduced because they behave favorably under image rotation. This property can be advantageously employed when constructing invariants with respect to rotation, as will be shown in the following chapter.

1.3.2 Orthogonal moments

If the polynomial basis $\{p_{kj}(x, y)\}$ is orthogonal, i.e. if its elements satisfy the condition of orthogonality

$$\iint\limits_{\Omega} p_{pq}(x, y) \cdot p_{mn}(x, y)\, dx\, dy = 0 \tag{1.6}$$

or weighted orthogonality

$$\iint\limits_{\Omega} w(x, y) \cdot p_{pq}(x, y) \cdot p_{mn}(x, y)\, dx\, dy = 0 \tag{1.7}$$

for any indexes $p \neq m$ or $q \neq n$, we speak about *orthogonal (OG) moments*. Ω is the area of orthogonality.

---

[2]While the proof of (1.4) is straightforward, the proof of (1.5) requires, first, x and y to be expressed as $x = ((x + iy) + (x - iy))/2$ and $y = ((x + iy) - (x - iy))/2i$.

In theory, all polynomial bases of the same degree are equivalent because they generate the same space of functions. Any moment with respect to a certain basis can be expressed in terms of moments with respect to any other basis. From this point of view, OG moments of any type are equivalent to geometric moments.

However, a significant difference appears when considering stability and computational issues in a discrete domain. Standard powers are nearly dependent both for small and large values of the exponent and increase rapidly in range as the order increases. This leads to correlated geometric moments and to the need for high computational precision. Using lower precision results in unreliable computation of geometric moments. OG moments can capture the image features in an improved, nonredundant way. They also have the advantage of requiring lower computing precision because we can evaluate them using recurrent relations, without expressing them in terms of standard powers.

Unlike geometric moments, OG moments are coordinates of f in the polynomial basis in the common sense used in linear algebra. Thanks to this, the image reconstruction from OG moments can be performed easily as

$$f(x, y) = \sum_{k, j} M_{kj} \cdot p_{kj}(x, y).$$

Moreover, this reconstruction is "optimal" because it minimizes the mean-square error when using only a finite set of moments. On the other hand, image reconstruction from geometric moments cannot be performed directly in the spatial domain. It is carried out in the Fourier domain using the fact that geometric moments form Taylor coefficients of the Fourier transform $F(u, v)$

$$F(u, v) = \sum_p \sum_q \frac{(-2\pi i)^{p+q}}{p!q!} m_{pq} u^p v^q.$$

(To prove this, expand the kernel of the Fourier transform $e^{-2\pi i (ux+vy)}$ into a power series.) Reconstruction of $f(x, y)$ is then achieved via inverse Fourier transform.

We will discuss various OG moments and their properties in detail in Chapter 6. Their usage for stable implementation of implicit invariants will be shown in Chapter 4 and practical applications will be demonstrated in Chapter 8.

1.4   Outline of the book

This book deals in general with moments and moment invariants of 2D and 3D images and with their use in object description, recognition, and in other applications.

Chapters 2–5 are devoted to four classes of moment invariant. In Chapter 2, we introduce moment invariants with respect to the simplest spatial transforms – translation, rotation, and scaling. We recall the classical Hu invariants first and then present a general method for constructing invariants of arbitrary orders by means of complex moments. We prove the existence of a relatively small basis of invariants that is complete and independent. We also show an alternative approach – constructing invariants via normalization. We discuss the difficulties which the recognition of symmetric objects poses and present moment invariants suitable for such cases.

Chapter 3 deals with moment invariants to the affine transform of spatial coordinates. We present three main approaches showing how to derive them – the graph method, the method of normalized moments, and the solution of the Cayley–Aronhold equation. Relationships between invariants from different methods are mentioned and the dependency of generated invariants is studied. We describe a technique used for elimination of reducible and dependent invariants. Finally, numerical experiments illustrating the performance of the affine moment invariants are carried out and a brief generalization to color images and to 3D images is proposed.

In Chapter 4, we introduce a novel concept of so-called implicit invariants to elastic deformations. Implicit invariants measure the similarity between two images factorized by admissible image deformations. For many types of image deformation traditional invariants do not exist but implicit invariants can be used as features for object recognition. We present implicit moment invariants with respect to the polynomial transform of spatial coordinates and demonstrate their performance in artificial as well as real experiments.

Chapter 5 deals with a completely different kind of moment invariant, with invariants to convolution/blurring. We derive invariants with respect to image blur regardless of the convolution kernel, provided that it has a certain degree of symmetry. We also derive so-called combined invariants, which are invariant to composite geometric and blur degradations. Knowing these features, we can recognize objects in the degraded scene without any restoration.

Chapter 6 presents a survey of various types of orthogonal moments. They are divided into two groups, the first being moments orthogonal on a rectangle and the second orthogonal on a unit disk. We review Legendre, Chebyshev, Gegenbauer, Jacobi, Laguerre, Hermite, Krawtchouk, dual Hahn, Racah, Zernike, Pseudo–Zernike and Fourier–Mellin polynomials and moments. The use of orthogonal moments on a disk in the capacity of rotation invariants is discussed. The second part of the chapter is devoted to image reconstruction from its moments. We explain why orthogonal moments are more suitable for reconstruction than geometric ones and a comparison of reconstructing power of different orthogonal moments is presented.

In Chapter 7, we focus on computational issues. Since the computing complexity of all moment invariants is determined by the computing complexity of moments, efficient algorithms for moment calculations are of prime importance. There are basically two major groups of methods. The first one consists of methods that attempt to decompose the object into nonoverlapping regions of a simple shape. These "elementary shapes" can be pixel rows or their segments, square and rectangular blocks, among others. A moment of the object is then calculated as a sum of moments of all regions. The other group is based on Green's theorem, which evaluates the double integral over the object by means of single integration along the object boundary.

We present efficient algorithms for binary and graylevel objects and for geometric as well as selected orthogonal moments.

Chapter 8 is devoted to various applications of moments and moment invariants in image analysis. We demonstrate their use in image registration, object recognition, medical imaging, content-based image retrieval, focus/defocus measurement, forensic applications, robot navigation and digital watermarking.

References

[1] Gonzalez, R. C. and Woods, R. E. (2007) *Digital Image Processing*. Prentice Hall, 3rd edn.

[2] Pratt, W. K. (2007) *Digital Image Processing.* New York: Wiley Interscience, 4th edn.

[3] Šonka, M., Hlaváč, V. and Boyle, R. (2007) *Image Processing, Analysis and Machine Vision.* Toronto: Thomson, 3rd edn.

[4] Duda, R. O., Hart, P. E. and Stork, D. G. (2001) *Pattern Classification.* New York: Wiley Interscience, 2nd edn.

[5] Kundur, D. and Hatzinakos, D. (1996) "Blind image deconvolution," *IEEE Signal Processing Magazine*, vol. 13, no. 3, pp. 43–64.

[6] Zitová, B. and Flusser, J. (2003) "Image registration methods: A survey," *Image and Vision Computing*, vol. 21, no. 11, pp. 977–1000.

[7] Lin, C. C. and Chellapa, R. (1987) "Classification of partial 2-D shapes using Fourier descriptors," *IEEE Transactions on Pattern Analysis and Machine Intelligence*, vol. 9, no. 5, pp. 686–90.

[8] Arbter, K., Snyder, W. E., Burkhardt, H. and Hirzinger, G. (1990) "Application of affine-invariant Fourier descriptors to recognition of 3-D objects," *IEEE Transactions Pattern Analysis and Machine Intelligence*, vol. 12, no. 7, pp. 640–47.

[9] Tieng, Q. M. and Boles, W. W. (1995) "An application of wavelet-based affine-invariant representation," *Pattern Recognition Letters*, vol. 16, no. 12, pp. 1287–96.

[10] Khalil, M. and Bayeoumi, M. (2001) "A dyadic wavelet affine invariant function for 2D shape recognition," *IEEE Transactions on Pattern Analysis and Machine Intelligence*, vol. 23, no. 10, pp. 1152–63.

[11] Mundy, J. L. and Zisserman, A. (1992) *Geometric Invariance in Computer Vision.* Cambridge, Massachusetts: MIT Press.

[12] Suk, T. and Flusser, J. (1996) "Vertex-based features for recognition of projectively deformed polygons," *Pattern Recognition*, vol. 29, no. 3, pp. 361–67.

[13] Lenz, R. and Meer, P. (1994) "Point configuration invariants under simultaneous projective and permutation transformations," *Pattern Recognition*, vol. 27, no. 11, pp. 1523–32.

[14] Rao, N. S. V., Wu, W. and Glover, C. W. (1992) "Algorithms for recognizing planar polygonal configurations using perspective images," *IEEE Transactions on Robotics and Automation*, vol. 8, no. 4, pp. 480–86.

[15] Wilczynski, E. (1906) *Projective Differential Geometry of Curves and Ruled Surfaces.* Leipzig: B. G. Teubner.

[16] Weiss, I. (1988) "Projective invariants of shapes," in *Proceedings of Computer Vision and Pattern Recognition CVPR'88* (Ann Arbor, Michigan), pp. 1125–34, IEEE Computer Society.

[17] Rothwell, C. A., Zisserman, A., Forsyth, D. A. and Mundy, J. L. (1992) "Canonical frames for planar object recognition," in *Proceedings of the Second European Conference on Computer Vision ECCV'92* (St. Margherita, Italy), LNCS vol. 588, pp. 757–72, Springer.

[18] Weiss, I. (1992) "Differential invariants without derivatives," in *Proceedings of the Eleventh International Conference on Pattern Recognition ICPR'92* (Hague, The Netherlands), pp. 394–8, IEEE Computer Society.

[19] Mokhtarian, F. and Abbasi, S. (2002) "Shape similarity retrieval under affine transforms," *Pattern Recognition*, vol. 35, no. 1, pp. 31–41.

[20] Ibrahim Ali, W. S. and Cohen, F. S. (1998) "Registering coronal histological 2-D sections of a rat brain with coronal sections of a 3-D brain atlas using geometric curve invariants and B-spline representation," *IEEE Transactions on Medical Imaging*, vol. 17, no. 6, pp. 957–66.

[21] Yang, Z. and Cohen, F. (1999) "Image registration and object recognition using affine invariants and convex hulls," *IEEE Transactions on Image Processing*, vol. 8, no. 7, pp. 934–46.

[22] Flusser, J. (2002) "Affine invariants of convex polygons," *IEEE Transactions on Image Processing*, vol. 11, no. 9, pp. 1117–18,

[23] Rothwell, C. A., Zisserman, A., Forsyth, D. A. and Mundy, J. L. (1992) "Fast recognition using algebraic invariants," in *Geometric Invariance in Computer Vision* (Mundy, J. L. and Zisserman, A., eds), pp. 398–407, MIT Press.

[24] Lamdan, Y., Schwartz, J. and Wolfson, H. (1988) "Object recognition by affine invariant matching," in *Proceedings of Computer Vision and Pattern Recognition CVPR'88* (Ann Arbor, Michigan), pp. 335–44, IEEE Computer Society.

[25] Krolupper, F. and Flusser, J. (2007) "Polygonal shape description for recognition of partially occluded objects," *Pattern Recognition Letters*, vol. 28, no. 9, pp. 1002–11.

[26] Horáček, O., Kamenický, J. and Flusser, J. (2008) "Recognition of partially occluded and deformed binary objects," *Pattern Recognition Letters*, vol. 29, no. 3, pp. 360–69.

[27] Hilbert, D. (1993) *Theory of Algebraic Invariants*. Cambridge: Cambridge University Press.

[28] Gurevich, G. B. (1964) *Foundations of the Theory of Algebraic Invariants*. Groningen, The Netherlands: Nordhoff.

[29] Schur, I. (1968) *Vorlesungen über Invariantentheorie*. Berlin: Springer (in German).

[30] Hu, M.-K. (1962) "Visual pattern recognition by moment invariants," *IRE Transactions on Information Theory*, vol. 8, no. 2, pp. 179–87.

2

Moment invariants to translation, rotation and scaling

2.1 Introduction

Translation, rotation and scaling (TRS) are the simplest transformations of spatial coordinates. TRS, sometimes called *similarity* transform, is a four-parameter transform, which can be described as

$$\mathbf{x}' = s\mathbf{R} \cdot \mathbf{x} + \mathbf{t},$$

where $\mathbf{t}$ is a translation vector, s is a positive scaling factor (note that here we consider *uniform* scaling only, i.e. s is the same, both in horizontal and vertical directions), and $\mathbf{R}$ is a rotation matrix

$$\mathbf{R} = \begin{pmatrix} \cos \alpha & -\sin \alpha \\ \sin \alpha & \cos \alpha \end{pmatrix}$$

where α is the angle of rotation.

Invariance with respect to TRS is required in almost all practical applications, because the object should be correctly recognized, regardless of its position and orientation in the scene and of the object-to-camera distance. On the other hand, the TRS model is a sufficient approximation of the actual image deformation if the scene is flat and (almost) perpendicular to the optical axis. Therefore, much attention has been paid to TRS invariants. While translation and scaling invariants can be derived in an intuitive way, derivation of invariants to rotation is far more complicated.

2.1.1 Invariants to translation

Invariance to translation can be achieved simply by seemingly shifting the object such that its centroid coincides with the origin of the coordinate system or, vice versa, by shifting the polynomial basis into the object centroid. In the case of geometric moments, we have so-called *central* geometric moments

$$\mu_{pq} = \int_{-\infty}^{\infty} \int_{-\infty}^{\infty} (x - x_c)^p (y - y_c)^q f(x, y) \, dx \, dy, \tag{2.1}$$

Moments and Moment Invariants in Pattern Recognition   Jan Flusser, Tomáš Suk and Barbara Zitová
© 2009 John Wiley & Sons, Ltd

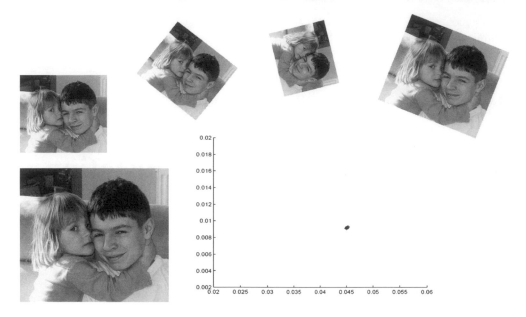

Figure 2.1 The desired behavior of TRS moment invariants – in all instances, the rotated and scaled image has approximately the same value as the invariant (depicted for two invariants).

where

$$x_c = m_{10}/m_{00}, \quad y_c = m_{01}/m_{00}$$

are the coordinates of the object centroid. Note that it always holds $\mu_{10} = \mu_{01} = 0$, and $\mu_{00} = m_{00}$. Translation invariance of the central moments is straightforward.

The central moments can be expressed in terms of geometric moments as

$$\mu_{pq} = \sum_{k=0}^{p} \sum_{j=0}^{q} \binom{p}{k} \binom{q}{j} (-1)^{k+j} x_c^k y_c^j m_{p-k,q-j}.$$

Although this relation has little importance for theoretic consideration, it is sometimes used when we want to calculate central moments by means of some fast algorithm for geometric moments.

2.1.2 Invariants to uniform scaling

Scaling invariance is obtained by proper normalization of each moment. In principle, any moment can be used as a normalizing factor provided that it is nonzero for all images in the experiment. Since low-order moments are more stable to noise and easier to calculate, we normalize most often by a proper power of μ_{00}

$$\nu_{pq} = \frac{\mu_{pq}}{\mu_{00}^w}, \tag{2.2}$$

where

$$w = \frac{p+q}{2} + 1. \tag{2.3}$$

The moment v_{pq} is called *normalized* central geometric moment.[1] After scaling by a factor s, central moments in the new coordinates change as

$$\mu'_{pq} = \int_{-\infty}^{\infty} \int_{-\infty}^{\infty} (x' - x'_c)^p (y' - y'_c)^q f'(x', y') \, dx' \, dy' \tag{2.4}$$

$$= \int_{-\infty}^{\infty} \int_{-\infty}^{\infty} s^p (x - x_c)^p s^q (y - y_c)^q f(x, y) s^2 \, dx \, dy = s^{p+q+2} \mu_{pq}. \tag{2.5}$$

In particular,

$$\mu'_{00} = s^2 \mu_{00}.$$

Consequently,

$$v'_{pq} = \frac{\mu'_{pq}}{(\mu'_{00})^w} = \frac{s^{p+q+2} \mu_{pq}}{(s^2 \mu_{00})^w} = v_{pq},$$

which proves the scaling invariance of the normalized moments.

The moment that was used for scaling normalization can no longer be used for recognition because the value of the corresponding normalized moment is always one (in the above normalization, $v_{00} = 1$). If we want to keep the zero-order moment valid, we can normalize by another moment, but other than for the zero-order scaling, normalization is used very rarely. The only meaningful different normalization is by $(\mu_{20} + \mu_{02})^{w/2}$.

2.1.3   Traditional invariants to rotation

Rotation moment invariants were first introduced in 1962 by Hu [1], who employed the results of the theory of algebraic invariants and derived his seven famous invariants to an in-plane rotation around the origin

$\phi_1 = m_{20} + m_{02}$,

$\phi_2 = (m_{20} - m_{02})^2 + 4m_{11}^2$,

$\phi_3 = (m_{30} - 3m_{12})^2 + (3m_{21} - m_{03})^2$,

$\phi_4 = (m_{30} + m_{12})^2 + (m_{21} + m_{03})^2$,

$\phi_5 = (m_{30} - 3m_{12})(m_{30} + m_{12})((m_{30} + m_{12})^2 - 3(m_{21} + m_{03})^2)$

$\qquad + (3m_{21} - m_{03})(m_{21} + m_{03})(3(m_{30} + m_{12})^2 - (m_{21} + m_{03})^2)$,

$\phi_6 = (m_{20} - m_{02})((m_{30} + m_{12})^2 - (m_{21} + m_{03})^2) + 4m_{11}(m_{30} + m_{12})(m_{21} + m_{03})$,

$\phi_7 = (3m_{21} - m_{03})(m_{30} + m_{12})((m_{30} + m_{12})^2 - 3(m_{21} + m_{03})^2)$

$\qquad - (m_{30} - 3m_{12})(m_{21} + m_{03})(3(m_{30} + m_{12})^2 - (m_{21} + m_{03})^2)$.

$$\tag{2.6}$$

---

[1]This normalization has already been used in Hu [1] but the exponent was stated incorrectly as $w = (p + q + 1)/2$. Many authors adopted this error and some other authors introduced a new one, claiming that $w = [(p + q)/2] + 1$, where [ ] denotes an integer part.

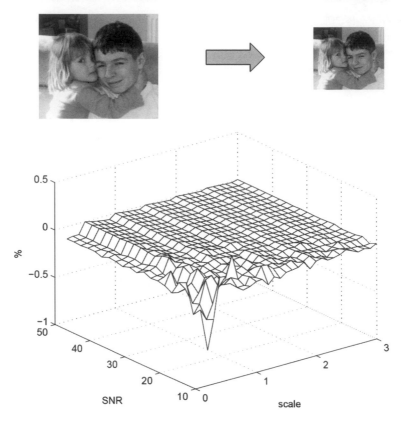

Figure 2.2 Numerical test of the normalized moment ν_{20}. Computer-generated scaling of the test image ranged form $s = 0.2$ to $s = 3$. To show robustness, each image was corrupted by additive Gaussian white noise. Signal-to-noise ratio (SNR) ranged from 50 dB (low noise) to 10 dB (heavy noise). Horizontal axes: scaling factor s and SNR, respectively. Vertical axis – relative deviation (in %) between ν_{20} of the original and that of the scaled and noisy image. The test proves the invariance of ν_{20} and illustrates its high robustness to noise.

If we replace geometric moments by central or normalized moments in these relations, we obtain invariants not only to rotation but also to translation and/or scaling, which also ensures invariance to rotation around an arbitrary point. The Hu derivation was rather complicated and that is why only these seven invariants were derived explicitly and no hint about how to derive invariants from higher-order moments was given. However, once we have the formulas, the proof of rotation invariance is easy. Let us demonstrate it for ϕ_1 and ϕ_2, both with central moments.

The second-order moments after rotation by angle α can be expressed as

$$\mu'_{20} = \cos^2 \alpha \cdot \mu_{20} + \sin^2 \alpha \cdot \mu_{02} - \sin 2\alpha \cdot \mu_{11},$$

$$\mu'_{02} = \sin^2 \alpha \cdot \mu_{20} + \cos^2 \alpha \cdot \mu_{02} + \sin 2\alpha \cdot \mu_{11},$$

$$\mu'_{11} = \tfrac{1}{2} \sin 2\alpha \cdot (\mu_{20} - \mu_{02}) + \cos 2\alpha \cdot \mu_{11}.$$

Thus,

$$\phi_1' = \mu_{20}' + \mu_{02}' = (\sin^2 \alpha + \cos^2 \alpha)(\mu_{20} + \mu_{02}) = \phi_1$$

and similarly for ϕ_2', applying the formula $\cos 2\alpha = \cos^2 \alpha - \sin^2 \alpha$.

Although the Hu invariants suffer from limited recognition power, mutual dependence and restriction to second- and third-order moments only, they have become classics and, inspite of their drawbacks, they have found numerous successful applications in various areas. The major weakness of Hu's theory is that it does not provide for the possibility of any generalization. Using it, we could not derive invariants from higher-order moments and invariants to more general transformations. These limitations were overcome 30 years later.

After Hu, various approaches to the theoretical derivation of moment-based rotation invariants were published. Li [2] used the Fourier–Mellin transform to derive invariants up to the order 9, Wong *et al.* [3] used complex monomials up to the fifth order that originate from the theory of algebraic invariants, and Jin and Tianxu [4] published another technique used to derive higher-order moment invariants. Teague [5] and Wallin *et al.* [6] proposed using Zernike moments. Mostafa and Psaltis [7] introduced the idea of using complex moments for deriving invariants but they focused on the evaluation of the invariants rather than on constructing higher-order systems. This approach was later followed by Flusser [8, 9], who proposed a general theory of constructing rotation moment invariants. His theory, presented in the next section, is based on complex moments. It is formally simple and transparent, allows invariants of any orders to be derived, and enables mutual dependence/independence of the invariants to be studied in a readable way.

All the above-mentioned approaches differ from each other formally in the mathematical tools and notation used, but they have common ground generally and, if applied correctly, yield similar or equivalent results.

2.2 Rotation invariants from complex moments

In this section, we present a general method to derive complete and independent rotation invariants of any orders. This method employs complex moments of the image.

2.2.1 Construction of rotation invariants

The complex moment c_{pq} of image function $f(x, y)$ was already defined by equation (1.3) as

$$c_{pq} = \int_{-\infty}^{\infty} \int_{-\infty}^{\infty} (x + iy)^p (x - iy)^q f(x, y) \, dx \, dy. \qquad (2.7)$$

It follows from the definition that only the indices $p \geq q$ are meaningful because $c_{pq} = c_{qp}^*$ (the asterisk denotes complex conjugate).

The complex moment can easily be expressed in *polar coordinates* (r, θ)

$$\begin{aligned} x &= r \cos \theta & r &= \sqrt{x^2 + y^2} \\ y &= r \sin \theta & \theta &= \arctan(y/x), \end{aligned} \qquad (2.8)$$

then, equation (2.7) takes the form

$$c_{pq} = \int_0^\infty \int_0^{2\pi} r^{p+q+1} e^{i(p-q)\theta} f(r,\theta)\, dr\, d\theta. \tag{2.9}$$

The following lemma describes an important rotation property of the complex moments that will be used later for invariant construction.

Lemma 2.1 *Let f' be a rotated version (around the origin) of f, i.e. $f'(r,\theta) = f(r, \theta + \alpha)$ where α is the angle of rotation. Let us denote the complex moments of f' as c'_{pq}. Then,*

$$c'_{pq} = e^{-i(p-q)\alpha} \cdot c_{pq}. \tag{2.10}$$

Proof. Using equation (2.9), the proof of Lemma 2.1 is straightforward.

$$c'_{pq} = \int_0^\infty \int_0^{2\pi} r^{p+q+1} e^{i(p-q)(\theta-\alpha)} f(r,\theta)\, dr\, d\theta = e^{-i(p-q)\alpha} \cdot c_{pq}. \tag{2.11}$$

$\square$

Equation (2.10) implies rotation invariance of the moment magnitude $|c_{pq}|$ while the phase is shifted by $(p - q)\alpha$ (recall the clear analogue with the Fourier shift theorem and Fourier descriptors). Any approach to the construction of rotation invariants must be based on a proper kind of phase cancelation. The simplest method proposed by many authors (see [10], for instance) is to use as invariants the moment magnitudes themselves. However, they do not generate a complete set of invariants – by taking the magnitudes only, we miss many useful invariants. In the following theorem, phase cancelation is achieved by multiplication of appropriate moment powers.

Theorem 2.2 *Let $n \geq 1$ and let k_i, p_i, and q_i $(i = 1, \ldots, n)$ be non-negative integers such that*

$$\sum_{i=1}^n k_i (p_i - q_i) = 0.$$

Then

$$I = \prod_{i=1}^n c_{p_i q_i}^{k_i} \tag{2.12}$$

is invariant to rotation.

Proof. Let us rotate the image by an angle α. Then

$$I' = \prod_{i=1}^n c_{p_i q_i}'^{k_i} = \prod_{i=1}^n e^{-ik_i(p_i-q_i)\alpha} \cdot c_{p_i q_i}^{k_i} = e^{-i\alpha \sum_{i=1}^n k_i(p_i-q_i)} \cdot I = I. \qquad \square$$

According to Theorem 2.2, some simple examples of rotation invariants are c_{11}, $c_{20}c_{02}$, $c_{20}c_{12}^2$, etc. Most invariants (2.12) are complex. If real-valued features are required, we consider real and imaginary parts of each of them separately. To also achieve translation invariance, we use central coordinates in the definition of the complex moments (2.7) and discard all invariants containing c_{10} and c_{01}. Scaling invariance can be achieved by the same normalization as described in the previous section.

2.2.2 Construction of the basis

In this section, we pay attention to the construction of a basis of invariants up to a given order. Theorem 2.2 allows us to construct an infinite number of invariants for any order of moments, but only a few of them are mutually independent. By the term *basis* we intuitively understand the smallest set required to express all other invariants. More precisely, the basis must be *independent*, which means that none of its elements can be expressed as a function of the other elements, and also *complete*, meaning that any rotation invariant can be expressed by means of the basis elements only.

The knowledge of the basis is a crucial point in all pattern recognition tasks because the basis provides the same discriminative power as the set of all invariants and thus minimizes the computational cost. For instance, the set

$$\{c_{20}c_{02}, \; c_{21}^2 c_{02}, \; c_{12}^2 c_{20}, \; c_{21}c_{12}, \; c_{21}^3 c_{02}c_{12}\}$$

is a dependent set whose basis is $\{c_{12}^2 c_{20}, \; c_{21}c_{12}\}$.

To formalize these terms, we first introduce the following definitions.

Definition 2.3 Let $k \geq 1$ and let $\mathcal{I} = \{I_1, \ldots, I_k\}$ be a set of rotation invariants. Also let J be a rotation invariant. J is said to be *dependent on* $\mathcal{I}$ if and only if there exists a function F of k variables such that

$$J = F(I_1, \ldots, I_k).$$

J is said to be *independent* on $\mathcal{I}$ otherwise.

Definition 2.4 Let $k > 1$ and let $\mathcal{I} = \{I_1, \ldots, I_k\}$ be a set of rotation invariants. The set $\mathcal{I}$ is said to be *dependent* if and only if there exists $k_0 \leq k$ such that I_{k_0} depends on $\mathcal{I} \dotminus \{I_{k_0}\}$. The set $\mathcal{I}$ is said to be *independent* otherwise.[2]

According to this definition, $\{c_{20}c_{02}, \; c_{20}^2 c_{02}^2\}$, $\{c_{21}^2 c_{02}, \; c_{21}c_{12}, \; c_{21}^3 c_{02}c_{12}\}$ and $\{c_{20}c_{12}^2, \; c_{02}c_{21}^2\}$ are examples of dependent invariant sets.

Definition 2.5 Let $\mathcal{I}$ be a set of rotation invariants and let $\mathcal{B}$ be its subset. $\mathcal{B}$ is called a *complete* subset if and only if any element of $\mathcal{I} \dotminus \mathcal{B}$ depends on $\mathcal{B}$. The set $\mathcal{B}$ is a *basis* of $\mathcal{I}$ if and only if it is independent and complete.

Now we can formulate a fundamental theorem that tells us how to construct an invariant basis of a given order.

Theorem 2.6 *Let us consider complex moments up to the order $r \geq 2$. Let a set of rotation invariants $\mathcal{B}$ be constructed as follows:*

$$\mathcal{B} = \{\Phi(p, q) \equiv c_{pq}c_{q_0 p_0}^{p-q} | p \geq q \wedge p + q \leq r\},$$

where p_0 and q_0 are arbitrary indices such that $p_0 + q_0 \leq r$, $p_0 - q_0 = 1$ and $c_{p_0 q_0} \neq 0$ for all admissible images. Then $\mathcal{B}$ is a basis of all rotation invariants created from the moments of any kind up to the order r.

---

[2]The symbol $\dotminus$ stands for a set minus.

Theorem 2.6 is very strong because it claims that $\mathcal{B}$ is a basis of *all possible* rotation moment invariants, not only of those constructed according to (2.12) and not only of those which are based on complex moments. In other words, $\mathcal{B}$ provides at least the same discrimination power as any other set of moment invariants up to the given order $r \geq 2$ because any possible invariant can be expressed in terms of $\mathcal{B}$. (Note that this is a theoretical property; it may be violated in the discrete domain where different polynomials have different numerical properties.)

Proof. Let us prove the independence of $\mathcal{B}$ first. Let us assume that $\mathcal{B}$ is dependent, i.e. there exists $\Phi(p, q) \in \mathcal{B}$, such that it depends on $\mathcal{B} \doteq \{\Phi(p, q)\}$. As follows from the linear independence of the polynomials $(x + iy)^p (x - iy)^q$ and, consequently, from independence of the complex moments themselves, it must hold that $p = p_0$ and $q = q_0$. This means, according to the above assumption, that there exist invariants $\Phi(p_1, q_1), \ldots, \Phi(p_n, q_n)$ and $\Phi(s_1, t_1), \ldots, \Phi(s_m, t_m)$ from $\mathcal{B} \doteq \{\Phi(p_0, q_0)\}$ and positive integers $k_1, \ldots, k_n$ and $\ell_1, \ldots, \ell_m$ such that

$$\Phi(p_0, q_0) = \frac{\prod_{i=1}^{n_1} \Phi(p_i, q_i)^{k_i} \cdot \prod_{i=n_1+1}^{n} \Phi^*(p_i, q_i)^{k_i}}{\prod_{i=1}^{m_1} \Phi(s_i, t_i)^{\ell_i} \cdot \prod_{i=m_1+1}^{m} \Phi^*(s_i, t_i)^{\ell_i}}. \tag{2.13}$$

Substituting into (2.13) and grouping the factors $c_{p_0 q_0}$ and $c_{q_0 p_0}$ together, we obtain

$$\Phi(p_0, q_0) = \frac{c_{q_0 p_0}^{\sum_{i=1}^{n_1} k_i(p_i - q_i)} \cdot c_{p_0 q_0}^{\sum_{i=n_1+1}^{n} k_i(q_i - p_i)}}{c_{q_0 p_0}^{\sum_{i=1}^{m_1} \ell_i(s_i - t_i)} \cdot c_{p_0 q_0}^{\sum_{i=m_1+1}^{m} \ell_i(t_i - s_i)}} \cdot \frac{\prod_{i=1}^{n_1} c_{p_i q_i}^{k_i} \cdot \prod_{i=n_1+1}^{n} c_{q_i p_i}^{k_i}}{\prod_{i=1}^{m_1} c_{s_i t_i}^{\ell_i} \cdot \prod_{i=m_1+1}^{m} c_{t_i s_i}^{\ell_i}}. \tag{2.14}$$

Comparing the exponents of $c_{p_0 q_0}$ and $c_{q_0 p_0}$ on both sides, we obtain the constraints

$$K_1 = \sum_{i=1}^{n_1} k_i(p_i - q_i) - \sum_{i=1}^{m_1} \ell_i(s_i - t_i) = 1 \tag{2.15}$$

and

$$K_2 = \sum_{i=n_1+1}^{n} k_i(q_i - p_i) - \sum_{i=m_1+1}^{m} \ell_i(t_i - s_i) = 1. \tag{2.16}$$

Since the rest of the right-hand side of equation (2.14) must be equal to 1 and since the moments themselves are mutually independent, the following constraints must be fulfilled for any index i:

$$n_1 = m_1, \qquad n = m, \qquad p_i = s_i, \qquad q_i = t_i, \qquad k_i = \ell_i.$$

Introducing these constraints into (2.15) and (2.16), we obtain $K_1 = K_2 = 0$ which is a contradiction.

To prove the completeness of $\mathcal{B}$, it is sufficient to resolve the so-called *inverse problem*, thereby recovering all complex moments (and, consequently, all geometric moments) up to the order r when the elements of $\mathcal{B}$ are known. Thus, the following nonlinear system of equations must be resolved for the c_{pq}'s:

$$\Phi(p_0, q_0) = c_{p_0 q_0} c_{q_0 p_0},$$

$$\Phi(0, 0) = c_{00},$$

$$\Phi(1, 0) = c_{10} c_{q_0 p_0},$$

$$\Phi(2, 0) = c_{20} c_{q_0 p_0}^2,$$

$$\Phi(1, 1) = c_{11},$$

$$\Phi(3, 0) = c_{30} c_{q_0 p_0}^3,$$

$$\cdots$$

$$\Phi(r, 0) = c_{r0} c_{q_0 p_0}^r,$$

$$\Phi(r - 1, 1) = c_{r-1,1} c_{q_0 p_0}^{r-2},$$

$$\cdots \tag{2.17}$$

Since $\mathcal{B}$ is a set of rotation invariants, it does not reflect the orientation of the object. Thus, there is one degree of freedom when recovering the object moments that corresponds to the choice of the object orientation. Without loss of generality, we can choose such orientation in which $c_{p_0 q_0}$ is real and positive. As can be seen from equation (2.10), if $c_{p_0 q_0}$ is nonzero, then such orientation always exists. The first equation of (2.17) can then be immediately resolved for $c_{p_0 q_0}$:

$$c_{p_0 q_0} = \sqrt{\Phi(p_0, q_0)}.$$

Consequently, using the relationship $c_{q_0 p_0} = c_{p_0 q_0}$, we obtain the solutions

$$c_{pq} = \frac{\Phi(p, q)}{c_{q_0 p_0}^{p-q}}$$

and

$$c_{pp} = \Phi(p, p)$$

for any p and q. Recovering the geometric moments is straightforward from equation (1.5). Since any polynomial is a linear combination of standard powers $x^p y^q$, any moment (and any moment invariant) can be expressed as a function of geometric moments, which completes the proof. □

Theorem 2.6 not only enables the basis to be created but also the number of its elements to be calculated in advance. Let us denote it as $|\mathcal{B}|$. If r is odd, then

$$|\mathcal{B}| = \tfrac{1}{4}(r + 1)(r + 3),$$

if r is even, then

$$|\mathcal{B}| = \tfrac{1}{4}(r + 2)^2.$$

(These numbers refer to complex-valued invariants.) We can see that the authors who used the moment magnitudes only actually lost about one half of the information contained in basis $\mathcal{B}$.

The basis defined in Theorem 2.6 is generally not unique. It depends on the particular choice of p_0 and q_0 which is very important. How shall we select these indices in practice?

On one hand, we want to keep p_0 and q_0 as small as possible because lower-order moments are less sensitive to noise than the higher-order ones. On the other hand, a close-to-zero value of $c_{p_0 q_0}$ may cause numerical instability of the invariants. Thus, we propose the following algorithm. We start with $p_0 = 2$ and $q_0 = 1$ and check whether $|c_{p_0 q_0}|$ exceeds a pre-defined threshold for all objects (in practice, this means for all given training samples or database elements). If this condition is met, we accept the choice; if not, we increase both p_0 and q_0 by one and repeat the above procedure.

2.2.3   Basis of invariants of the second and third orders

In this section, we present a basis of the rotation invariants composed of the moments of the second and third orders, that is constructed according to Theorem 2.6 by choosing $p_0 = 2$ and $q_0 = 1$. The basis is

$$\Phi(1, 1) = c_{11},$$

$$\Phi(2, 1) = c_{21}c_{12},$$

$$\Phi(2, 0) = c_{20}c_{12}^2,$$

$$\Phi(3, 0) = c_{30}c_{12}^3. \tag{2.18}$$

In this case, the basis is determined unambiguously and contains six real-valued invariants. It is worth noting that formally, according to Theorem 2.6, the basis should also contain invariants $\Phi(0, 0) = c_{00}$ and $\Phi(1, 0) = c_{10}c_{12}$. We did not include these two invariants in the basis because $c_{00} = \mu_{00}$ is in most recognition applications used for normalization to scaling and $c_{10} = m_{10} + im_{01}$ is used to achieve translation invariance. Then $\Phi(0, 0) = 1$ and $\Phi(1, 0) = 0$ for any object and it is useless to consider them.

2.2.4   Relationship to the Hu invariants

In this section, we highlight the relationship between Hu invariants (2.6) and the proposed invariants (2.18). We show the Hu invariants to be incomplete and mutually dependent, which can explain some practical problems connected with their use.

It can be seen that the Hu invariants are nothing other than particular representatives of the general form (2.12)

$$\phi_1 = c_{11},$$

$$\phi_2 = c_{20}c_{02},$$

$$\phi_3 = c_{30}c_{03},$$

$$\phi_4 = c_{21}c_{12},$$

$$\phi_5 = \mathcal{R}e(c_{30}c_{12}^3),$$

$$\phi_6 = \mathcal{R}e(c_{20}c_{12}^2),$$

$$\phi_7 = \mathcal{I}m(c_{30}c_{12}^3) \tag{2.19}$$

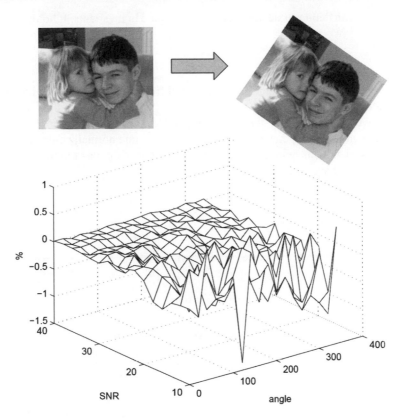

Figure 2.3 Numerical test of the basic invariant $\Phi(2, 0)$. Computer-generated rotation of the test image ranged form 0 to 360°. To show robustness, each image was corrupted by additive Gaussian white noise. SNR ranged from 40 dB (low noise) to 10 dB (heavy noise). Horizontal axes: rotation angle and SNR, respectively. Vertical axis – relative deviation (in %) between $\mathcal{Re}(\Phi(2, 0))$ of the original and that of the rotated and noisy image. The test proves the invariance of $\mathcal{Re}(\Phi(2, 0))$ and illustrates its high robustness to noise.

and that they can be expressed also in terms of the basis (2.18)

$$\phi_1 = \Phi(1, 1),$$

$$\phi_2 = \frac{|\Phi(2, 0)|^2}{\Phi(2, 1)^2},$$

$$\phi_3 = \frac{|\Phi(3, 0)|^2}{\Phi(2, 1)^3},$$

$$\phi_4 = \Phi(2, 1),$$

$$\phi_5 = \mathcal{Re}(\Phi(3, 0)),$$

$$\phi_6 = \mathcal{Re}(\Phi(2, 0)),$$

$$\phi_7 = \mathcal{Im}(\Phi(3, 0)).$$

Using (2.19), we can demonstrate the dependency of the Hu invariants. It holds

$$\phi_3 = \frac{\phi_5^2 + \phi_7^2}{\phi_4^3},$$

which means that ϕ_3 is useless and can be excluded from the Hu's system without any loss of discrimination power.

Moreover, we show the Hu system to be incomplete. Let us try to recover complex and geometric moments knowing that $\phi_1, \ldots, \phi_7$ under the same normalization constraint as in the previous case, i.e. c_{21} is required to be real and positive. Complex moments $c_{11}, c_{21}, c_{12}, c_{30}$ and c_{03} can be recovered in a straightforward way:

$$c_{11} = \phi_1,$$

$$c_{21} = c_{12} = \sqrt{\phi_4},$$

$$Re(c_{30}) = Re(c_{03}) = \frac{\phi_5}{c_{12}^3},$$

$$Im(c_{30}) = -Im(c_{03}) = \frac{\phi_7}{c_{12}^3}.$$

Unfortunately, c_{20} cannot be fully recovered. We have, for its real part,

$$Re(c_{20}) = \frac{\phi_6}{c_{12}^2}$$

but for its imaginary part, we obtain

$$(Im(c_{20}))^2 = |c_{20}|^2 - (Re(c_{20}))^2 = \phi_2 - \left(\frac{\phi_6}{c_{12}^2}\right)^2.$$

There is no way of determining the sign of $Im(c_{20})$. In terms of geometric moments, it means that the sign of m_{11} cannot be recovered.

The incompleteness of the Hu invariants implies their lower discrimination power compared to the proposed invariants of the same order. Let us consider two objects, $f(x, y)$ and $g(x, y)$, both in the normalized central positions, having the same geometric moments up to the third order except m_{11}, for which $m_{11}^{(f)} = -m_{11}^{(g)}$. In the case of artificial data, such object $g(x, y)$ exists for any given $f(x, y)$ and can be designed as

$$g(x, y) = (f(x, y) + f(-x, y) + f(x, -y) - f(-x, -y))/2, \qquad (2.20)$$

see Figure 2.4 for an example. It is easy to prove that under (2.20) the moment constraints are always fulfilled. While the basic invariants (2.18) distinguish these two objects by the imaginary part of $\Phi(2, 0)$, the Hu invariants are not able to do so (see Table 2.1), even if the objects are easy to discriminate visually.

This property can also be demonstrated on real data. In Figure 2.5(a), one can see the photograph of a pan. A picture of a virtual "two-handle" pan (Figure 2.5(b)) was created from the original image according to (2.20). Although these two objects are apparently different, all their Hu invariants are exactly the same. On the other hand, the new invariant $\Phi(2, 0)$ distinguishes these two objects clearly, owing to the opposite signs of its imaginary part.

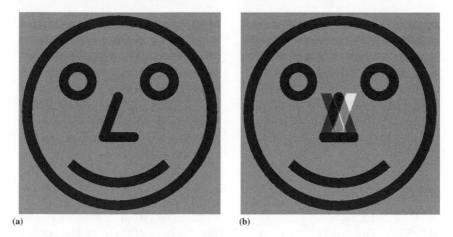

(a)   (b)

Figure 2.4 The smiles: (a) original, and (b) another figure created from the original according to equation (2.20). For the values of the respective invariants, see Table 2.1.

Table 2.1 The values of the Hu invariants and the basic invariants (2.18) of "The smiles" in Figure 2.4. The only invariant discriminating them is $Im(\Phi(2, 0))$. (The values shown here were scaled to the range from -10 to 10.)

Hu	Figure 2.4(a)	Figure 2.4(b)	Basic	Figure 2.4(a)	Figure 2.4(b)
ϕ_1	4.2032	4.2032	$\Phi(1, 1)$	4.2032	4.2032
ϕ_2	1.7563	1.7563	$\Phi(2, 1)$	6.5331	6.5331
ϕ_3	3.6395	3.6395	$Re(\Phi(2, 0))$	8.6576	8.6576
ϕ_4	6.5331	6.5331	$\boldsymbol{Im(\Phi(2, 0))}$	**7.6437**	**−7.6437**
ϕ_5	2.5749	2.5749	$Re(\Phi(3, 0))$	1.0074	1.0074
ϕ_6	8.6576	8.6576	$Im(\Phi(3, 0))$	−6.9850	−6.9850
ϕ_7	−6.9850	−6.9850			

This ambiguity cannot be avoided by a different choice of normalization constraint when recovering the moments. Let us, for instance, consider another normalization constraint which requires c_{20} to be real and positive. In terms of geometric moments, it corresponds to the requirement $m_{11} = 0$ and $m_{20} > m_{02}$. Then, similarly to the previous case, the signs of $(m_{30} + m_{12})$ and $(m_{03} + m_{21})$ cannot be unambiguously determined.

Note also, that other earlier sets of rotation invariants are dependent and/or incomplete; this was not mentioned in the original papers, partly because the authors did not pay attention to these issues and partly because their methods do not allow the dependencies among the invariants to be discovered in a systematic way. Li [2] published a set of invariants from moments up to the ninth order. Unfortunately, his system includes the Hu invariants and therefore neither can it be a basis. Jin and Tianxu [4] derived 12 invariants in explicit form but only eight of them are independent. Wong [3] presented a set of 16 invariants from moments

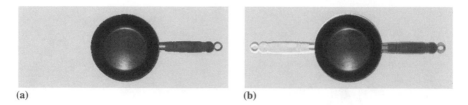

(a)   (b)

Figure 2.5 (a) Original image of a pan, and (b) a virtual "two-handle" pan. These objects are distinguishable by the basic invariants but not by the Hu invariants.

up to the third order and a set of "more than 49" invariants from moments up to the fourth order. It follows immediately from Theorem 2.6 that the basis of the third-order invariants has only six elements and the basis of the fourth-order invariants has nine elements. Thus, most of Wong's invariants are dependent and of no practical significance. Even the Mukundan monograph [11] presents a dependent and incomplete set of 22 rotation invariants up to the sixth order (see [11, p. 125]).

2.3   Pseudoinvariants

In this section, we investigate the behavior of rotation invariants under mirror reflection of the image. Mirror reflection is not a special case of TRS transform and can occur in some applications only, but it is useful to know how the rotation invariants behave in that case. We show that they cannot change arbitrarily. There are only two possibilities – a (real-valued) invariant can either stay constant or change its sign. The invariants, which do not change the sign under reflection are traditionally called *true invariants*, while the others are called *pseudoinvariants* [5] or, misleadingly, *skew invariants* [1]. Pseudoinvariants distinguish between mirrored images of the same object which is useful in some applications but may be undesirable in other cases. We show which invariants from the basis introduced in Theorem 2.6 are pseudoinvariants and which are the true ones.

Let us consider a basic invariant and investigate its behavior under a reflection across an arbitrary line. Due to the rotation and shift invariance, we can limit ourselves to the reflection across the x-axis.

Let $\overline{f}(x, y)$ be a mirrored version of $f(x, y)$, i.e. $\overline{f}(x, y) = f(x, -y)$. Clearly,

$$\overline{c_{pq}} = c_{qp} = c_{pq}^*.$$

Thus, it holds for any basic invariant $\Phi(p, q)$

$$\overline{\Phi(p, q)} = \overline{c_{pq} c_{q_0 p_0}^{p-q}} = c_{pq}^* \cdot (c_{q_0 p_0}^*)^{p-q} = \Phi^*(p, q).$$

This proves that the real parts of the basic invariants are true invariants. On the other hand, the imaginary parts of them are pseudoinvariants, because they change their signs under reflection.

Figure 2.6 The test image and its mirrored version. Basic invariants of the mirrored image are complex conjugates of those of the original.

2.4 Combined invariants to TRS and contrast changes

So far, we have considered invariants to spatial transformations only. However, in practice the features used in a recognition system should also be invariant to graylevel or color changes. In Chapter 5 we describe invariants to graylevel degradations caused by linear filtering. In this section we consider contrast stretching only, which is a very simple graylevel transform given by

$$f'(x, y) = a \cdot f(x, y),$$

where a is a positive stretching factor.[3] We will demonstrate how to combine invariance to contrast stretching and to TRS.

Since $\mu'_{pq} = a \cdot \mu_{pq}$ and $c'_{pq} = a \cdot c_{pq}$, pure contrast (and translation) invariance can be achieved simply by normalizing each central or complex moment by μ_{00}. Now consider also

---

[3]In case of digital images, $f'(x, y)$ might overflow the 8-bit intensity range. In such cases, the following consideration might become invalid.

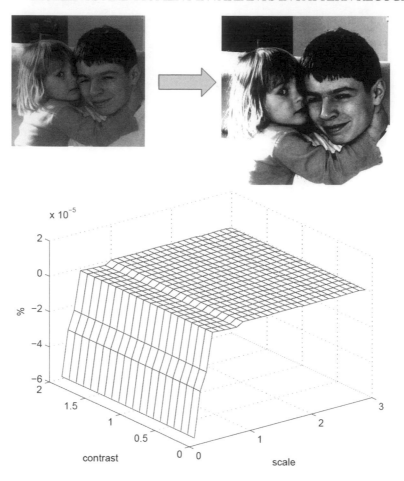

Figure 2.7   Numerical test of the contrast and TRS invariant $\Psi(2, 0)$ for $p_0 = 2$ and $q_0 = 1$. Computer-generated scaling of the test image ranged from $s = 0.2$ to $s = 3$ and the contrast stretching factor ranged from $a = 0.1$ to $a = 2$. Horizontal axes: scaling factor s and contrast stretching factor a, respectively. Vertical axis – relative deviation (in %) between $\mathcal{R}e(\Psi(2, 0))$ of the original and that of the scaled and stretched image. The test proves the invariance of $\mathcal{R}e(\Psi(2, 0))$ with respect to both factors. However, for downscaling with $s = 0.2$ and $s = 0.3$, the resampling effect leads to slightly higher relative errors.

rotation. Then, it holds for any basic invariant $\Phi(p, q)$

$$\Phi'(p, q) = a^{p-q+1}\Phi(p, q)$$

and it is sufficient to use the normalization

$$\frac{\Phi(p, q)}{\mu_{00}^{p-q+1}}.$$

This approach unfortunately cannot be further extended to scaling and contrast invariance. If μ_{00} is used for normalization to scaling, it cannot be at the same time used for contrast normalization. We have to normalize by some other invariant, preferably by $\Phi(p_0, q_0)$, because of simplicity.

Denote scale-normalized complex moments as $\widetilde{c}_{pq}$

$$\widetilde{c}_{pq} = \frac{c_{pq}}{\mu_{00}^{((p+q)/2)+1}}$$

and scale-normalized basic invariants as $\widetilde{\Phi(p, q)}$

$$\widetilde{\Phi(p, q)} = \widetilde{c}_{pq}\widetilde{c}_{q_0 p_0}{}^{p-q}.$$

Under contrast stretching, these quantities change as

$$\widetilde{c}_{pq}{}' = a^{-(p+q)/2}\widetilde{c}_{pq}$$

and

$$\widetilde{\Phi(p, q)}{}' = a^{-((p_0+q_0)(p-q)+(p+q))/2}\widetilde{\Phi(p, q)}.$$

In particular,

$$\widetilde{\Phi(p_0, q_0)}{}' = a^{-(p_0+q_0)}\widetilde{\Phi(p_0, q_0)}.$$

Eliminating the contrast factor a, we obtain combined TRS and contrast invariants $\Psi(p, q)$ in the form

$$\Psi(p, q) = \frac{\widetilde{\Phi(p, q)}}{|\widetilde{c}_{pq}| \cdot \widetilde{\Phi(p_0, q_0)}{}^{(p-q)/2}}.$$

Note that due to the normalization used, $\Psi(p_0, q_0) = 1$ for any object.

2.5 Rotation invariants for recognition of symmetric objects

In many applied tasks we want to classify man-made objects or natural shapes from their silhouettes (i.e. from binary images) only. In most cases such shapes have some kind of symmetry. Different classes may or may not have the same symmetry that makes the task even more difficult. While humans can use the symmetry as a cue that helps them to recognize objects, moment-based classifiers suffer from the fact that some moments of symmetric shapes are zero and corresponding invariants do not provide any discrimination power. This is why we must pay attention to these situations and why it is necessary to design special invariants for each type of symmetry [12].

For example, all odd-order moments of a centrosymmetric object are identically equal to zero. If an object is circularly symmetric, all its complex moments, whose indices are different, vanish. Thus, Theorem 2.6 either cannot be applied at all or many rotation invariants might be useless. Let us imagine an illustrative example. We want to recognize three shapes – square, cross and circle – independently of their orientation. Because of symmetry, all complex moments of the second and third orders except c_{11} are zero. If the shapes are appropriately scaled, c_{11} can be the same for all of them. Consequently, neither the Hu invariants nor the basic invariants from Theorem 2.6 provide any discrimination power, even

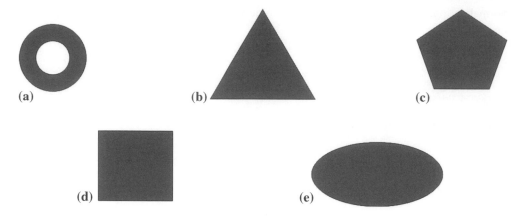

Figure 2.8 Objects having N-fold rotation symmetry. (a) $N = \infty$, (b) $N = 3$, (c) $N = 5$, (d) $N = 4$, and (e) $N = 2$.

if the shapes can be readily recognized visually. Appropriate invariants in this case would be c_{22}, $c_{40}c_{04}$, $c_{51}c_{04}$, c_{33}, $c_{80}c_{04}^2$, $c_{62}c_{04}$, c_{44}, etc.

The above simple example shows the necessity of having different systems of invariants for objects with different types of symmetry. In this section we consider the so-called *N-fold rotation symmetry* (*N*-FRS), which is the most frequent symmetry met in practice. An object is said to have N-FRS if it repeats itself when it rotates around its centroid by $2\pi j/N$ for all $j = 1, \ldots, N$. We use this definition not only for finite N but also for $N = \infty$. Thus, in our terminology, objects having *circular symmetry* $f(x, y) = f(\sqrt{x^2 + y^2})$ are said to have ∞-FRS.

In addition to rotation symmetry, axial symmetry also appears often in the images and contributes to vanishing of some moments, too. There is a close connection between axial and rotation symmetry – if an object has K axes of symmetry ($K > 0$), then it is also rotationally symmetric and K is exactly its number of folds [13]. Thus, we will not discuss the choice of invariants for axially symmetric objects separately.

Lemma 2.7 *If $f(x, y)$ has an N-fold rotation symmetry (N finite) and if $(p - q)/N$ is not an integer, then $c_{pq} = 0$.*

Proof. Let us rotate f around the origin by $2\pi/N$. Due to its symmetry, the rotated object f' must be the same as the original. In particular, it must hold $c'_{pq} = c_{pq}$ for any p and q. On the other hand, it follows from equation (2.10) that

$$c'_{pq} = e^{-2\pi i(p-q)/N} \cdot c_{pq}.$$

Since $(p - q)/N$ is assumed not to be an integer, this equation can be fulfilled only if $c_{pq} = 0$.   □

Lemma 2.7a *If $f(x, y)$ has an ∞-fold rotation symmetry and if $p \neq q$, then $c_{pq} = 0$.*

Proof. Let us rotate f around the origin by an arbitrary angle α. The rotated version f' must be the same as the original for any α and, consequently, its moments cannot change under rotation. Equation (2.10) implies that

$$c'_{pq} = e^{-i(p-q)\alpha} \cdot c_{pq}.$$

Since $p \neq q$ and α may be arbitrary, this equation can be fulfilled only if $c_{pq} = 0$. $\square$

In order to derive invariants for recognition of objects with N-fold rotation symmetry, Theorem 2.6 can be generalized in the following way.

Theorem 2.8 *Let $N \geq 1$ be a finite integer. Let us consider complex moments up to the order $r \geq N$. Let a set of rotation invariants $\mathcal{B}_N$ be constructed as follows:*

$$\mathcal{B}_N = \{\Phi_N(p, q) \equiv c_{pq}c_{q_0 p_0}^k \mid p \geq q \wedge p + q \leq r \wedge k \equiv (p - q)/N \text{ is integer}\},$$

where p_0 and q_0 are arbitrary indices such that $p_0 + q_0 \leq r$, $p_0 - q_0 = N$, and $c_{p_0 q_0} \neq 0$ for all admissible objects. Then $\mathcal{B}_N$ is a basis of all nontrivial[4] rotation invariants for objects with N-FRS, created from the moments up to the order r.

Proof. Rotation invariance of all elements of $\mathcal{B}_N$ follows immediately from Theorem 2.2. The independence and completeness of $\mathcal{B}_N$ can be proven exactly in the same way as in Theorem 2.6, when only nonzero moments are recovered. $\square$

For $N = 1$, which means no rotation symmetry, Theorem 2.8 is reduced exactly to Theorem 2.6; $\mathcal{B}_1 = \mathcal{B}$ and $\Phi_1(p, q) = \Phi(p, q)$. The following modification of Theorem 2.8 deals with the case of $N = \infty$.

Theorem 2.8a *Let us consider complex moments up to the order $r \geq 2$. Then the basis $\mathcal{B}_\infty$ of all nontrivial rotation invariants for objects with ∞-FRS is*

$$\mathcal{B}_\infty = \{c_{pp} \mid p \leq r/2\}.$$

The proof of Theorem 2.8a follows immediately from Theorem 2.2 and Lemma 2.4a.

Theorems 2.8 and 2.8a have several interesting consequences. Some of them are summarized in the following Lemma.

Lemma 2.9 *Let us denote all rotation invariants that can be expressed by means of elements of basis $\mathcal{B}$ as $\langle \mathcal{B} \rangle$. Then it holds for any order r:*

1. *If M and N are finite and L is their least common multiple, then*

$$\langle \mathcal{B}_M \rangle \cap \langle \mathcal{B}_N \rangle = \langle \mathcal{B}_L \rangle.$$

In particular, if M/N is integer, then $\langle \mathcal{B}_M \rangle \subset \langle \mathcal{B}_N \rangle$.

2.

$$\bigcap_{N=1}^{\infty} \langle \mathcal{B}_N \rangle = \langle \mathcal{B}_\infty \rangle.$$

---

[4]The invariants which are constant for any image are called trivial. The others are nontrivial. Trivial invariants provide no recognition power at all.

3. *If N is finite, the number of elements of $\mathcal{B}_N$ is*

$$|\mathcal{B}_N| = \sum_{j=0}^{[r/N]} \left[\frac{r - jN + 2}{2}\right],$$

where [a] means an integer part of a. For $N = \infty$ it holds

$$|\mathcal{B}_\infty| = \left[\frac{r + 2}{2}\right].$$

From the last point of Lemma 2.9 we can see that the higher the number of folds, the fewer nontrivial invariants exist. Their number ranges from approximately $r^2/4$ for nonsymmetric objects to approximately $r/2$ for circularly symmetric ones.

In practical pattern recognition experiments, the number of folds N may not be known a priori. In that case we can apply a fold detector (see references [14–16] for algorithms detecting the number of folds) to all elements of the training set before we choose an appropriate system of moment invariants. In the case of equal fold numbers of all classes, proper invariants can be chosen directly according to Theorems 2.8 or 2.8a. Usually we start selecting from the lowest orders and the number of invariants is determined to provide a sufficient discriminability of the training set. However, it is not realistic to meet such a simple situation in practice. Different shape classes typically have different numbers of folds. The previous theory does not provide a solution to this problem.

As can be seen from Lemma 2.9, we cannot simply choose one of the number of folds detected as the appropriate N for constructing an invariant basis according to Theorem 2.8 (although one could intuitively expect the highest number of folds to be a good choice, this is not the case here). A more sophisticated choice is to take the least common multiple of all finite fold numbers and then to apply Theorem 2.8. Unfortunately, taking the least common multiple often leads to high-order instable invariants. This is why in practice one may prefer to decompose the problem into two steps – first, preclassification into "groups of classes" according to the number of folds performed and then final classification by means of moment invariants, which are defined separately in each group. This decomposition can be performed explicitly in a separate preclassification stage or implicitly during the classification. The word "implicitly" here means that the number of folds of an unknown object is not explicitly tested; however, at the start, we must test the numbers of folds in the training set. We explain the latter version.

Let us have C classes altogether such that C_k classes have N_k folds of symmetry; $k = 1, \ldots, K$; $N_1 > N_2 > \cdots N_K$. The set of proper invariants can be chosen as follows. Starting from the highest symmetry, we iteratively select those invariants providing (jointly with the invariants already selected) a sufficient discriminability between the classes with the fold numbers N_F and higher, but which may equal zero for some (or all) other classes. Note that for some F, the algorithm need not select any invariant because the discriminability can be assured by the invariants selected previously or because $C_F = 1$.

In order to illustrate the importance of a careful choice of invariant in pattern recognition tasks, we carried out the following experimental study.

2.5.1   Logo recognition

In the first experiment, we tested the capability of recognizing objects having the same number of folds, particularly $N = 3$. As a test set we used three logos of major companies

(Mercedes-Benz, Mitsubishi and Fischer) and two commonly used symbols ("recycling" and "woolen product"). All logos were downloaded from the respective websites, resampled to 128×128 pixels and binarized. We decided to use logos as the test objects because most logos have a certain degree of symmetry and all commercial logo/trademark recognition systems face the problem of symmetry. A comprehensive case study on trademark recognition and retrieval [17] used the Hu moment invariants as a preselector; here, we show that Theorem 2.8 yields more discriminative features.

As can be seen in Figure 2.9, all our test logos have threefold rotation symmetry. Each logo was rotated ten times by randomly generated angles. Since the spatial resolution of the images was relatively high, the discretization effect was negligible. Moment invariants from Theorem 2.8 ($N = 3$, $p_0 = 3$ and $q_0 = 0$) provide an excellent discrimination power even if only the two simplest are used (see Figure 2.10), while the invariants from Theorem 2.6 are not able to distinguish the marks at all (see Figure 2.11).

Figure 2.9 The test logos (from left to right): Mercedes-Benz, Mitsubishi, Recycling, Fischer, and Woolen product. (Reprinted from [12] with permission, © 2006 IEEE.)

2.5.2 Recognition of simple shapes

In the second experiment, we used nine simple binary patterns with various numbers of folds: capitals F and L ($N = 1$), rectangle and diamond ($N = 2$), equilateral triangle and tripod ($N = 3$), cross ($N = 4$), and circle and ring ($N = \infty$) (see Figure 2.12). As in the previous case, each pattern was ten times rotated by ten random angles.

First, we applied rotation invariants according to Theorem 2.6 choosing $p_0 = 2$ and $q_0 = 1$. The positions of our test patterns in the feature space are plotted in Figure 2.13. Although only a 2D subspace showing the invariants $c_{21}c_{12}$ and $\mathcal{Re}(c_{20}c_{12}^2)$ is visualized here, we can easily observe that the patterns form one dense cluster around the origin (the only exception being the tripod, which is slightly biased because of its nonsymmetry caused by the quantization effect). Two nonsymmetric objects – letters F and L – are far from the origin, beyond the displayed area. The only source of nonzero variance of the cluster are spatial quantization errors. All other invariants of the form $c_{pq}c_{12}^{p-q}$ behave in the same way. Thus, according to our theoretical expectation, we cannot discriminate among symmetric objects (even if they are very different) by means of invariants defined in Theorem 2.6.

Second, we employed the invariants introduced in Theorem 2.8, choosing $N = 4$ (the highest finite number of folds among the test objects), $p_0 = 4$ and $q_0 = 0$ to resolve the above recognition experiment. The situation in the feature space appears to be different from the previous case (see the plot of the two simplest invariants $c_{40}c_{04}$ and $\mathcal{Re}(c_{51}c_{04})$ in Figure 2.14). Five test patterns formed their own very compact clusters that are well separated from each other. However, the patterns circle, ring, triangle and tripod still made a mixed

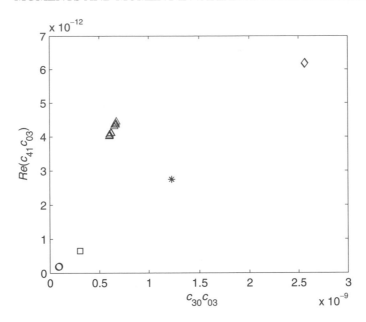

Figure 2.10 The logo positions in the space of two invariants $c_{30}c_{03}$ and $\mathcal{R}e(c_{41}c_{03})$ showing good discrimination power. The symbols: □ – Mercedes-Benz, ◇ – Mitsubishi, △ – Recycling, ∗ – Fischer, and ○ – Woolen product. (Reprinted from [12] with permission, © 2006 IEEE.)

cluster around the origin and remained nonseparable. This result is fully in accordance with the theory, because the number of folds used here is not optimal for our test set.

Third, we repeated this experiment again with invariants according to Theorem 2.8 but selecting N as the least common multiple of all finite fold numbers involved, i.e. $N = 12$. One can learn from Figure 2.15 that now all clusters are well separated (because of the high dynamic range, logarithmic scale was used for visualization purposes). The only exception are two patterns having circular symmetry – circle and ring – that still created a mixed cluster. If, also, these two patterns were to be separated from one another, we could use the invariants c_{pp}. On the other hand, using *only* these invariants for the whole experiment is not a good choice from the practical point of view – since there is only one such invariant for each order, we would be pushed into using high-order noise-sensitive moments.

Finally, we used the algorithm described at the end of section 2.5. In this case, two invariants $c_{30}c_{03}$ and $c_{40}c_{04}$ are sufficient to separate all classes (excepting, of course, the circle and the ring), see Figure 2.16. Compared to the previous case, note less correlation of the invariants, their higher robustness and lower dynamic range. On the other hand, neither $c_{30}c_{03}$ nor $c_{40}c_{04}$ provide enough discrimination power when used individually while the twelfth-order invariants are able to distinguish all classes.

2.5.3   Experiment with a baby toy

We demonstrate the performance of invariants in an object-matching task. We used a popular baby toy (see Figure 2.17) that is also commonly used in testing computer vision algorithms

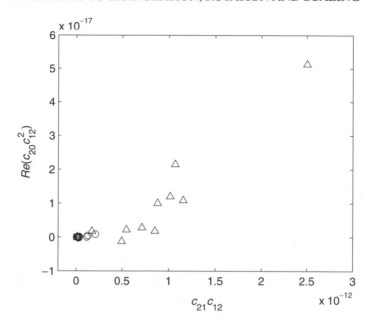

Figure 2.11 The logo positions in the space of two invariants $c_{21}c_{12}$ and $\mathcal{R}e(c_{20}c_{12}^2)$ introduced in Theorem 2.6. These invariants have no discrimination power with respect to this logo set. The symbols: $\square$ – Mercedes-Benz, $\diamond$ – Mitsubishi, $\triangle$ – Recycling, $*$ – Fischer, and $\circ$ – Woolen product. (Reprinted from [12] with permission, © 2006 IEEE.)

Figure 2.12 The test patterns: capital L, rectangle, equilateral triangle, circle, capital F, diamond, tripod, cross and ring. (Reprinted from [12] with permission, © 2006 IEEE.)

and robotic systems. The toy consists of a hollow sphere with 12 holes and of 12 objects of various shapes. Each object matches one particular hole. The baby (or the algorithm) is supposed to assign the objects to the corresponding holes and insert them into the sphere. The baby can employ both the color and shape information; however, in our experiment we completely disregarded the colors to make the task more difficult.

First, we binarized the pictures of the holes (one picture per each hole) by simple thresholding. Binarization was the only preprocessing; no sphere-to-plane corrections were applied.

To select proper invariants, we applied the algorithm from section 2.5 on the images of the holes. As a discriminability measure we took weighted Euclidean distance, where the weights were set up to normalize the dynamic range of the invariants. As one can observe, the highest

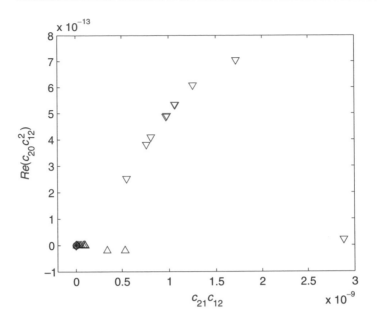

Figure 2.13 The space of two invariants $c_{21}c_{12}$ and $\mathcal{R}e(c_{20}c_{12}^2)$ introduced in Theorem 2.6. The symbols: $\times$ – rectangle, $\Diamond$ – diamond, $\triangle$ – equilateral triangle, $\triangledown$ – tripod $+$ – cross, $\cdot$ – circle and $\bigcirc$ – ring. (Reprinted from [12] with permission, © 2006 IEEE.)

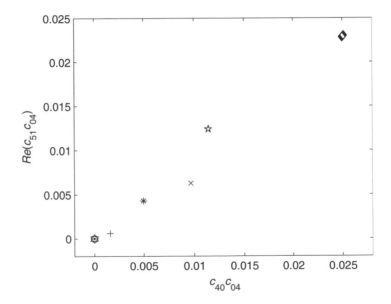

Figure 2.14   The space of two invariants $c_{40}c_{04}$ and $\mathcal{R}e(c_{51}c_{04})$ introduced in Theorem 2.8, $N = 4$. The symbols: $\times$ – rectangle, $\Diamond$ – diamond, $\triangle$ – equilateral triangle, $\triangledown$ – tripod $+$ – cross, $\cdot$ – circle, and $\bigcirc$ – ring, $*$ – capital F and $\star$ – capital L. (Reprinted from [12] with permission, © 2006 IEEE.)

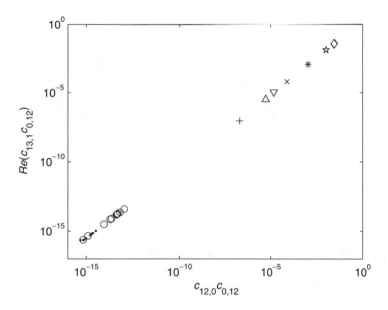

Figure 2.15 The space of two invariants $c_{12,0}c_{0,12}$ and $\mathcal{R}e(c_{13,1}c_{0,12})$ introduced in Theorem 2.8, $N = 12$ (logarithmic scale). The symbols: $\times$ – rectangle, $\diamond$ – diamond, $\triangle$ – equilateral triangle, ∇ – tripod + – cross, $\cdot$ – circle, and $\bigcirc$ – ring, $\ast$ – capital F and $\star$ – capital L. (Reprinted from [12] with permission, © 2006 IEEE.)

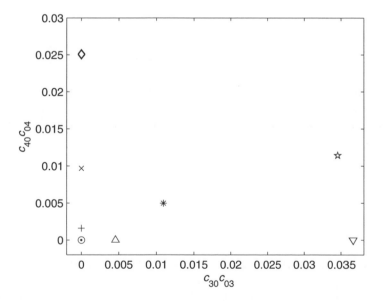

Figure 2.16 The space of two invariants $c_{3,0}c_{0,3}$ and $c_{40}c_{04}$. The symbols: $\times$ – rectangle, $\diamond$ – diamond, $\triangle$ – equilateral triangle, ∇ – tripod + – cross, $\cdot$ – circle, and $\bigcirc$ – ring, $\ast$ – capital F and $\star$ – capital L. Compared to Figure 2.15, note less correlation of the invariants and a lower dynamic range. (Reprinted from [12] with permission, © 2006 IEEE.)

Figure 2.17 The toy set used in the experiment. (Reprinted from [12] with permission, © 2006 IEEE.)

finite number of folds is 6. The algorithm terminated after passing three loops and selected the three following invariants: $c_{60}c_{06}$, $c_{50}c_{05}$ and $c_{40}c_{04}$.

Then we took ten pictures of each object with random rotations, binarized them and ran the classification. This task is not as easy as it might appear because the holes are a little larger than the objects but this relation is morphological rather than linear and does not preserve the shapes exactly. Fortunately, all 120 unknown objects were recognized correctly and assigned to proper holes. It should be emphasized that only three invariants without any other information yielded 100% recognition rate for 12 classes, which is a very good result even though the shapes are relatively simple.

We repeated the classification once again with invariants constructed according to Theorem 2.6 setting $p_0 = 2$, $q_0 = 1$. Only three objects having one-fold symmetry were recognized correctly; the others were classified randomly.

2.6 Rotation invariants via image normalization

TRS invariants can also be derived in a way other than by means of Theorem 2.6. This alternative method called *image normalization* first brings the image into some "standard" or "canonical" position, defined by setting certain image moments to predefined values (usually 0 or 1). Then, the moments of the normalized image (except those that were used to define normalization constraints) are used as TRS invariants of the original image. It should be noted that no actual geometric transformation and resampling of the image is required; the moments of the normalized image can be calculated directly from the original image by means of normalization constraints.

While normalization to translation and scaling is trivial by setting $m'_{10} = m'_{01} = 0$ and $m'_{00} = 1$, there are many possibilities of how to define the normalization constraints to rotation. The traditional one known as the *principal axis* method [1] requires the constraint $\mu'_{11} = 0$ for the normalized image. It leads to diagonalization of the second-order moment matrix

$$\mathbf{M} = \begin{pmatrix} \mu_{20} & \mu_{11} \\ \mu_{11} & \mu_{02} \end{pmatrix}. \tag{2.21}$$

It is well known that each symmetric matrix can be diagonalized on the basis of its (orthogonal) eigenvectors. After finding the eigenvalues of $\mathbf{M}$ (they are guaranteed to be real) by solving

$$|M - \lambda I| = 0$$

we obtain the diagonal form

$$\mathbf{M'} = \mathbf{G}^T \mathbf{M} \mathbf{G} = \begin{pmatrix} \lambda_1 & 0 \\ 0 & \lambda_2 \end{pmatrix} = \begin{pmatrix} \mu'_{20} & 0 \\ 0 & \mu'_{02} \end{pmatrix},$$

where $\mathbf{G}$ is an orthogonal matrix composed of the eigenvectors of $\mathbf{M}$. It is easy to show that for the eigenvalues, we obtain

$$\mu'_{20} \equiv \lambda_1 = ((\mu_{20} + \mu_{02}) + \sqrt{(\mu_{20} - \mu_{02})^2 + 4\mu_{11}^2})/2,$$

$$\mu'_{02} \equiv \lambda_2 = ((\mu_{20} + \mu_{02}) - \sqrt{(\mu_{20} - \mu_{02})^2 + 4\mu_{11}^2})/2,$$

and that the angle between the first eigenvector and the x-axis, which is actually the normalization angle, is[5]

$$\alpha = \frac{1}{2} \cdot \arctan\left(\frac{2\mu_{11}}{\mu_{20} - \mu_{02}}\right).$$

To remove ambiguity, we add the constraints $\mu'_{20} \geq \mu'_{02}$ and $\mu'_{30} > 0$. (This method cannot be used if $\mu_{11} = 0$ and $\mu_{20} = \mu_{02}$.) Now each moment of the normalized image μ'_{pq} is a rotation invariant of the original one. For the second-order moments, we can see a clear correspondence with the Hu invariants

$$\mu'_{20} = (\phi_1 + \sqrt{\phi_2})/2,$$

$$\mu'_{02} = (\phi_1 - \sqrt{\phi_2})/2.$$

The principal axis normalization is linked with a notion of the *reference ellipse*, which is a (binary) ellipse having the same first- and second-order moments as the image (the zero-order moments may differ from one another). After the principal axis normalization, the reference

---

[5]Some authors define the normalization angle with an opposite sign. Both ways are, however, equivalent – we may "rotate" either the image or the coordinates.

ellipse is in axial position and we have, for its moments,

$$\mu_{11}^{(e)} = 0,$$

$$\mu_{20}^{(e)} = \frac{\pi a^3 b}{4},$$

$$\mu_{02}^{(e)} = \frac{\pi a b^3}{4},$$

where a is the length of the major semiaxis and b of the minor semiaxis. All images having the same first- and second-order moments have identical reference ellipses. Conversely, by knowing the first- and second-order image moments, we can reconstruct the reference ellipse only, without any finer image details.

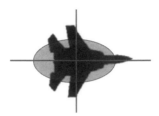

Figure 2.18 Principal axis normalization to rotation – an object in the normalized position along with its superimposed reference ellipse.

Abu-Mostafa and Psaltis [18] proposed a general normalization scheme by means of complex moments. Their normalization constraint requires one complex moment c_{st}, where $s \neq t$, of the image to be real and positive. As follows from equation (2.10), this is always possible to achieve provided that $c_{st} \neq 0$. Then, the constraint leads to a normalization angle

$$\alpha = \frac{1}{s-t} \cdot \arctan \left(\frac{\mathcal{Im}(c_{st})}{\mathcal{Re}(c_{st})} \right). \tag{2.22}$$

This normalization is unambiguous only if $s - t = 1$. For $s - t > 1$, additional constraints are required. Note that the principal axis normalization is just a special case of (2.22) having $s = 2$ and $t = 0$.

The selection of the normalizing moment c_{st} is a critical step and in practice it must be done very carefully. If we calculated the normalization angle with an error, this error would propagate and affect all invariants. Even a very small error in the normalizing angle can influence some invariants significantly. The nonzero constraint is easy to check in a continuous domain but in a discrete case some moments that are in fact zero due to the object symmetry may appear in certain positions of the object as nonzero because of the quantization effect. We have to identify such moments and avoid their usage in the normalization. At the same time, we want to keep the moment order $s + t$ as low as possible in order to make the normalization robust.

The normalizing moment can be found as follows. We sort all complex moments (except those that were already used for translation and scaling normalization) up to the given order r

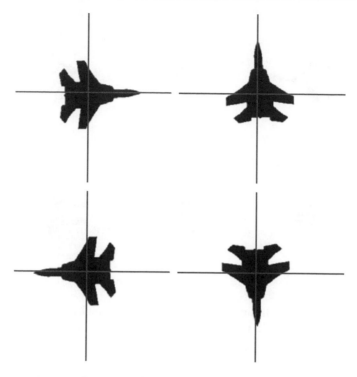

Figure 2.19 Ambiguity of the principal axis normalization. These four positions of the object satisfy $\mu'_{11} = 0$. Additional constraints $\mu'_{20} \geq \mu'_{02}$ and $\mu'_{30} > 0$ make the normalization unique.

according to the index difference $p - q$ and, those moments with the same index differences, according to the order. We obtain an ordered sequence $c_{21}, c_{32}, c_{43}, \ldots, c_{20}, c_{31}, c_{42},$ $c_{53}, \ldots$, etc. The first nonzero moment in this sequence is then used for normalization. To test if the moment is nonzero, we compare its value with a predefined threshold. The choice of the threshold must reflect the magnitudes of the moments in question and, especially when considering a high r, it may be useful to define different thresholds for different moment orders. If all moments in the sequence are zero, we consider the object circularly symmetric and no normalization is necessary.

One can, of course, use another ordering scheme when looking for the normalizing moment. An alternative is to sort the moments the other way round – first according to their orders and the moments of the same order according to the index differences. Both methods have certain advantages and drawbacks. For nonsymmetric objects there is no difference because we always choose c_{21}. For symmetric and close-to-symmetric objects, when we sort the moments according to their orders first, then the order of the chosen moment is kept as low as possible. This is a favorable property because low-order moments are more robust to noise than higher-order ones. On the other hand, one can find objects where this approach fails. Sorting moments according to the index difference first generally leads to a higher-order normalization constraint that is numerically less stable but from the theoretical point of view works for any object.

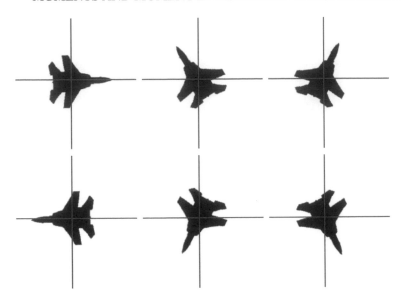

Figure 2.20 An example of the ambiguity of the normalization by complex moments. In all of these six positions, c_{60}' is real and positive, as required.

There have been many discussions on the advantages and drawbacks of both approaches – moment invariants and image normalization. Here we show that, from the theoretical point of view, they are fully equivalent and that the normalizing moment c_{st} and the basic moment $c_{p_0 q_0}$ play a very similar role.

Let us consider the basis of rotation invariants designed according to Theorem 2.6. We can apply the Mostafa and Psaltis normalization method with $c_{st} = c_{p_0 q_0}$. Since each $\Phi(p, q)$ is a rotation invariant, it holds

$$\Phi(p, q) = c_{pq}' c_{q_0 p_0}'^{p-q},$$

where c_{pq}' denotes the moments of the normalized image. Consequently, since $c_{q_0 p_0}'$ is real and positive, we obtain

$$\Phi(p, q) = c_{pq}' |c_{p_0 q_0}|^{p-q}. \qquad (2.23)$$

Equation (2.23) shows that there is a one-to-one mapping between moment invariants $\Phi(p, q)$ of the original image and complex moments c_{pq}' of the normalized image. An analogous result can be obtained for symmetric objects and the invariants from Theorem 2.8. Thus, the normalization approach and the invariant-based approach are equivalent (at least on the theoretical level; computational issues in particular applications might be different).

2.7   Invariants to nonuniform scaling

Nonuniform scaling is a transform beyond the TRS framework that maps a unit square onto a rectangle. It is defined as

$$x' = ax,$$

$$y' = by,$$

where $a \neq b$ are positive scaling factors. Invariants to nonuniform scaling are sometimes called *aspect-ratio invariants*.

While complex moments change in a rather complicated way under this transform, geometric and central moments change simply as

$$\mu'_{pq} = \int_{-\infty}^{\infty} \int_{-\infty}^{\infty} (x' - x'_c)^p (y' - y'_c)^q f'(x', y') \, dx' \, dy'$$

$$= \int_{-\infty}^{\infty} \int_{-\infty}^{\infty} a^p (x - x_c)^p b^q (y - y_c)^q f(x, y) ab \, dx \, dy = a^{p+1} b^{q+1} \mu_{pq}.$$

To eliminate the scaling factors a and b, we need at least two normalizing moments. When using, for instance, μ_{00} and μ_{20}, we obtain the invariants

$$A_{pq} = \frac{\mu_{pq}}{\mu_{00}^{\alpha} \mu_{20}^{\beta}},$$

where

$$\alpha = \frac{3q - p}{2} + 1$$

and

$$\beta = \frac{p - q}{2}.$$

One can derive many different invariants to nonuniform scaling. For instance, Pan and Keane [19] proposed more "symmetric" normalization by three moments μ_{00}, μ_{20} and μ_{02}, which leads to

$$S_{pq} = \frac{\mu_{00}^{(p+q+2)/2}}{\mu_{20}^{(p+1)/2} \cdot \mu_{02}^{(q+1)/2}} \cdot \mu_{pq}.$$

Invariance to nonuniform scaling cannot be combined in a simple way with rotation invariance. The reason is that translation, rotation and nonuniform scaling are not closed operations. When applied repeatedly, they generate affine group of transformations, implying that if we want to have invariance to nonuniform scaling and rotation, we must use affine invariants (they will be introduced in the next chapter). There is no transformation group "between" the TRS and affine groups and, consequently, no special set of invariants "between" TRS and affine moment invariants. This is why the invariants to nonuniform scaling described above are of only little importance in practical applications.

2.8   TRS invariants in 3D

So far, we have been dealing with 2D invariants because most images are 2D, even if they depict a projection of a 3D scene. However, in some applications (when using active range finders and in medicine when analyzing CT and MRI images) we can encounter true 3D images – image cubes. In those cases, 3D moment invariants may help to classify such objects.

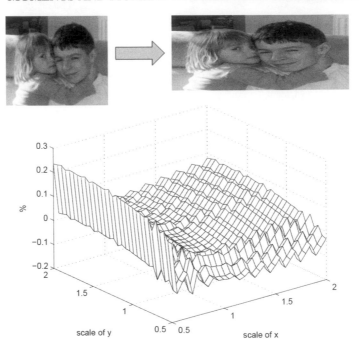

Figure 2.21 Numerical test of the aspect-ratio invariant A_{22}. Computer-generated scaling of the test image ranged from 0.5 to 2 in both directions independently. Horizontal axes: scaling factors a (the scale of x) and b (the scale of y), respectively. Vertical axis – relative deviation (in %) between A_{22} of the original and that of the scaled image. The test illustrates the invariance of A_{22}. Higher relative errors for low scaling factors and the typical jagged surface of the graph are the consequences of the image resampling.

In comparison with a huge number of papers on 2D moment invariants, only few papers on 3D and n-D invariants have been published. The first attempt to extend 2D moment invariants to 3D was carried out by Sadjadi and Hall [20]. Their results were later rediscovered (with some modifications) by Guo [21], who derived only three invariants without any possibility of further extension. Probably the first systematic approach to derivation of 3D moment invariants to rotation was published by Lo and Don [22]. It was based on group representation theory. Galvez and Canton [23] derived essentially the same invariants by the normalization approach.

Since rotation in 3D is more complicated than in 2D (spherical harmonics are used instead of complex moments) we do not present a comprehensive derivation of 3D rotation invariants here. The reader can find derivation of 3D affine invariants in the next chapter, rotation invariants being their special cases. Here, we present three very simple 3D TRS invariants for illustration.

Geometric and central moments in 3D are defined analogously as in 2D

$$m_{pqr} = \int\limits_{-\infty}^{\infty} \int\limits_{-\infty}^{\infty} \int\limits_{-\infty}^{\infty} x^p y^q z^r f(x, y, z) \, \mathrm{d}x \, \mathrm{d}y \, \mathrm{d}z \qquad (2.24)$$

and

$$\mu_{pqr} = \int\limits_{-\infty}^{\infty} \int\limits_{-\infty}^{\infty} \int\limits_{-\infty}^{\infty} (x - x_c)^p (y - y_c)^q (z - z_c)^r f(x, y, z) \, dx \, dy \, dz. \qquad (2.25)$$

Invariance to uniform scaling is also achieved by a very similar normalization

$$\nu_{pqr} = \frac{\mu_{pqr}}{\mu_{000}^w}, \qquad (2.26)$$

where

$$w = \frac{p + q + r}{3} + 1. \qquad (2.27)$$

Rotation invariants of the second order are

$$\psi_1 = \mu_{200} + \mu_{020} + \mu_{002},$$

$$\psi_2 = \mu_{020}\mu_{002} + \mu_{200}\mu_{002} + \mu_{200}\mu_{020} - (\mu_{011}^2 + \mu_{101}^2 + \mu_{110}^2), \qquad (2.28)$$

$$\psi_3 = \mu_{200}\mu_{020}\mu_{002} + 2\mu_{110}\mu_{101}\mu_{011} - \mu_{200}\mu_{011}^2 - \mu_{020}\mu_{101}^2 - \mu_{002}\mu_{110}^2.$$

As in 2D, when using ν_{pqr} instead of μ_{pqr}, we obtain TRS invariants in 3D.

2.9 Conclusion

In this chapter, we introduced moment invariants with respect to the TRS transform. We described a general theory showing how to generate invariants of any order. We defined the "basis of invariants" as the smallest complete subset, we showed its importance and described an explicit method of how to find it. As a consequence of this theory, we proved that some traditional invariant sets are dependent and/or incomplete, which explains certain failures reported in the literature. It was shown that the proposed invariants outperform the widely used Hu moment invariants both in discrimination power and dimensionality requirements.

Furthermore, we described invariants to transforms beyond the TRS – nonuniform scaling and contrast stretching. We discussed the difficulties with recognition of symmetric objects, where most moment invariants vanish, and we presented moment invariants suitable for such cases. Finally, we briefly reviewed an alternative approach to constructing invariants via normalization and showed that both methods are theoretically equivalent.

References

[1] Hu, M.-K. (1962) "Visual pattern recognition by moment invariants," *IRE Transactions on Information Theory*, vol. 8, no. 2, pp. 179–87.

[2] Li, Y. (1992) "Reforming the theory of invariant moments for pattern recognition," *Pattern Recognition*, vol. 25, no. 7, pp. 723–30.

[3] Wong, W.-H., Siu, W.-C. and Lam, K.-M. (1995) "Generation of moment invariants and their uses for character recognition," *Pattern Recognition Letters*, vol. 16, no. 2, pp. 115–23.

[4] Jin, L. and Tianxu, Z. (2004) "Fast algorithm for generation of moment invariants," *Pattern Recognition*, vol. 37, no. 8, pp. 1745–56.

[5] Teague, M. R. (1980) "Image analysis via the general theory of moments," *Journal of the Optical Society of America*, vol. 70, no. 8, pp. 920–30.

[6] Wallin, A. and Kübler, O. (1995) "Complete sets of complex Zernike moment invariants and the role of the pseudoinvariants," *IEEE Transactions on Pattern Analysis and Machine Intelligence*, vol. 17, no. 11, pp. 1106–10.

[7] Abu-Mostafa, Y. S. and Psaltis, D. (1984) "Recognitive aspects of moment invariants," *IEEE Transactions on Pattern Analysis and Machine Intelligence*, vol. 6, no. 6, pp. 698–706.

[8] Flusser, J. (2000) "On the independence of rotation moment invariants," *Pattern Recognition*, vol. 33, no. 9, pp. 1405–10.

[9] Flusser, J. (2002) "On the inverse problem of rotation moment invariants," *Pattern Recognition*, vol. 35, no. 12, pp. 3015–17.

[10] Bhattacharya, D. and Sinha, S. (1997) "Invariance of stereo images via theory of complex moments," *Pattern Recognition*, vol. 30, no. 9, pp. 1373–86.

[11] Mukundan, R. and Ramakrishnan, K. R. (1998) *Moment Functions in Image Analysis*. Singapore: World Scientific.

[12] Flusser, J. and Suk, T. (2006) "Rotation moment invariants for recognition of symmetric objects," *IEEE Transactions on Image Processing*, vol. 15, no. 12, pp. 3784–90.

[13] Shen, D., Ip, H. H.-S. and Teoh, E. K. (2000) "A novel theorem on symmetries of 2D images," in *Proceedings of the 15th International Conference on Pattern Recognition ICPR'00*, vol. 3, pp. 1014–17, IEEE Computer Society.

[14] Lin, J., Tsai, W. and Chen, J. (1994) "Detecting number of folds by a simple mathematical property," *Pattern Recognition Letters*, vol. 15, no. 11, pp. 1081–88.

[15] Lin, J. (1996) "A simplified fold number detector for shapes with monotonic radii," *Pattern Recognition*, vol. 29, no. 6, pp. 997–1005.

[16] Shen, D., Ip, H. H.-S., Cheung, K. K. T. and Teoh, E. K. (1999) "Symmetry detection by generalized complex (GC) moments: A close-form solution," *IEEE Transactions on Pattern Analysis and Machine Intelligence*, vol. 21, no. 5, pp. 466–76.

[17] Jain, A. and Vailaya, A. (1998) "Shape-based retrieval: A case study with trademark image databases," *Pattern Recognition*, vol. 31, no. 9, pp. 1369–90.

[18] Abu-Mostafa, Y. S. and Psaltis, D. (1985) "Image normalization by complex moments," *IEEE Transactions on Pattern Analysis and Machine Intelligence*, vol. 7, no. 1, pp. 46–55.

[19] Pan, F. and Keane, M. (1994) "A new set of moment invariants for handwritten numeral recognition," in *Proceedings of the International Conference on Image Processing ICIP'94* (Austin, Texas), pp. 154–8, IEEE Computer Society.

[20] Sadjadi, F. A. and Hall, E. L. (1980) "Three dimensional moment invariants," *IEEE Transactions on Pattern Analysis and Machine Intelligence*, vol. 2, no. 2, pp. 127–36.

[21] Guo, X. (1993) "Three dimensional moment invariants under rigid transformation," in *Proceedings of the Fifth International Conference on Computer Analysis of Images and Patterns CAIP'93* (Budapest, Hungary), LNCS vol. 719 (D. Chetverikov and W. Kropatsch, eds), pp. 518–22, Springer.

[22] Lo, C.-H. and Don, H.-S., (1989) "3-D moment forms: Their construction and application to object identification and positioning," *IEEE Transactions on Pattern Analysis and Machine Intelligence*, vol. 11, no. 10, pp. 1053–64.

[23] Galvez, J. M. and Canton, M. (1993) "Normalization and shape recognition of three-dimensional objects by 3D moments," *Pattern Recognition*, vol. 26, no. 5, pp. 667–81.

3

Affine moment invariants

3.1 Introduction

This chapter is devoted to *affine moment invariants* (AMIs), which play a very important role in moment-based theories. They are invariant with respect to affine transform of the spatial coordinates. We explain why the AMIs are important in image analysis and we present three different methods that allow a systematic (and automatic) derivation of AMIs of any order – a graph method, a normalization method and a method based on the Cayley–Aronhold equation. We discuss the differences between them and use instruments to illustrate the numerical properties of the AMIs.

3.1.1 Projective imaging of a 3D world

In a vast majority of practical cases, pictures of 3D scenes or planar scenes are taken placed arbitrarily in a 3D environment. In this way 3D objects and structures are represented by their projections onto a 2D plane because photography is a 2D medium (here, we deal neither with intrinsically 3D imaging modalities such as computer tomography, magnetic resonance imaging and other medical imaging devices, nor with a coupling of camera and rangefinder, that actually yield 3D information). In such cases, we often face object deformations that are beyond the translation-rotation-scaling model (see Figure 3.1).

An exact model of photographing a planar scene by a pinhole camera whose optical axis is not perpendicular to the scene is a *projective transform* (sometimes also called *perspective projection*)

$$
\begin{aligned}
x' &= \frac{a_0 + a_1 x + a_2 y}{1 + c_1 x + c_2 y}, \\
y' &= \frac{b_0 + b_1 x + b_2 y}{1 + c_1 x + c_2 y}.
\end{aligned}
\tag{3.1}
$$

Jacobian J of the projective transform is

$$J = \frac{(a_1b_2 - a_2b_1) + (a_2b_0 - a_0b_2)c_1 + (a_0b_1 - a_1b_0)c_2}{(1 + c_1x + c_2y)^3}$$

$$= \frac{\begin{vmatrix} a_0 & a_1 & a_2 \\ b_0 & b_1 & b_2 \\ 1 & c_1 & c_2 \end{vmatrix}}{(1 + c_1x + c_2y)^3}. \tag{3.2}$$

Figure 3.1 The projective deformation of a 3D scene due to a nonperpendicular view. Square tiles appear as quadrilaterals; the transformation preserves straight lines but does not preserve their collinearity.

The projective transform is not a linear transform but exhibits certain linear properties. It maps a square onto a general quadrilateral (see Figure 3.2) and, importantly, it preserves straight lines. However, it does not preserve their collinearity.

3.1.2   Projective moment invariants

Projective invariants are in great demand, especially in computer vision and robotics. Unfortunately, construction of moment invariants to projective transform is practically impossible. Major difficulties with projective moment invariants originate from the fact that projective transform is not linear; its Jacobian is a function of spatial coordinates and the transform does not preserve the center of gravity of the object. The tools that have been

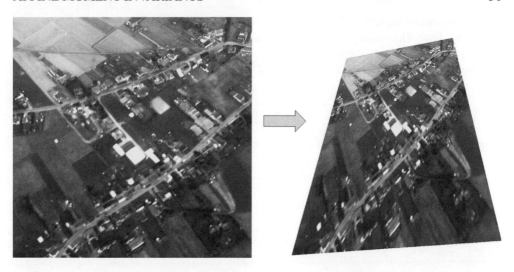

Figure 3.2 The projective transform maps a square onto a quadrilateral (computer-generated example).

successfully applied to derive moment invariants to various linear transformations cannot be exploited in the case of projective invariants.[1]

Van Gool *et al.* [9] employed the Lie group theory to prove the nonexistence of projective invariants from a finite set of moments. Here, we present another proof that provides a better insight and does not use the group theory tools.

Let us decompose the projective transform (3.1) into eight one-parametric transformations:

horizontal and vertical translations

$$(a) \quad \begin{aligned} x' &= x + \alpha \\ y' &= y \end{aligned} \qquad\qquad (b) \quad \begin{aligned} x' &= x \\ y' &= y + \beta, \end{aligned}$$

scaling and stretching

$$(c) \quad \begin{aligned} x' &= \omega x \\ y' &= \omega y \end{aligned} \qquad\qquad (d) \quad \begin{aligned} x' &= r x \\ y' &= \tfrac{1}{r} y, \end{aligned}$$

horizontal and vertical skewing

$$(e) \quad \begin{aligned} x' &= x + t_1 y \\ y' &= y \end{aligned} \qquad\qquad (f) \quad \begin{aligned} x' &= x \\ y' &= t_2 x + y, \end{aligned}$$

and horizontal and vertical pure projections

$$(g) \quad \begin{aligned} x' &= \frac{x}{1 + c_1 x} \\ y' &= \frac{y}{1 + c_1 x} \end{aligned} \qquad\qquad (h) \quad \begin{aligned} x' &= \frac{x}{1 + c_2 y} \\ y' &= \frac{y}{1 + c_2 y}. \end{aligned}$$

---

[1]Moments are not the only possibility to construct projective invariants. Differential invariants based on local shape properties were described in [1–3] and various invariants defined by means of salient points [4–8]. Since they are of different nature and have different usage than global moment invariants, we will not discuss them further.

Any projective invariant F would have to be invariant to all elementary transformations $(a) - (h)$. Each elementary transform imposes special constraints on F. Particularly, from the invariance to horizontal pure projection (g) it follows that the derivative of F with respect to parameter c_1 must be zero (assuming that all derivatives exist):

$$\frac{\mathrm{d}F}{\mathrm{d}c_1} = \sum_p \sum_q \frac{\partial F}{\partial m_{pq}} \frac{\mathrm{d}m_{pq}}{\mathrm{d}c_1} = 0. \tag{3.3}$$

The derivatives of moments can be expressed as

$$\frac{\mathrm{d}m_{pq}}{\mathrm{d}c_1} = \frac{\mathrm{d}}{\mathrm{d}c_1} \iint_S \frac{x^p y^q}{(1+c_1 x)^{p+q}} \frac{1}{|1+c_1 x|^3} f(x, y) \, \mathrm{d}x \, \mathrm{d}y$$

$$= \iint_S (-p - q - 3) \frac{x^{p+1} y^q}{(1+c_1 x)^{p+q+1}} \frac{1}{|1+c_1 x|^3} f(x, y) \, \mathrm{d}x \, \mathrm{d}y \tag{3.4}$$

$$= -(p + q + 3)m_{p+1,q}.$$

Thus, equation (3.3) becomes

$$-\sum_p \sum_q (p + q + 3)m_{p+1,q} \frac{\partial F}{\partial m_{pq}} = 0. \tag{3.5}$$

The constraint (3.5) must be fulfilled for any image. Assuming that F contains only a finite number of moments, we denote their maximum order as r. However, equation (3.5) contains moments up to the order $r + 1$. If equation (3.5) was satisfied for a certain image, we could always construct another image with identical moments up to the order r and different moments of the order $r + 1$ such that equation (3.5) does not hold for this new image. Thus, finite projective moment invariants cannot exist.[2]

This result can be extended to prove the nonexistence of invariants, that would have a form of infinite series having each term in a form of a finite product of moments of nonnegative indices.

Suk and Flusser [12] found projective moment invariants in a form of infinite series of products of moments with both positive and negative indices. However, practical applicability of these invariants is very limited because of their high dynamic range and slow convergence.

3.1.3 Affine transformation

Affine transformation is a general linear transform of spatial coordinates of the image, which can – under certain circumstances – approximate the projective transform. Due to this property, it is of prime importance in image analysis and has been thoroughly studied.[3] Unlike projective moment invariants, AMIs from a finite set of moments do exist and can be used in practice.

---

[2]Some authors proved that finite moment-like projective invariants do exist for certain particular projective transforms (see, for instance, reference [10] for the case $a_0 = a_2 = b_0 = b_1 = 0$ and reference [11] for the case $a_0 = a_2 = b_0 = 0$). Unfortunately, those special forms of invariant cannot be generalized to the projective transform (3.1).

[3]Pure affine transform appears rarely in practice. An example is a skewing of satellite images taken by line scanners, caused by Earth rotation.

Affine transformation can be expressed as

$$x' = a_0 + a_1 x + a_2 y$$
$$y' = b_0 + b_1 x + b_2 y$$

(3.6)

and in a matrix form

$$\mathbf{x}' = \mathbf{A}\mathbf{x} + \mathbf{b},$$

where $\mathbf{A} = \begin{pmatrix} a_1 & a_2 \\ b_1 & b_2 \end{pmatrix}$ and $\mathbf{b} = \begin{pmatrix} a_0 \\ b_0 \end{pmatrix}$.

(3.7)

Affine transform maps a square onto a parallelogram and preserves collinearity (see Figure 3.3). Apparently, TRS is just a special case of the affine transform when $a_1 = b_2$ and $a_2 = -b_1$. The Jacobian of the affine transform is $J = a_1 b_2 - a_2 b_1$. Contrary to the projective transform, J does not depend on the spatial coordinates x and y, which makes searching for invariants easier. Affine transform is a particular case of projective transform with $c_1 = c_2 = 0$. If the object is small compared to the camera-to-scene distance, then both c_1 and c_2 approach zero, the perspective effect becomes negligible and the affine model is a reasonable approximation of the projective model. This is why the affine transform and affine invariants are so important in computer vision.

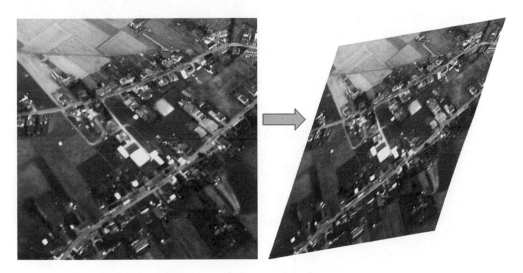

Figure 3.3 The affine transform maps a square to a parallelogram.

3.1.4 AMIs

For the reasons mentioned above, AMIs play an important role in view-independent object recognition and have been widely used not only in tasks where image deformation is intrinsically affine but also commonly substitute projective invariants.

The history of AMIs began almost 50 years ago and is rather complicated. It is unclear who published the first correct and usable set of AMIs. The first attempt to derive moment invariants to affine transform had already been presented in Hu's first paper [13] but the *Fundamental theorem of affine invariants* was stated incorrectly there. Thirty years later, Reiss [14] and Flusser and Suk [15, 16] independently discovered and corrected this mistake, published new sets of AMIs and proved their applicability in simple recognition tasks. Here, the derivation of the AMIs originated from the traditional theory of algebraic invariants from the nineteenth century, e.g. references [17–21].[4] The same results achieved by a slightly different approach were later published by Mamistvalov [22].[5]

The AMIs can be derived in several ways that differ from each other in the mathematical tools used. Apart from the above-mentioned algebraic invariants, one may use graph theory, tensor algebra, direct solution of proper partial differential equations and derivation via image normalization. In this chapter, we review the most important approaches and show that the invariants created by different methods are equivalent. A common point of all automatic methods is that they can generate as many invariants as one wishes but only a few of them are independent. Since dependent invariants are useless in practice, much effort has been spent on their elimination; we mention the most successful methods here.

3.2   AMIs derived from the Fundamental theorem

The theory of AMIs is closely connected to the theory of algebraic invariants and the Fundamental theorem describes this connection. The algebraic invariant is a polynomial of coefficients of a binary form, whose value remains the same after an affine transform of the coordinates. In the theory of algebraic invariants, only the transforms without translation (i.e. $\mathbf{b} = 0$) are considered.

The *binary algebraic form* of order p is defined as a homogeneous polynomial

$$\sum_{k=0}^{p} \binom{p}{k} a_k x^{p-k} y^k \tag{3.8}$$

of the coordinates $\mathbf{x}$. The numbers a_k are called coefficients of the binary form. After the affine transform

$$\mathbf{x}' = \mathbf{A}\mathbf{x} \tag{3.9}$$

the coefficients are changed and the binary form becomes

$$\sum_{k=0}^{p} \binom{p}{k} a_k' x'^{p-k} y'^k. \tag{3.10}$$

Hilbert [21] defined an *algebraic invariant* as a polynomial function of the coefficients a, $b, \ldots$ of binary forms that satisfies the equation

$$I(a_0', a_1', \ldots, a_{p_a}'; b_0', b_1', \ldots, b_{p_b}'; \ldots) = J^w I(a_0, a_1, \ldots, a_{p_a}; b_0, b_1, \ldots, b_{p_b}; \ldots). \tag{3.11}$$

---

[4]The fundamental book [21] is based on the original notes of the course held by David Hilbert in 1897 in Gottingen, Germany. The notes were discovered 50 years after Hilbert's death, translated into English and first published in this book in 1993.

[5]The author had probably already published the same results in 1970 [23] but in Russian and not commonly available.

The w is called the *weight* of the invariant. If there is only one binary form with coefficients a, the invariant is called *homogeneous*, otherwise it is called *simultaneous*. If $w = 1$, the invariant is called *absolute*, otherwise it is called *relative*. The Fundamental theorem can be formulated as follows.

Theorem 3.1 (Fundamental theorem of AMIs) *If the binary forms of orders* p_a, p_b, $\ldots$ *have an algebraic invariant of weight w and degree r*

$$I(a'_0, a'_1, \ldots, a'_{p_a}; b'_0, b'_1, \ldots, b'_{p_b}; \ldots) = J^w I(a_0, a_1, \ldots, a_{p_a}; b_0, b_1, \ldots, b_{p_b}; \ldots),$$

then the moments of the same orders have the same invariant but with the additional factor $|J|^r$:

$$I(\mu'_{p_a 0}, \mu'_{p_a-1,1}, \ldots, \mu'_{0p_a}; \mu'_{p_b 0}, \mu'_{p_b-1,1}, \ldots, \mu'_{0p_b}; \ldots)$$
$$= J^w |J|^r I(\mu_{p_a 0}, \mu_{p_a-1,1}, \ldots, \mu_{0p_a}; \mu_{p_b 0}, \mu_{p_b-1,1}, \ldots, \mu_{0p_b}; \ldots).$$

A traditional proof of this theorem[6] can be found in reference [14]. A 3D version of this theorem along with its proof was published in reference [22]. In Appendix 3.A, we present another, much simpler proof using the Fourier transform.

The power of the Fundamental theorem depends on our ability to find algebraic invariants of binary forms. Having an algebraic invariant, we can construct a corresponding AMI easily just by interchanging the coefficients and the central moments and normalizing the invariant by μ_{00}^{w+r} to eliminate the factor $J^w \cdot |J|^r$ (if $J < 0$ and w is odd, then after this normalization we obtain only a pseudoinvariant, i.e. $I' = -I$). However, the theorem does not provide any instructions on how to find algebraic invariants.

3.3 AMIs generated by graphs

The simplest and most transparent way for AMIs of any orders and weights to be generated systematically is probably based on representing invariants by graphs. This *graph method* was proposed in reference [24] and is similar to an earlier *tensor method* [25]. One of the main advantages of the graph method is that it provides an insight into the structure of the invariants and allows the dependencies among them to be eliminated.

3.3.1 The basic concept

Let us consider an image f and two arbitrary points (x_1, y_1), (x_2, y_2) from its support. Let us denote the "cross-product" of these points as C_{12}:

$$C_{12} = x_1 y_2 - x_2 y_1.$$

After an affine transform (as in the previous section, we consider no shift) it holds $C'_{12} = J \cdot C_{12}$, which means that C_{12} is a relative affine invariant.[7] We consider various numbers of

---

[6]Probably, the first version of this theorem appeared in Hu's original paper [13] but was stated incorrectly without the factor $|J|^r$. That is why the Hu affine invariants are not actually invariant to a general affine transform; they perform well only if $J = 1$ or $r = 0$.

[7]The geometric meaning of C_{12} is the oriented double area of the triangle, whose vertices are (x_1, y_1), (x_2, y_2) and $(0, 0)$.

points and we integrate their cross-products (or some powers of their cross-products) on the support of f. These integrals can be expressed in terms of moments and, after eliminating the Jacobian by proper normalization, they yield affine invariants.

More precisely, having r points ($r \geq 2$) we define functional I depending on r and on non-negative integers n_{kj} as

$$I(f) = \int_{-\infty}^{\infty} \int_{-\infty}^{\infty} \prod_{k,j=1}^{r} C_{kj}^{n_{kj}} \cdot \prod_{i=1}^{r} f(x_i, y_i) \, dx_i \, dy_i. \tag{3.12}$$

Note that it is meaningful to consider only $j > k$, because $C_{kj} = -C_{jk}$ and $C_{kk} = 0$. After an affine transform, $I(f)$ becomes

$$I(f)' = J^w |J|^r \cdot I(f),$$

where $w = \sum_{k,j} n_{kj}$ is called the *weight* of the invariant and r is called the *degree* of the invariant. If $I(f)$ is normalized by μ_{00}^{w+r}, we obtain a desirable absolute affine invariant

$$\left(\frac{I(f)}{\mu_{00}^{w+r}} \right)' = \left(\frac{I(f)}{\mu_{00}^{w+r}} \right)$$

(if w is odd and $J < 0$, there is an additional factor -1).

The maximum order of moments of which the invariant is composed is called the *order* of the invariant. The order is always less than or equal to the weight. Another important characteristic of the invariant is its *structure*. The structure of the invariant is defined by an integer vector $s = (k_2, k_3, \ldots, k_s)$, where s is the invariant order and k_j is the total number of moments of the jth order contained in each term of the invariant (all terms have the same structure).[8]

We illustrate the general formula (3.12) on two simple invariants. First, let $r = 2$ and $w = n_{12} = 2$. Then

$$I(f) = \int_{-\infty}^{\infty} \int_{-\infty}^{\infty} (x_1 y_2 - x_2 y_1)^2 f(x_1, y_1) f(x_2, y_2) \, dx_1 \, dy_1 \, dx_2 \, dy_2 = 2(m_{20} m_{02} - m_{11}^2). \tag{3.13}$$

When replacing the geometric moments by corresponding central moments and normalizing the invariant by μ_{00}^4 we obtain a "full" invariant to general affine transform

$$I_1 = (\mu_{20} \mu_{02} - \mu_{11}^2)/\mu_{00}^4.$$

This is the simplest affine invariant, uniquely containing the second-order moments only. Its structure is (2). In this form it is commonly referred to in the literature regardless of the method used for derivation.

---

[8] Since always $k_0 = k_1 = 0$, these two quantities are not included in the structure vector.

Similarly, for $r = 3$ and $n_{12} = 2$, $n_{13} = 2$, $n_{23} = 0$ we obtain

$$I(f) = \int_{-\infty}^{\infty} \int_{-\infty}^{\infty} (x_1 y_2 - x_2 y_1)^2 (x_1 y_3 - x_3 y_1)^2$$

$$\times f(x_1, y_1) f(x_2, y_2) f(x_3, y_3) \, dx_1 \, dy_1 \, dx_2 \, dy_2 \, dx_3 \, dy_3$$

$$= m_{20}^2 m_{04} - 4 m_{20} m_{11} m_{13} + 2 m_{20} m_{02} m_{22} + 4 m_{11}^2 m_{22} - 4 m_{11} m_{02} m_{31} + m_{02}^2 m_{40}$$

$$(3.14)$$

and the normalizing factor is in this case μ_{00}^7. The weight of this invariant is $w = 4$, the order $s = 4$ and its structure is $(2, 0, 1)$.

3.3.2 Representing the invariants by graphs

Each invariant generated by the formula (3.12) can be represented by a connected graph, where each point (x_k, y_k) corresponds to one node and each cross-product C_{kj} corresponds to one edge of the graph. If $n_{kj} > 1$, the respective term $C_{kj}^{n_{kj}}$ corresponds to n_{kj} edges connecting kth and jth nodes (in graph theory, such graphs with multiple edges are called *multigraphs*). Thus, the number of nodes equals the degree of the invariant and the total number of the graph edges equals the weight w of the invariant. From the graph one can also learn about the orders of the moments of which the invariant is composed and about its structure. The number of edges originating from each node equals the order of the moments involved. Each invariant of the form (3.12) is in fact a sum where each term is a product of a certain number of moments. This number, the degree of the invariant, is constant for all terms of one particular invariant and is equal to the total number of graph nodes.

Now one can see that the problem of derivation of the AMIs up to the given weight w is equivalent to generating all connected graphs with at least two nodes and at most w edges. Let us denote this set of graphs as G_w. Generating all graphs from G_w is a combinatorial task with exponential complexity but formally easy to implement. An example of the graphs is in Figure 3.4.

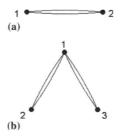

Figure 3.4 The graphs corresponding to the invariants (a) (3.13) and (b) (3.14).

3.3.3   Independence of the AMIs

We already pointed out that dependent invariants do not increase the discrimination power of the recognition system at all while increasing the dimensionality of the problem, which leads to growth of the complexity and even to misclassifications. Using dependent features in recognition tasks is a serious mistake. Thus, identifying and discarding dependent invariants is highly desirable. In the case of AMIs generated by the graph method, this is extremely important because, as we will see later, most of them are dependent on some other invariants.

The number of independent invariants

Before we proceed to the selection of independent invariants, it is worth analyzing how many invariants may exist. An intuitive rule suggests that the number n of independent invariants created from m independent measurements (i.e. moments in our case) is

$$n = m - p, \tag{3.15}$$

where p is the number of independent constraints that must be satisfied (see, e.g., reference [9]). This is true but the number of independent constraints is often hard to determine. The dependencies both in constraints and in measurements may be hidden. It may be hard to identify those measurements and constraints that are actually independent and those that are not.

In most cases we estimate p as the number of transformation parameters. This works perfectly in the case of rotation but for affine transform this estimate is not true in general. An affine transform has six parameters, so we would not expect any second-order affine invariant (6 moments − 6 parameters = 0). However, in the previous section we proved the existence of the second-order invariant I_1. On the other hand, for invariants up to the third order we have 10 moments − 6 parameters = 4, which is actually the correct number of independent invariants. The same estimation works well for the fourth order (15 − 6 = 9 invariants) and for all higher orders where the actual number of independent invariants is known. It is a common belief (although not exactly proven) that the order $r = 2$ is the only exception to this rule.

A detailed analysis of the number of independent AMIs regarding their weights, degrees and structures can be found in reference [26]. It is based on classic results of the theory of algebraic invariants, particularly on the Cayley–Sylvester theorem [19, 21].

Possible dependencies among the AMIs

There might be various kinds of dependency in the set of all AMIs (i.e. in the set G_w of all graphs). Let us categorize them into five groups and explain how they can be eliminated.

1. *Zero invariants.* Some AMIs may be identically zero regardless of the image from which they are calculated. If there are one or more nodes with one adjacent edge only, then all terms of the invariants contain first-order moment(s). When using central moments, they are zero by definition and, consequently, such an invariant is zero, too. However, some other graphs may also generate zero invariants, for instance the graph

in Figure 3.5 leads to

$$I(f) = \int_{-\infty}^{\infty} (x_1 y_2 - x_2 y_1)^3 f(x_1, y_1) f(x_2, y_2) \, dx_1 \, dy_1 \, dx_2 \, dy_2$$

$$= m_{30} m_{03} - 3m_{21} m_{21} + 3m_{21} m_{21} - m_{30} m_{03} = 0.$$

(3.16)

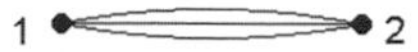

Figure 3.5 The graph leading to the zero invariant (3.16).

2. *Identical invariants.* All isomorphic graphs (and also some nonisomorphic graphs) generate identical invariants. Elimination can be done by comparing the invariants termwise.

3. *Products.* Some invariants may be products of other invariants.

4. *Linear combinations.* Some invariants may be linear combinations of other invariants.

5. *Polynomial dependencies.* If there exists a finite sum of products of invariants (including their integer powers) that equals zero, the invariants involved are polynomially dependent.

The invariants suffering from dependencies 1–4 are called *reducible* invariants. After eliminating all of them, we obtain a set of so-called *irreducible* invariants. However, irreducibility does not mean independence, as we will see later. In the following two sections we show how to detect reducible and polynomially dependent invariants.

Removing reducible invariants

The dependencies 1 and 2 are trivial and easy to find. To remove the products (type 3), we perform an incremental exhaustive search. All possible pairs of the admissible invariants (the sum of their individual weights must not exceed w) are multiplied and the independence of the result is checked.

To remove the dependencies of type 4 (linear combinations), all possible linear combinations of the admissible invariants (the invariants must have the same structure) are considered and their independence is checked. The algorithm should not miss any possible combination and can be implemented as follows.

Since only invariants of the same structure can be linearly dependent, we sort the invariants (including all products) according to the structure first. For each group of the same structure we construct a matrix of coefficients of all invariants. The "coefficient" here means the multiplicative constant of each term; if the invariant does not contain all possible terms, the corresponding coefficients are zero. Thanks to this, all coefficient vectors are of the same length and we can actually arrange them into a matrix. If this matrix has a full rank, all invariants are linearly independent. To calculate the rank of the matrix, we use singular value decomposition (SVD). The rank of the matrix is given by the number of nonzero singular values and equals the number of linearly independent invariants of the given structure. Each zero singular value corresponds to a linearly dependent invariant that can be removed.

At the end of the above procedure we obtain a set of all irreducible invariants up to the given weight or order. However, it is of exponential complexity and thus is a very expensive procedure even for low weights. On the other hand, this procedure is run only once. As soon as the invariants are known in explicit forms they can be used in future experiments without repeating the derivation.

For example, for $w \leq 12$ we generated $2\,533\,942\,752$ graphs altogether, $2\,532\,349\,394$ of them being zero. $1\,575\,126$ invariants were equal to some other invariants, 2105 were products of other invariants and $14\,538$ invariants were linearly dependent. After eliminating all these reducible invariants, we obtained 1589 irreducible invariants. Here we present the first ten examples of irreducible invariants in explicit forms along with their generating graphs and structures. Other important invariants are listed in Appendix 3.B and a complete list of all 1589 irreducible invariants along with their MATLAB implementation is available on the accompanying website.

1.

$$I_1 = (\mu_{20}\mu_{02} - \mu_{11}^2)/\mu_{00}^4$$

$w = 2$, $s = (2)$

2.

$$I_2 = (-\mu_{30}^2\mu_{03}^2 + 6\mu_{30}\mu_{21}\mu_{12}\mu_{03} - 4\mu_{30}\mu_{12}^3 - 4\mu_{21}^3\mu_{03} + 3\mu_{21}^2\mu_{12}^2)/\mu_{00}^{10}$$

$w = 6$, $s = (0, 4)$

3.

$$I_3 = (\mu_{20}\mu_{21}\mu_{03} - \mu_{20}\mu_{12}^2 - \mu_{11}\mu_{30}\mu_{03} + \mu_{11}\mu_{21}\mu_{12} + \mu_{02}\mu_{30}\mu_{12}$$

$$- \mu_{02}\mu_{21}^2)/\mu_{00}^7$$

$w = 4$, $s = (1, 2)$

4.

$$I_4 = (-\mu_{20}^3\mu_{03}^2 + 6\mu_{20}^2\mu_{11}\mu_{12}\mu_{03} - 3\mu_{20}^2\mu_{02}\mu_{12}^2 - 6\mu_{20}\mu_{11}^2\mu_{21}\mu_{03} - 6\mu_{20}\mu_{11}^2\mu_{12}^2$$

$$+ 12\mu_{20}\mu_{11}\mu_{02}\mu_{21}\mu_{12} - 3\mu_{20}\mu_{02}^2\mu_{21}^2 + 2\mu_{11}^3\mu_{30}\mu_{03} + 6\mu_{11}^3\mu_{21}\mu_{12}$$

$$- 6\mu_{11}^2\mu_{02}\mu_{30}\mu_{12} - 6\mu_{11}^2\mu_{02}\mu_{21}^2 + 6\mu_{11}\mu_{02}^2\mu_{30}\mu_{21} - \mu_{02}^3\mu_{30}^2)/\mu_{00}^{11}$$

$w = 6, \mathbf{s} = (3, 2)$

5.

$$
\begin{aligned}
I_5 = (&\mu_{20}^3\mu_{30}\mu_{03}^3 - 3\mu_{20}^3\mu_{21}\mu_{12}\mu_{03}^2 + 2\mu_{20}^3\mu_{12}^3\mu_{03} - 6\mu_{20}^2\mu_{11}\mu_{30}\mu_{12}\mu_{03}^2 \\
&+ 6\mu_{20}^2\mu_{11}\mu_{21}^2\mu_{03}^2 + 6\mu_{20}^2\mu_{11}\mu_{21}\mu_{12}^2\mu_{03} - 6\mu_{20}^2\mu_{11}\mu_{12}^4 \\
&+ 3\mu_{20}^2\mu_{02}\mu_{30}\mu_{12}^2\mu_{03} - 6\mu_{20}^2\mu_{02}\mu_{21}^2\mu_{12}\mu_{03} + 3\mu_{20}^2\mu_{02}\mu_{21}\mu_{12}^3 \\
&+ 12\mu_{20}\mu_{11}^2\mu_{30}\mu_{12}^2\mu_{03} - 24\mu_{20}\mu_{11}^2\mu_{21}^2\mu_{12}\mu_{03} + 12\mu_{20}\mu_{11}^2\mu_{21}\mu_{12}^3 \\
&- 12\mu_{20}\mu_{11}\mu_{02}\mu_{30}\mu_{12}^3 + 12\mu_{20}\mu_{11}\mu_{02}\mu_{21}^3\mu_{03} - 3\mu_{20}\mu_{02}^2\mu_{30}\mu_{21}^2\mu_{03} \\
&+ 6\mu_{20}\mu_{02}^2\mu_{30}\mu_{21}\mu_{12}^2 - 3\mu_{20}\mu_{02}^2\mu_{21}^3\mu_{12} - 8\mu_{11}^3\mu_{30}\mu_{12}^3 + 8\mu_{11}^3\mu_{21}^3\mu_{03} \\
&- 12\mu_{11}^2\mu_{02}\mu_{30}\mu_{21}^2\mu_{03} + 24\mu_{11}^2\mu_{02}\mu_{30}\mu_{21}\mu_{12}^2 - 12\mu_{11}^2\mu_{02}\mu_{21}^3\mu_{12} \\
&+ 6\mu_{11}\mu_{02}^2\mu_{30}\mu_{21}\mu_{03} - 6\mu_{11}\mu_{02}^2\mu_{30}\mu_{12}^2 - 6\mu_{11}\mu_{02}^2\mu_{30}\mu_{21}^2\mu_{12} \\
&+ 6\mu_{11}\mu_{02}^2\mu_{21}^4 - \mu_{02}^3\mu_{30}^3\mu_{03} + 3\mu_{02}^3\mu_{30}^2\mu_{21}\mu_{12} - 2\mu_{02}^3\mu_{30}\mu_{21}^3)/\mu_{00}^{16}
\end{aligned}
$$

$w = 9, \mathbf{s} = (3, 4)$

6.

$$I_6 = (\mu_{40}\mu_{04} - 4\mu_{31}\mu_{13} + 3\mu_{22}^2)/\mu_{00}^6$$

$w = 4, \mathbf{s} = (0, 0, 2)$

7.

$$I_7 = (\mu_{40}\mu_{22}\mu_{04} - \mu_{40}\mu_{13}^2 - \mu_{31}^2\mu_{04} + 2\mu_{31}\mu_{22}\mu_{13} - \mu_{22}^3)/\mu_{00}^9$$

$w = 6, \mathbf{s} = (0, 0, 3)$

8.

$$
\begin{aligned}
I_8 = (&\mu_{20}^2\mu_{04} - 4\mu_{20}\mu_{11}\mu_{13} + 2\mu_{20}\mu_{02}\mu_{22} + 4\mu_{11}^2\mu_{22} \\
&- 4\mu_{11}\mu_{02}\mu_{31} + \mu_{02}^2\mu_{40})/\mu_{00}^7
\end{aligned}
$$

$w = 4$, $\mathbf{s} = (2, 0, 1)$

9.

$$I_9 = (\mu_{20}^2\mu_{22}\mu_{04} - \mu_{20}^2\mu_{13}^2 - 2\mu_{20}\mu_{11}\mu_{31}\mu_{04} + 2\mu_{20}\mu_{11}\mu_{22}\mu_{13}$$
$$+ \mu_{20}\mu_{02}\mu_{40}\mu_{04} - 2\mu_{20}\mu_{02}\mu_{31}\mu_{13} + \mu_{20}\mu_{02}\mu_{22}^2 + 4\mu_{11}^2\mu_{31}\mu_{13} - 4\mu_{11}^2\mu_{22}^2$$
$$- 2\mu_{11}\mu_{02}\mu_{40}\mu_{13} + 2\mu_{11}\mu_{02}\mu_{31}\mu_{22} + \mu_{02}^2\mu_{40}\mu_{22} - \mu_{02}^2\mu_{31}^2)/\mu_{00}^{10}$$

$w = 6$, $\mathbf{s} = (2, 0, 2)$

10.

$$I_{10} = (\mu_{20}^3\mu_{31}\mu_{04}^2 - 3\mu_{20}^3\mu_{22}\mu_{13}\mu_{04} + 2\mu_{20}^3\mu_{13}^3 - \mu_{20}^2\mu_{11}\mu_{40}\mu_{04}^2$$
$$- 2\mu_{20}^2\mu_{11}\mu_{31}\mu_{13}\mu_{04} + 9\mu_{20}^2\mu_{11}\mu_{22}^2\mu_{04} - 6\mu_{20}^2\mu_{11}\mu_{22}\mu_{13}^2$$
$$+ \mu_{20}^2\mu_{02}\mu_{40}\mu_{13}\mu_{04} - 3\mu_{20}^2\mu_{02}\mu_{31}\mu_{22}\mu_{04} + 2\mu_{20}^2\mu_{02}\mu_{31}\mu_{13}^2$$
$$+ 4\mu_{20}\mu_{11}^2\mu_{40}\mu_{13}\mu_{04} - 12\mu_{20}\mu_{11}^2\mu_{31}\mu_{22}\mu_{04} + 8\mu_{20}\mu_{11}^2\mu_{31}\mu_{13}^2$$
$$- 6\mu_{20}\mu_{11}\mu_{02}\mu_{40}\mu_{13}^2 + 6\mu_{20}\mu_{11}\mu_{02}\mu_{31}^2\mu_{04} - \mu_{20}\mu_{02}^2\mu_{40}\mu_{31}\mu_{04}$$
$$+ 3\mu_{20}\mu_{02}^2\mu_{40}\mu_{22}\mu_{13} - 2\mu_{20}\mu_{02}^2\mu_{31}^2\mu_{13} - 4\mu_{11}^3\mu_{40}\mu_{13}^2 + 4\mu_{11}^3\mu_{31}^2\mu_{04}$$
$$- 4\mu_{11}^2\mu_{02}\mu_{40}\mu_{31}\mu_{04} + 12\mu_{11}^2\mu_{02}\mu_{40}\mu_{22}\mu_{13} - 8\mu_{11}^2\mu_{02}\mu_{31}^2\mu_{13}$$
$$+ \mu_{11}\mu_{02}^2\mu_{40}^2\mu_{04} + 2\mu_{11}\mu_{02}^2\mu_{40}\mu_{31}\mu_{13} - 9\mu_{11}\mu_{02}^2\mu_{40}\mu_{22}^2$$
$$+ 6\mu_{11}\mu_{02}^2\mu_{31}^2\mu_{22} - \mu_{02}^3\mu_{40}^2\mu_{13} + 3\mu_{02}^3\mu_{40}\mu_{31}\mu_{22} - 2\mu_{02}^3\mu_{31}^3)/\mu_{00}^{15}$$

$w = 9$, $\mathbf{s} = (3, 0, 3)$

Removing polynomial dependencies

Polynomial dependencies of orders higher than one among irreducible invariants pose a serious problem because the dependent invariants cannot be easily identified and removed. Unfortunately we cannot ignore them. Out of 1589 irreducible invariants of the weight of

12 and less, only 85 invariants at most can be independent, which means that at least 1504 of them must be polynomially dependent. Although an algorithm for a complete removal of dependent invariants is known in principle, it would be so difficult to implement and so computationally demanding that even for low weights it lies beyond the capacity of current computers. Nevertheless, many dependencies were discovered. For instance, among the ten invariants from the previous section, there are two polynomial dependencies

$$-4I_1^3 I_2^2 + 12I_1^2 I_2 I_3^2 - 12I_1 I_3^4 - I_2 I_4^2 + 4I_3^3 I_4 - I_5^2 = 0, \tag{3.17}$$

and

$$-16I_1^3 I_7^2 - 8I_1^2 I_6 I_7 I_8 - I_1 I_6^2 I_8^2 + 4I_1 I_6 I_9^2 + 12I_1 I_7 I_8 I_9 + I_6 I_8^2 I_9 - I_7 I_8^3 - 4I_9^3 - I_{10}^2 = 0. \tag{3.18}$$

This proves that the invariants I_5 and I_{10} are dependent. To obtain a complete and independent set of AMIs up to the fourth order, I_5 and I_{10} are omitted and one of I_{11}, I_{19} or I_{25} is added to $\{I_1, I_2, I_3, I_4, I_6, I_7, I_8, I_9\}$ (for explicit forms of I_{11}, I_{19} and I_{25}, see Appendix 3.B).

Looking for polynomial dependencies is carried out by a constrained exhaustive search of all possibilities, which is extremely expensive. However, even if we found (all or some) polynomial dependencies among the irreducible invariants, selecting independent invariants would not be straightforward. In the previous case of linear dependencies, an invariant that has been proven to be a linear combination of other invariants was simply omitted. This cannot be done in the case of polynomial dependencies because the identified dependencies among invariants may not be independent. Let us illustrate this in a hypothetical example. Assume I_a, I_b, I_c and I_d to be irreducible invariants with the three dependencies:

$$
\begin{aligned}
S_1: && I_a^2 + I_b I_c &= 0, \\
S_2: && I_d^2 - I_b I_c^2 &= 0, \\
S_3: && I_a^4 + 2I_a^2 I_b I_c + I_d^2 I_b &= 0.
\end{aligned}
$$

If we claimed three invariants to be dependent and we omitted them, it would be a mistake because the third dependency is a combination of the first and the second ones

$$S_1^2 + I_b S_2 - S_3 = 0$$

and does not bring any new information. Among these four invariants, only two of them are dependent and two are independent. This is an example of a *second-order dependency*, i.e. "dependency among dependencies", while S_1, S_2 and S_3 are first-order dependencies. The second-order dependencies may be of the same kind as the first-order dependencies – identities, products, linear combinations and polynomials. They can be found in the same way; the algorithm from the previous section requires only minor modifications.

This consideration can be further extended and we can define kth order dependencies. The number n of independent invariants is then

$$n = n_0 - n_1 + n_2 - n_3 + \cdots, \tag{3.19}$$

where n_0 is the number of irreducible invariants, n_1 is the number of first-order dependencies, n_2 is the number of second-order dependencies, etc. If we consider only the invariants of a certain finite order, this chain is always finite (the proof of finiteness for algebraic invariants originates from Hilbert [21], the proof for moment invariants is essentially the same).

3.3.4   The AMIs and tensors

There is an analogy between the graph method and the derivation of the AMIs by means of tensors. The tensor method is one of the traditional approaches to the AMIs, employed for instance by Reiss [25]. In this section, we briefly show the link between the graph and tensor methods. We recall a few basic terms of the tensor theory first. For more details about tensors we refer to reference [20] or to the Russian edition [27] of this famous book.

A tensor is an r-dimensional array of numbers that expresses the coordinates of a certain geometric figure in n-dimensional space together with the rule, how they change under an affine transform. These rules are of two kinds, covariant and contravariant.

A *contravariant vector* is given by n numbers $x^1, x^2, \ldots, x^n$, that are transformed as

$$x^i = \sum_{\alpha=1}^{n} p_\alpha^i \hat{x}^\alpha, \qquad i = 1, 2, \ldots, n \tag{3.20}$$

which is in the Einstein tensor notation

$$x^i = p_\alpha^i \hat{x}^\alpha. \tag{3.21}$$

The 2D affine transform p_α^i is given by its matrix

$$\mathbf{A} = \begin{pmatrix} p_1^1 & p_2^1 \\ p_1^2 & p_2^2 \end{pmatrix}. \tag{3.22}$$

If we denote the inverse transform

$$\mathbf{A}^{-1} = \begin{pmatrix} q_1^1 & q_2^1 \\ q_1^2 & q_2^2 \end{pmatrix}, \tag{3.23}$$

we can write

$$\hat{x}^\alpha = q_i^\alpha x^i. \tag{3.24}$$

A *covariant vector* is given by n numbers $u_1, u_2, \ldots, u_n$, that are transformed as

$$\hat{u}_\alpha = p_\alpha^i u_i. \tag{3.25}$$

A *covariant tensor* of order r is given by n^r numbers $a_{i_1,i_2,\ldots,i_r}$, that are transformed by an affine transformation as

$$\hat{a}_{\alpha_1,\alpha_2,\ldots,\alpha_r} = p_{\alpha_1}^{i_1} p_{\alpha_2}^{i_2} \cdots p_{\alpha_r}^{i_r} a_{i_1,i_2,\ldots,i_r}, \qquad i_1, i_2, \ldots, i_r, \alpha_1, \alpha_2, \ldots, \alpha_r = 1, 2, \ldots, n. \tag{3.26}$$

A *contravariant tensor* of order r is given by n^r numbers $a^{i_1,i_2,\ldots,i_r}$, that are transformed as

$$a^{i_1,i_2,\ldots,i_r} = p_{\alpha_1}^{i_1} p_{\alpha_2}^{i_2} \cdots p_{\alpha_r}^{i_r} \hat{a}^{\alpha_1,\alpha_2,\ldots,\alpha_r}. \tag{3.27}$$

A *mixed tensor* of the covariant order r_1, the contravariant order r_2 and the general order $r = r_1 + r_2$ has the transformation rule

$$\hat{a}_{\alpha_1,\alpha_2,\ldots,\alpha_{r_1}}^{\beta_1,\beta_2,\ldots,\beta_{r_2}} = q_{j_1}^{\beta_1} q_{j_2}^{\beta_2} \cdots q_{j_{r_2}}^{\beta_{r_2}} p_{\alpha_1}^{i_1} p_{\alpha_2}^{i_2} \cdots p_{\alpha_{r_1}}^{i_{r_1}} a_{i_1,i_2,\ldots,i_{r_1}}^{j_1,j_2,\ldots,j_{r_2}},$$

$$i_1, i_2, \ldots, i_{r_1}, j_1, j_2, \ldots, j_{r_2}, \alpha_1, \alpha_2, \ldots, \alpha_{r_1}, \beta_1, \beta_2, \ldots, \beta_{r_2} = 1, 2, \ldots, n. \tag{3.28}$$

In tensor multiplication, we perform addition over indices that are used once as upper ones and once as lower ones and mere enumeration over indices used only once. This means that mixed tensors with covariant order one and contravariant order one are multiplied in the same way as matrices:

$$a^i_j b^j_k = c^i_k, \qquad i, j, k = 1, 2, \ldots, n, \tag{3.29}$$

we perform addition over j and enumeration over i and k.

The *relative tensor* of the weight g is transformed as

$$\hat{a}^{\beta_1,\beta_2,\ldots,\beta_{r_2}}_{\alpha_1,\alpha_2,\ldots,\alpha_{r_1}} = J^g q^{\beta_1}_{i_1} q^{\beta_2}_{i_2} \cdots q^{\beta_{r_2}}_{i_{r_2}} p^{i_1}_{\alpha_1} p^{i_2}_{\alpha_2} \cdots p^{i_{r_1}}_{\alpha_{r_1}} a^{i_1,i_2,\ldots,i_{r_2}}_{i_1,i_2,\ldots,i_{r_1}} \tag{3.30}$$

where J is a determinant of $\mathbf{A}$.

The moments themselves do not behave under affine transform like tensors, but we can define a *moment tensor* [28]

$$M^{i_1 i_2 \cdots i_r} = \int_{-\infty}^{\infty} \int_{-\infty}^{\infty} x^{i_1} x^{i_2} \cdots x^{i_r} f(x^1, x^2)\, dx^1\, dx^2, \tag{3.31}$$

where $x^1 = x$ and $x^2 = y$. If p indices equal 1 and q indices equal 2, then $M^{i_1 i_2 \cdots i_r} = m_{pq}$. The behavior of the moment tensor under an affine transform

$$M^{i_1 i_2 \cdots i_r} = |J| p^{i_1}_{\alpha_1} p^{i_2}_{\alpha_2} \cdots p^{i_r}_{\alpha_r} \hat{M}^{\alpha_1 \alpha_2 \cdots \alpha_r} \tag{3.32}$$

or

$$\hat{M}^{i_1 i_2 \cdots i_r} = |J|^{-1} q^{i_1}_{\alpha_1} q^{i_2}_{\alpha_2} \cdots q^{i_r}_{\alpha_r} M^{\alpha_1 \alpha_2 \cdots \alpha_r}, \qquad i_1, i_2, \ldots, i_r, \alpha_1, \alpha_2, \ldots, \alpha_r = 1, 2, \ldots, n.$$

It means that the moment tensor is a relative contravariant tensor with the weight $g = -1$.

For an explanation of the method of tensors, it is necessary to introduce a concept of unit polyvectors. $\varepsilon_{i_1 i_2 \cdots i_n}$ is a *covariant unit polyvector*, if it is a skew-symmetric tensor over all indices and $\varepsilon_{12 \cdots n} = 1$. The term *skew-symmetric* means that the tensor element changes its sign and preserves its magnitude when interchanging two indices.

In two dimensions, it means that $\varepsilon_{12} = 1$, $\varepsilon_{21} = -1$, $\varepsilon_{11} = 0$ and $\varepsilon_{22} = 0$. Under an affine transform, a covariant unit polyvector is changed as

$$\hat{\varepsilon}_{\alpha_1 \alpha_2 \cdots \alpha_n} = J^{-1} p^{i_1}_{\alpha_1} p^{i_2}_{\alpha_2} \cdots p^{i_n}_{\alpha_n} \varepsilon_{i_1 i_2 \cdots i_n}. \tag{3.33}$$

Then, if we multiply a proper number of moment tensors and unit polyvectors such that the number of upper indices at the moment tensors equals the number of lower indices at polyvectors, we obtain a real-valued relative affine invariant, e.g.

$$M^{ij} M^{klm} M^{nop} \varepsilon_{ik} \varepsilon_{jn} \varepsilon_{lo} \varepsilon_{mp}$$

$$= 2(m_{20}(m_{21}m_{03} - m_{12}^2) - m_{11}(m_{30}m_{03} - m_{21}m_{12}) + m_{02}(m_{30}m_{12} - m_{21}^2)). \tag{3.34}$$

This method is analogous to the method of graphs. Each moment tensor corresponds to a node of the graph and each unit polyvector corresponds to an edge. The indices indicate which edge connects which nodes. The graph corresponding to the invariant (3.34) is on Figure 3.6.

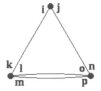

Figure 3.6 The graph corresponding to the invariant from (3.34).

3.3.5   Robustness of the AMIs

In this section, we illustrate the robustness (and vulnerability) of the AMIs on simple computer-simulated examples.

The invariance property of I_1 is experimentally verified in Figure 3.7, the behavior of other invariants is quite similar.

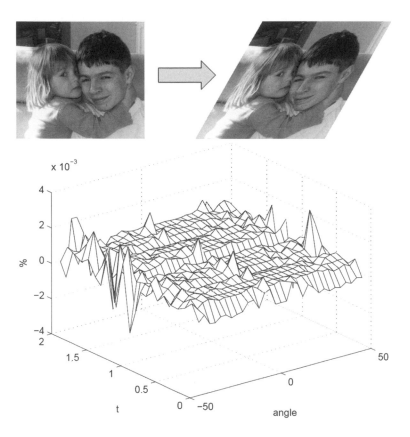

Figure 3.7 Numerical test of invariance of I_1. Horizontal axes: horizontal skewing and rotation angle, respectively. Vertical axis – relative deviation (in %) between I_1 of the original and that of the transformed image. The test proves the invariance of I_1.

We also investigated the robustness of the AMIs with respect to projective distortion. The traditional Lena and Lisa images (Figure 3.8) were projectively distorted with gradually increasing degree of distortion (Figure 3.9) and their affine moment invariants were computed. The values of the AMIs I_1 and I_6 are plotted in the graph on Figure 3.10(a), those of I_3 and I_4 are on Figure 3.10(b). One can see that the even-order invariants I_1 and I_6 are in this case more robust to the projective distortion than the odd-order ones I_3 and I_4, but generally, the robustness of the affine moment invariants with respect to projective distortion is limited.

(a)   (b)

Figure 3.8 (a) The Lena image, and (b) the Lisa image.

We performed a similar experiment except for using real images and real projective distortions. We took five pictures of a comb lying on a black background by a digital camera (see Figure 3.11). The images differ from each other by viewing angles. We used five viewing angles ranging from a perpendicular view (0°) to approximately 75°, that yielded perspective deformations of the image to various extents. To eliminate the influence of varying lighting conditions, the images were normalized to the same contrast and the background was thresholded.

The values of the 30 independent AMIs up to the seventh order were computed for each image (see Table 3.1 for eight independent invariants). Observe that they change a little when the viewing angle is changed. This is because the invariants are invariant to affine transform but not to perspective projection, which occurs in this experiment. Observe also the loss of invariance property when taking the picture from sharp angles. On the other hand, notice reasonable stability in the case of weak perspective projections.

3.4 AMIs via image normalization

In Chapter 2, we demonstrated that the rotation moment invariants can be derived not only in the "direct" way but also by means of image normalization. The normalization approach

Figure 3.9 The Lena image with gradually increasing projective distortion.

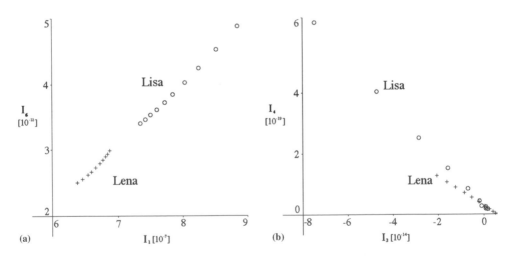

Figure 3.10 The space of the AMIs (a) I_1 and I_6, and (b) I_3 and I_4 of the Lena and Lisa images. While I_1 and I_6 created two separate clusters, the values of I_3 and I_4 fall into one mixed cluster.

can also be applied in the case of affine transform, with similar advantages and drawbacks as in the case of rotation. Recall that in image normalization the object is brought into certain "normalized" (also "standard" or "canonical") positions, which is independent of the actual position, rotation, scale and skewing of the original object. In this way, the influence of

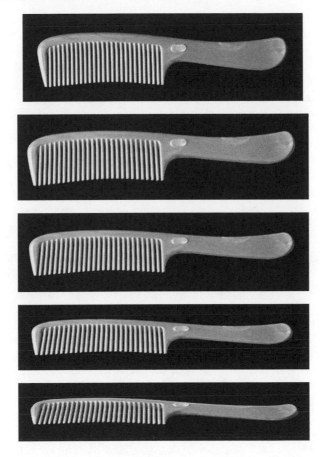

Figure 3.11 The comb. The viewing angle increases from 0° (top) to 75° (bottom).

Table 3.1 The values of the AMIs of the comb; γ is the approximate viewing angle.

$\gamma[°]$	$I_1[10^{-2}]$	$I_2[10^{-6}]$	$I_3[10^{-4}]$	$I_4[10^{-5}]$	$I_6[10^{-3}]$	$I_7[10^{-5}]$	$I_8[10^{-3}]$	$I_9[10^{-4}]$
0	2.802	1.538	−2.136	−1.261	4.903	6.498	4.526	1.826
30	2.679	1.292	−1.927	−1.109	4.466	5.619	4.130	1.589
45	2.689	1.328	−1.957	−1.129	4.494	5.684	4.158	1.605
60	2.701	1.398	−2.011	−1.162	4.528	5.769	4.193	1.626
75	2.816	1.155	−1.938	−1.292	5.033	6.484	4.597	1.868

an affine transform is eliminated. Since the normalized position is the same for all objects differing from each other only by an affine transform, the moments of the normalized object are in fact affine invariants of the original. The normalized objects can be viewed as representatives of "shape classes".

We emphasize that no actual spatial transformation of the original object is necessary. Such a transformation would slow down the process and would introduce resampling errors. Instead, the moments of the normalized object can be calculated directly from the original object using the normalization constraints. These constraints are formulated by means of low-order moments.

Image normalization to affine transform is always based on proper decomposition of an affine transform into several simpler transformations (usually into six elementary one-parametric transformations). The normalization methods differ from each other by the type of decomposition used and by elementary normalization constraints.

The idea of affine normalization via decomposition was introduced by Rothe *et al.* [29], where two different decompositions were used – so-called XSR decomposition to skew, nonuniform scaling and rotation, and XYS decomposition to two skews and nonuniform scaling. However, this approach led to some ambiguities, which were later studied in reference [30] in detail. Pei and Lin [31] presented a method based on decomposition to two rotations and a nonuniform scaling between them. Their method uses eigenvalues of the covariance matrix and tensors for the computation of the normalization coefficients. However, their method fails on symmetric objects. This is a serious weakness because in many applications we have to classify man-made or specific natural objects that are commonly symmetrical. Since many moments of symmetrical objects are zero, the normalization constraints may not be well defined. A similar decomposition scheme with normalization constraints adapted also for symmetric objects was used by Suk and Flusser [32]. Shen and Ip [33] used the so-called generalized complex moments for normalization constraints. Their approach combines moments with the Fourier transform in polar coordinates and is relatively stable also in the case of symmetric objects, yet exceedingly complicated. Sprinzak [34] and Heikkilä [35] used Cholesky factorization of the second-order moment matrix for derivation of the normalization constraints.

In this section we introduce a proper affine decomposition and show how the normalization constraints can be formulated. We also demonstrate that one can consider the moments of the normalized image as another type of AMI, that are, however, theoretically equivalent to the AMIs derived by the graph method. We call them *normalized moments* because of the distinction from the AMIs obtained by other methods.

The normalized moments have certain advantages in comparison with the AMIs, namely an easy creation of complete and independent sets. There is no need of time-consuming algorithms for identification and removal of dependent invariants. On the other hand, normalized moments are less numerically stable than AMIs. If some of the constrained moments are calculated with errors, these errors propagate into all normalized moments. Even worse, the normalization may be "discontinuous" on nearly symmetric objects. A small, almost invisible change of the input object may cause a significant change in normalized position and may, consequently, change all normalized moments in a substantial way.

3.4.1   Decomposition of the affine transform

Among various possible decompositions of the affine transform, we prefer the following one. The normalization constraints to the elementary transforms are formulated in a simple way by geometric and complex moments of low orders and the method is well defined also for objects having an N-FRS. This decomposition avoids skewing and replaces it by two rotations and stretching between them.

Horizontal translation:

$$x' = x + \alpha,$$
$$y' = y. \tag{3.35}$$

Vertical translation:

$$x' = x,$$
$$y' = y + \beta. \tag{3.36}$$

Uniform scaling:

$$x' = \omega x,$$
$$y' = \omega y. \tag{3.37}$$

First rotation:

$$x' = x \cos \alpha - y \sin \alpha,$$
$$y' = x \sin \alpha + y \cos \alpha. \tag{3.38}$$

Stretching:

$$x' = \delta x,$$
$$y' = \frac{1}{\delta} y. \tag{3.39}$$

Second rotation:

$$x' = x \cos \varrho - y \sin \varrho,$$
$$y' = x \sin \varrho + y \cos \varrho. \tag{3.40}$$

Possible mirror reflection:

$$x' = x,$$
$$y' = \pm y. \tag{3.41}$$

These partial transforms are not commutative; stretching must always be between the two rotations.

In the following sections, we show how to normalize an image with respect to individual partial transforms.

Normalization to translation

Normalization to translation is analogous to that in the previous methods; we can easily translate the image in such a way that its centroid coincides with the origin. This is ensured by the constraint $m'_{10} = m'_{01} = 0$, which is equivalent to using central moments instead of geometric ones.

Normalization to uniform scaling

Normalization to scaling is expressed by the constraint $\mu'_{00} = 1$, which can be assured by using scale-normalized moments in exactly the same way as in Chapter 2:

$$\nu_{pq} = \frac{\mu_{pq}}{\mu_{00}^{((p+q)/2)+1}}. \tag{3.42}$$

Normalization to the first rotation

Normalization to rotation was already described in Chapter 2. To eliminate the first rotation, we use traditional "principal axis" normalization constrained by $\mu'_{11} = 0$ (i.e. $c'_{20} > 0$). As shown in Chapter 2, this constraint leads to the normalizing angle

$$\alpha = \frac{1}{2} \arctan \left(\frac{2\mu_{11}}{\mu_{20} - \mu_{02}} \right). \tag{3.43}$$

If it holds that $\mu_{11} = 0$ and $\mu_{20} = \mu_{02}$, then we consider the image already normalized to rotation and set $\alpha = 0$. After normalization to the first rotation, we have for the geometric moments

$$\mu'_{pq} = \sum_{k=0}^{p} \sum_{j=0}^{q} \binom{p}{k} \binom{q}{j} (-1)^{p+j} \sin^{p-k+j} \alpha \cos^{q+k-j} \alpha \, \nu_{k+j,p+q-k-j}. \tag{3.44}$$

Normalization to stretching

Under stretching, the moments are transformed as

$$\mu''_{pq} = \delta^{p-q} \mu'_{pq}$$

(note that the Jacobian of a stretching equals one). Normalization to stretching can be done by imposing a constraint $\mu''_{20} = \mu''_{02}$, which can be fulfilled by choosing

$$\delta = \sqrt[4]{\frac{\mu'_{02}}{\mu'_{20}}}.$$

In terms of the moments of the original image, we have

$$\delta = \sqrt{\frac{\mu_{20} + \mu_{02} - \sqrt{(\mu_{20} - \mu_{02})^2 + 4\mu_{11}^2}}{2\sqrt{\mu_{20}\mu_{02} - \mu_{11}^2}}} \tag{3.45}$$

(this is well defined because $\mu_{20}\mu_{02} \geq \mu_{11}^2$ always and, moreover, for nondegenerated 2D objects it holds $\mu_{20}\mu_{02} > \mu_{11}^2$) and, for the moments normalized both to the first rotation and stretching,

$$\mu''_{pq} = \delta^{p-q} \sum_{k=0}^{p} \sum_{j=0}^{q} \binom{p}{k} \binom{q}{j} (-1)^{p+j} \sin^{p-k+j} \alpha \cos^{q+k-j} \alpha \, \nu_{k+j,p+q-k-j}. \tag{3.46}$$

After this normalization, c''_{20} becomes zero and cannot be further used for another normalization.

Normalization to the second rotation

For normalization to the second rotation we adopt the Abu-Mostafa and Psaltis [36] normalization scheme by means of complex moments that was used in Chapter 2. The normalized position is defined in such a way that a certain nonzero moment c_{st} is required to be real and positive. As proven in Chapter 2, it is always possible but ambiguous for $s - t > 1$.

Normalization to the second rotation is a critical step. While small errors and inaccuracies in the previous stages can be compensated in the later steps, there is no chance to fix the errors in normalization to the second rotation. The selection of the normalizing moment must be done very carefully. If c_{21}'' was nonzero, it would be an optimal choice. However, as we saw in Chapter 2, many complex moments including c_{21}'' are zero for images having a certain degree of symmetry. Note that we are speaking about symmetry after normalization of the image with respect to stretching, which may be different from the symmetry of the original. For instance, a rectangle has a twofold rotational symmetry and fourfold symmetry after the normalization to stretching, and ellipse has circular symmetry after the stretching normalization.

The normalizing moment can be found in the same way as described in Chapter 2 – we sort the moments and look for the first moment whose magnitude exceeds a given threshold. The only difference here is that we cannot consider c_{20}'' because, due to previous normalization, $c_{20}'' = 0$.

Once the normalizing moment is determined, the normalizing angle ϱ is

$$\varrho = \frac{1}{s - t} \arctan\left(\frac{\mathcal{I}m(c_{st}'')}{\mathcal{R}e(c_{st}'')}\right). \tag{3.47}$$

Finally, the moments of the object normalized to the second rotation are calculated by means of a similar formula to that in the first rotation

$$\tau_{pq} = \sum_{k=0}^{p}\sum_{j=0}^{q} \binom{p}{k}\binom{q}{j}(-1)^{p+j}\,\sin^{p-k+j}\varrho\,\cos^{q+k-j}\varrho\,\mu_{k+j,p+q-k-j}''. \tag{3.48}$$

The normalized moments τ_{pq} are new AMIs of the original object. Some of them have "prescribed" values due to normalization, regardless of image:

$$\tau_{00} = 1,\ \tau_{10} = 0,\ \tau_{01} = 0,\ \tau_{02} = \tau_{20},\ \tau_{11} = 0,\ \tau_{03} = -\tau_{21} \tag{3.49}$$

(the last equality follows from the normalization to the second rotation and holds only if the normalizing moment was c_{21}'' or if $c_{21}'' = 0$ because of symmetry; otherwise it is replaced by an analogous condition for other moments).

Normalization to mirror reflection

Although the general affine transform may contain a mirror reflection, normalization to the mirror reflection should be done separately from the other partial transformations in equations (3.35)–(3.40) for practical reasons. In most affine deformations occurring in practice, no mirror reflection can in principle be present and it is necessary to classify mirrored images into different classes (in character recognition we certainly want to distinguish capital S from a question mark, for instance). Normalization to mirror reflection is not desirable in those cases.

If we still want to normalize objects to mirror reflection, we can do that, after all normalizations mentioned above, as follows. We find the first nonzero moment τ_{pq} with an odd second index q. If it is negative, then we change the signs of all moments with odd second indices. If it is positive or if all normalized moments up to the chosen order with odd second indices are zero, no action is necessary.

3.4.2   Violation of stability

In this simulation, we illustrate the potential vulnerability of the normalized moments. Observe that under certain circumstances the normalization to the second rotation might become unstable and "discontinuous". Even if it is probably a rare case in real applications, one should be aware of this danger.

Let us consider a binary fourfold cross inscribed in a square with a side a. Such a cross has only one parameter – the thickness b, for which $0 < b \le a$ (see Figure 3.12(a)). Since the cross has fourfold symmetry, one would expect c''_{40} to be chosen for the normalization to the second rotation. It holds for the cross

$$c''_{40} = (3a^5 b + 3ab^5 + 2b^6 - 10a^3 b^3)/120 \tag{3.50}$$

which is generally nonzero but if $b \to 0.593a$, then $c''_{40} \to 0$. If we take two very similar crosses – one with b slightly less than $0.593a$ and the other one with b slightly greater – then c''_{40} changes its sign and, consequently, the standard position changes from horizontal-vertical "+" shape to diagonal "×" shape (see Figure 3.12 for illustration). This leads to discontinuities in the normalized moments and finally, to false results of shape recognition (see Figure 3.13 showing the values of the normalized moments). The AMIs are not plagued by these problems because they are continuous in the neighborhood of the critical point.

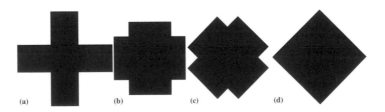

(a)   (b)   (c)   (d)

Figure 3.12   The standard positions of the cross with varying thickness b: (a) thin cross, $b \ll a$, (b) b slightly less than $0.593a$, (c) b slightly greater than $0.593a$, and (d) $b = a$. Note the difference between the two middle positions.

3.4.3   Relation between the normalized moments and the AMIs

In this section we show the close link between the normalized moments and the AMIs derived by the graph method.

The normalization is nothing other than a special affine transform, so the AMIs must be invariant to it. If we substitute τ_{pq} for μ_{pq} in the AMIs, considering the constraints (3.49), we

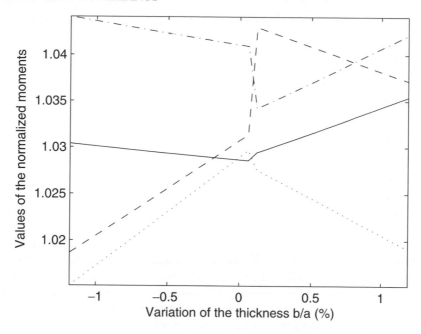

Figure 3.13 The behavior of the normalized moments in the neighborhood of the critical point. The horizontal axis shows the deviation of the line thickness from the critical value $b = 0.593a$, the vertical axis shows the values of the following normalized moments (with magnitude scaling to the same interval): τ_{40} (solid line), τ_{22} (dotted line), τ_{60} (dash-and-dot line) and τ_{42} (dashed line). The moments τ_{22} and τ_{40} are continuous because they can be obtained without normalization to the second rotation. The moments τ_{60} and τ_{42} have discontinuity in the critical point, where the normalizing moment c''_{40} changes its sign.

obtain for the second and third orders

$$I_1 = \tau_{20}^2,$$

$$I_2 = -\tau_{30}^2\tau_{21}^2 - 6\tau_{30}\tau_{21}^2\tau_{12} - 4\tau_{30}\tau_{12}^3 + 4\tau_{21}^4 + 3\tau_{21}^2\tau_{12}^2,$$

$$I_3 = \tau_{20}(-2\tau_{21}^2 - \tau_{12}^2 + \tau_{30}\tau_{12}),$$

$$I_4 = -\tau_{20}^3(4\tau_{21}^2 + 3\tau_{12}^2 + \tau_{30}^2).$$

(3.51)

For the fourth order, we obtain

$$I_6 = \tau_{40}\tau_{04} - 4\tau_{31}\tau_{13} + 3\tau_{22}^2,$$

$$I_7 = \tau_{40}\tau_{22}\tau_{04} - \tau_{40}\tau_{13}^2 - \tau_{31}^2\tau_{04} + 2\tau_{31}\tau_{22}\tau_{13} - \tau_{22}^3,$$

$$I_8 = \tau_{20}^2(\tau_{04} + 2\tau_{22} + \tau_{40}),$$

(3.52)

$$I_9 = \tau_{20}^2(\tau_{40}\tau_{04} + \tau_{40}\tau_{22} + \tau_{22}\tau_{04} + \tau_{22}^2 - \tau_{13}^2 - 2\tau_{31}\tau_{13} - \tau_{31}^2)$$

$$I_{25} = \tau_{20}^3(\tau_{12} + \tau_{30})^2(\tau_{04} + \tau_{22}).$$

Similar relations can also be obtained for higher-order invariants.

An inverse problem – calculating the normalized moments from the AMIs – can be resolved by solving the above equations for τ_{pq}. The system is not linear and does not have an analytical solution for orders higher than four. It must be solved numerically in those cases. Doing that, we can find one-to-one mapping between the two sets of invariants. It is naturally impractical to calculate the normalized moments in this way but this is a suitable illustration of the theoretical equivalency of both sets.

3.4.4   Affine invariants via half normalization

Half normalization is a "hybrid" approach to constructing invariants, trying to keep the positive properties of normalization as well as direct construction while avoiding the negative ones. This idea was first introduced by Heikkilä [35].

The first stage of the method is the same as the normalization described previously. The image is normalized to translation, uniform scaling, the first rotation and stretching, which ends up with the normalized moments μ''_{pq} and c''_{pq}. The normalization to the second rotation is not performed. Instead, the partially normalized image is described by proper rotation invariants as described in Chapter 2 (potential symmetry of the normalized image must be taken into account). Thus, we obtain affine invariants, for instance, in the form $c''_{pq}c''^{p-q}_{12}$.

The half normalization method is, of course, theoretically equivalent to both AMIs and normalized moments but may have certain favorable numerical properties. Avoiding normalization to the second rotation, we are rid of the danger of instability of the normalized position.

3.4.5   Affine invariants from complex moments

In normalization (including half normalization), some normalization constraints were formulated by means of complex moments and the resulting invariants can be considered both in terms of geometric as well as complex moments. In Chapter 2, we saw that complex moments provide an excellent insight into the behavior of invariants on symmetric images. For images having an N-FRS, we know precisely that complex moments are zero and can select only nontrivial invariants.

In contrast, the AMIs have been expressed as functions of geometric moments only, thus not providing any hints as to which invariants should be avoided when recognizing symmetric objects. To overcome this hindrance, we show how AMIs can be expressed by means of complex moments. Such relations also show a link between the AMIs and rotation invariants from Chapter 2.

Theorem 3.2 *Let us denote the value of an invariant I computed from the geometric moments as $I(\mu_{pq})$ and the value obtained when replacing μ_{pq} by c_{pq} as $I(c_{pq})$, respectively. Then, it holds*

$$I(c_{pq}) = (-2i)^w I(\mu_{pq}), \tag{3.53}$$

where w is the weight of the invariant I.

Proof. Consider an affine transform

$$x' = x + iy$$
$$y' = x - iy. \tag{3.54}$$

Under this transform, the geometric moments are transformed as

$$\mu'_{pq} = \int\limits_{-\infty}^{\infty} \int\limits_{-\infty}^{\infty} (x + iy)^P (x - iy)^q |J| f(x, y) \, dx \, dy = |J| c_{pq}, \tag{3.55}$$

where the Jacobian $J = -2i$. It follows from Theorem 3.1 that

$$I(\mu'_{pq}) = J^w |J|^r I(\mu_{pq}), \tag{3.56}$$

where r is a degree of I. On the other hand, from (3.55) we have

$$I(\mu'_{pq}) = I(|J| c_{pq}) = |J|^r I(c_{pq}). \tag{3.57}$$

Comparing (3.56) and (3.57), we immediately obtain (3.53). □

In particular, we obtain for I_1 and I_2

$$(-2i)^2 I_1 = c_{20} c_{02} - c_{11}^2 \tag{3.58}$$

and

$$(-2i)^6 I_2 = -c_{30}^2 c_{03}^2 + 6 c_{30} c_{21} c_{12} c_{03} - 4 c_{30} c_{12}^3 - 4 c_{21}^3 c_{03} + 3 c_{21}^2 c_{12}^2. \tag{3.59}$$

Similar relations can be derived for all the AMIs.

These formulas are useful not only for choosing proper invariants in case of a certain symmetry but also demonstrate the connection between AMIs and basic rotation invariants $\Phi(p, q)$.

This link can be advantageously used to prove the independency and completeness of a set of AMIs. Since the basis of rotation invariants is complete and independent, it is sufficient to find a one-to-one mapping between AMIs and rotation invariants.

Let us assume that the image has undergone a half normalization. The half normalization reduces an affine transform to a rotation and imposes $c_{00} = 1$, $c_{10} = c_{01} = 0$ and $c_{20} = c_{02} = 0$.[9] Under these constraints we obtain from Theorem 3.2

$$(-2i)^2 I_1 = -c_{11}^2,$$

$$(-2i)^6 I_2 = -c_{30}^2 c_{03}^2 + 6 c_{30} c_{21} c_{12} c_{03} - 4 c_{30} c_{12}^3 - 4 c_{21}^3 c_{03} + 3 c_{21}^2 c_{12}^2,$$

$$(-2i)^4 I_3 = -c_{11} c_{30} c_{03} + c_{11} c_{21} c_{12},$$

$$(-2i)^6 I_4 = 2 c_{11}^3 c_{30} c_{03} + 6 c_{11}^3 c_{21} c_{12},$$

$$(-2i)^4 I_6 = c_{40} c_{04} - 4 c_{31} c_{13} + 3 c_{22}^2,$$

$$(-2i)^6 I_7 = c_{40} c_{22} c_{04} - c_{40} c_{13}^2 - c_{31}^2 c_{04} + 2 c_{31} c_{22} c_{13} - c_{22}^3,$$

$$(-2i)^4 I_8 = 4 c_{11}^2 c_{22},$$

$$(-2i)^6 I_9 = 4 c_{11}^2 c_{31} c_{13} - 4 c_{11}^2 c_{22}^2,$$

$$(-2i)^8 I_{25} = -8 c_{11}^3 c_{21}^2 c_{13} + 16 c_{11}^3 c_{21} c_{12} c_{22} - 8 c_{11}^3 c_{12}^2 c_{31}. \tag{3.60}$$

---

[9] To be precise, we should write $c_{00}'' = 1$, $c_{10}'' = c_{01}'' = 0$, etc. For the sake of simplicity, we omit the double-primes in this section.

These equations can be inverted and resolved for rotation invariants in a much easier way than similar equations for normalizing moments (3.52). We obtain

$$c_{11} = 2\sqrt{I_1},$$

$$c_{21}c_{12} = \frac{1}{\sqrt{I_1}}\left(2I_3 - \frac{I_4}{I_1}\right),$$

$$c_{30}c_{03} = -\frac{1}{\sqrt{I_1}}\left(6I_3 + \frac{I_4}{I_1}\right),$$

$$Re(c_{30}c_{12}^3) = 8I_2 - 12\frac{I_3^2}{I_1} + \frac{I_4^2}{I_1^3},$$

$$c_{22} = \frac{I_8}{I_1},$$

$$c_{31}c_{13} = \frac{I_8^2}{I_1^2} - 4\frac{I_9}{I_1},$$

$$c_{40}c_{04} = 16I_6 + \frac{I_8^2}{I_1^2} - \frac{I_9}{I_1},$$

$$Re(c_{40}c_{13}^2) = 32I_7 + 8\frac{I_6 I_8}{I_1} - 12\frac{I_8 I_9}{I_1^2} + \frac{I_8^3}{I_1^3},$$

$$Re(c_{31}c_{12}^2) = \frac{1}{\sqrt{I_1}}\left(2\frac{I_3 I_8}{I_1} - \frac{I_4 I_8}{I_1^2} - 2\frac{I_{25}}{I_1}\right). \tag{3.61}$$

The set of rotation invariants $\{c_{11}, c_{21}c_{12}, c_{30}c_{03} \, Re(c_{30}c_{12}^3), c_{22}, c_{31}c_{13}, c_{40}c_{04}, Re(c_{40}c_{13}^2), Re(c_{31}c_{12}^2)\}$ is independent, which implies that the set of AMIs $\{I_1, I_2, I_3, I_4, I_6, I_7, I_8, I_9, I_{25}\}$ must be independent, too.

For higher-order invariants, it is difficult to resolve the equations with respect to the rotation invariants in general. As proposed in references [26] and [37], for the independence test it is sufficient to analyze one particular point in the space of invariants, in which most rotation invariants are ± 1 and many AMIs equal zero. A detailed description of this method is beyond the scope of the book but it provides the most successful method currently for independence tests of the AMIs. Thereby, the independence of the following set of AMIs of the weight $w \le 12$ has been proven: $\{I_1, I_2, I_3, I_4, I_6, I_7, I_8, I_9, I_{19}, I_{47}, I_{48}, I_{49}, I_{50}, I_{64}, I_{65}, I_{249}, I_{251}, I_{252}, I_{253}, I_{254}, I_{256}, I_{270}, I_{587}, I_{588}, I_{599}, I_{600}, I_{601}, I_{602}, I_{603}, I_{604}, I_{960}, I_{962}, I_{963}, I_{965}, I_{973}, I_{974}, I_{975}, I_{976}, I_{990}, I_{1241}, I_{1243}, I_{1249}, I_{1250}, I_{1251}, I_{1267}, I_{1268}, I_{1284}, I_{1285}, I_{1297}, I_{1423}, I_{1424}, I_{1425}, I_{1427}, I_{1429}, I_{1430}, I_{1434}, I_{1436}, I_{1437}, I_{1438}, I_{1456}, I_{1524}, I_{1525}, I_{1526}, I_{1528}, I_{1527}, I_{1531}, I_{1532}, I_{1536}, I_{1537}, I_{1538}, I_{1542}, I_{1545}, I_{1569}, I_{1570}, I_{1571}, I_{1574}, I_{1575}, I_{1576}, I_{1578}, I_{1579}\}$. This is the largest set of independent AMIs ever published; it contains 80 invariants. The explicit formulas of these AMIs can be found on the accompanying website.

3.5 Derivation of the AMIs from the Cayley–Aronhold equation

In this section, we present the last method for deriving AMIs. This method uses neither the theory of the algebraic invariants nor any normalization. It also does not generate "all" possibilities, as does the graph method. The basic idea is simple and its "manual" implementation for finding few low-order AMIs is easy (this approach has been used already in references [15] and [38]). The solution can be automated [39] which leads to a method of polynomial complexity as compared to the exponential complexity of the graph method. On the other hand, the implementation and programming of this method is much more difficult than that of the graph method.

3.5.1 Manual solution

The affine transformation (3.6) can be decomposed into seven partial transformations: horizontal translation:

$$x' = x + \alpha,$$
$$y' = y; \tag{3.62}$$

vertical translation:

$$x' = x,$$
$$y' = y + \beta; \tag{3.63}$$

scaling:

$$x' = \omega x,$$
$$y' = \omega y; \tag{3.64}$$

stretching:

$$x' = \delta x,$$
$$y' = \frac{1}{\delta} y; \tag{3.65}$$

horizontal skewing:

$$x' = x + ty,$$
$$y' = y; \tag{3.66}$$

vertical skewing:

$$x' = x,$$
$$y' = y + sx; \tag{3.67}$$

possible mirror reflection:

$$x' = x,$$
$$y' = \pm y. \tag{3.68}$$

This is a similar decomposition to that already used for image normalization but two rotations are replaced here by horizontal and vertical skewing.

Any function I of moments is invariant under these seven transformations if and only if it is invariant under the general affine transformation (3.6). Each of these transformations imposes one constraint on the invariants.

The invariance to translation, scaling and stretching can be achieved in exactly the same way as in the image normalization method; we do not repeat this here. In the following text, we explain how to derive constraints on the invariance to skewing.

If a function I should be invariant to the horizontal skew, then its derivative with respect to the skew parameter t must be zero

$$\frac{dI}{dt} = \sum_p \sum_q \frac{\partial I}{\partial \mu_{pq}} \frac{d\mu_{pq}}{dt} = 0. \qquad (3.69)$$

The derivative of the moment is

$$\frac{d\mu_{pq}}{dt} = \frac{d}{dt} \int_{-\infty}^{\infty} \int_{-\infty}^{\infty} (x + ty - x_c - ty_c)^p (y - y_c)^q f(x, y) \, dx \, dy$$

$$= \int_{-\infty}^{\infty} \int_{-\infty}^{\infty} p(x - x_c + t(y - y_c))^{p-1} (y - y_c)^{q+1} f(x, y) \, dx \, dy \qquad (3.70)$$

$$= p\mu_{p-1,q+1}.$$

If we substitute this into (3.69), we obtain

$$\sum_p \sum_q p\mu_{p-1,q+1} \frac{\partial I}{\partial \mu_{pq}} = 0. \qquad (3.71)$$

This equation is called the *Cayley–Aronhold differential equation*. We can derive a similar equation for the vertical skew

$$\sum_p \sum_q q\mu_{p+1,q-1} \frac{\partial I}{\partial \mu_{pq}} = 0. \qquad (3.72)$$

From the mirror reflection we can derive the condition of symmetry. If we combine it with interchanging of the indices

$$x' = y,$$
$$y' = -x \qquad (3.73)$$

then the moment after this transform is

$$\mu'_{pq} = (-1)^q \mu_{qp}. \qquad (3.74)$$

If there are two symmetric terms

$$c_1 \prod_{\ell=1}^{r} \mu_{p_{j\ell}, q_{j\ell}} \text{ and } c_2 \prod_{\ell=1}^{r} \mu_{q_{j\ell}, p_{j\ell}} \qquad (3.75)$$

in the invariant, then they become after the transform (3.73)

$$c_1(-1)^w \prod_{\ell=1}^{r} \mu_{q_{j\ell},p_{j\ell}} \text{ and } c_2(-1)^w \prod_{\ell=1}^{r} \mu_{p_{j\ell},q_{j\ell}}, \tag{3.76}$$

therefore $c_1 = (-1)^w c_2$. We can write a symbolic formula

$$I(\mu'_{pq}) = (-1)^w I(\mu_{qp}). \tag{3.77}$$

This means that the coefficients of the symmetric terms can be computed together. The invariants with odd weights change their signs under the mirror reflection, they are pseudo-invariants, while the invariants with even weights remain constant, they are true invariants. We can either use magnitudes of the invariants with odd weights or suppose the affine transform always has a positive Jacobian.

The Cayley–Aronhold differential equation leads to a solution of a system of linear equations and the AMIs can be derived in this way.

The invariants (i.e. the solutions of the Cayley–Aronhold equation) are usually supposed to have a form of a linear combination of moment products

$$I = \left(\sum_{j=1}^{n_t} c_j \prod_{\ell=1}^{r} \mu_{p_{j\ell},q_{j\ell}} \right) \Big/ \mu_{00}^{r+w}. \tag{3.78}$$

In reference [15], four solutions representing four low-order AMIs were found manually. According to the notation used in this book, they were I_1, I_2, I_3 and $-(I_4 + 6I_1 I_3)$.[10]

It is natural to ask the question whether or not all the AMIs generated by the graph method are combinations of the AMIs found by means of the Cayley–Aronhold equation, and vice versa. Although a positive answer seems to be intuitive (and is actually correct), to prove this assertion is rather difficult. We refer to the Gurevich proof of a similar theorem [20] dealing with tensors, which can be adapted for moments.

3.5.2 Automatic solution

The need for higher-order AMIs has led to development of an automatic method for solving the Cayley–Aronhold equation. The input parameters of the algorithm are the order s and the structure $(k_2, k_3, \ldots, k_s)$ of the desired invariant. Then, the degree $r = k_2 + k_3 + \cdots + k_s$ and the weight $w = (2k_2 + 3k_3 + \cdots + sk_s)/2$.

First, we generate moment indices of all possible terms of the invariant, meaning that we generate all possible partitions of the number w into sums of k_2 integers from 0 to 2, k_3 integers from 0 to 3 up to k_s integers from 0 to s. The generation of the partitions must be divided into two stages. In the first stage, we partition w among orders, in the second stage, we partition the quotas of specific moment orders among individual moments. The elements of the partitions are used as the first indices of the moments. When we complete the computation of the second indices from the moment orders, we have nearly generated the invariant; only the coefficients of the terms remain unknown. The number of all possible terms of the invariant is labeled n_t.

We sort both moments in a term and all terms in an invariant, because we need to find a term in an invariant or to detect identical invariants quickly.

---

[10]In reference [15], historical names of the invariants, such as "Apolar" and "Hankel determinant" are used.

The next step of the algorithm is to look for the symmetric terms. The first and second indices of the moments in each term are interchanged and the counterpart is sought. If it is found, then we compute coefficients of both terms together. The number of terms differing in another way is labeled n_s. Then we compute derivatives of all terms with respect to all moments (except μ_{0q}) and the terms of the Cayley–Aronhold equation from them.

Now we can construct the system of linear equations for unknown coefficients. The Cayley–Aronhold equation must hold for all values of the moments; therefore the sum of coefficients of identical terms must equal zero. The horizontal size of the matrix of the system equals the number of unknown coefficients, the vertical size equals the number of different derivative terms.

The right-hand side of the equation is zero, which means that the equation can have one zero solution or an infinite number of solutions. If it has zero solution only, it means that only invariants identically equaling zero can exist within the given structure. If it has infinite solutions, then the dimension of the space of the solutions equals the number of linearly independent invariants. If the dimension is n, and we have found n linearly independent solutions $I_1, I_2, \ldots, I_n$ of the equation, then each solution can be expressed as $k_1 I_1 + k_2 I_2 + \cdots + k_n I_n$, where $k_1, k_2, \ldots, k_n$ are arbitrary real numbers.

The base $I_1, I_2, \ldots, I_n$ of the solution can be found by SVD of the system matrix, which is decomposed to the product $\mathbf{U} \cdot \mathbf{W} \cdot \mathbf{V}^{\mathrm{T}}$, where $\mathbf{U}$ and $\mathbf{V}$ are orthonormal matrices and $\mathbf{W}$ is a diagonal matrix of the singular values. The solution of the system of linear equations is in each row of the matrix $\mathbf{V}$ that corresponds to a zero singular value in the matrix $\mathbf{W}$.

The solution from the matrix $\mathbf{V}$ does not have integer coefficients. Since it is always possible (and desirable) to obtain invariants with integer coefficients, we impose this requirement as an additional constraint on the solution.

An example with the structure (2,0,2)

An illustrative explanation of the above method can be seen from the following example. Let us consider the order $s = 4$ and the structure $\mathbf{s} = (2, 0, 2)$ as input parameters. Then the weight $w = 6$, and the degree $r = 4$.

Now we need to generate all partitions of the number 6 into a sum of two integers from 0 to 2 and two integers from 0 to 4; see Table 3.2. The first column contains partitions at the first stage among different orders, the second column shows partitions at the second stage inside the orders. One can see that the partitions $4 + 2$ and $2 + 4$ in the first stage are different; hence we must consider both cases. If we have partitions $3 + 2 + 1 + 0, 3 + 2 + 0 + 1, 2 + 3 + 1 + 0$ and $2 + 3 + 0 + 1$ in the second stage, we can see that it would be only interchanging of the factors in the term, so dealing with only one of these partitions is sufficient.

The moments in the terms are then sorted, so there are moments of the second order first in the fourth column of the table. The terms are also sorted, so the unknown coefficients in the third column are numbered according to this sorted order of terms. The coefficients are unknown at this phase of the computation, but their correct labeling is important. We can see that the theoretical number of terms $n_t = 14$. Then, the symmetric counterparts of each term are searched. The coefficients at the symmetric terms are labeled by the same number, as can be seen in the third column of the table; the number of different coefficients is $n_s = 10$.

Now, we need to compose the Cayley–Aronhold equation. The derivatives with respect to particular moments are

$$D(\mu_{13}) = 2c_2\mu_{20}^2\mu_{13}\mu_{04} + c_4\mu_{20}\mu_{11}\mu_{22}\mu_{04} + c_6\mu_{20}\mu_{02}\mu_{31}\mu_{04} + c_9\mu_{11}^2\mu_{31}\mu_{04}$$
$$+ c_3\mu_{11}\mu_{02}\mu_{40}\mu_{04}$$

Table 3.2 All possible terms of the invariant.

First stage	Second stage	Coefficient	Term
$6+0$	$4+2+0+0$	c_1	$\mu_{02}^2\mu_{40}\mu_{22}$
	$3+3+0+0$	c_2	$\mu_{02}^2\mu_{31}^2$
$5+1$	$4+1+1+0$	c_3	$\mu_{11}\mu_{02}\mu_{40}\mu_{13}$
	$3+2+1+0$	c_4	$\mu_{11}\mu_{02}\mu_{31}\mu_{22}$
$4+2$	$4+0+2+0$	c_5	$\mu_{20}\mu_{02}\mu_{40}\mu_{04}$
	$4+0+1+1$	c_8	$\mu_{11}^2\mu_{40}\mu_{04}$
	$3+1+2+0$	c_6	$\mu_{20}\mu_{02}\mu_{31}\mu_{13}$
	$3+1+1+1$	c_9	$\mu_{11}^2\mu_{31}\mu_{13}$
	$2+2+2+0$	c_7	$\mu_{20}\mu_{02}\mu_{22}^2$
	$2+2+1+1$	c_{10}	$\mu_{11}^2\mu_{22}^2$
$3+3$	$3+0+2+1$	c_3	$\mu_{20}\mu_{11}\mu_{31}\mu_{04}$
	$2+1+2+1$	c_4	$\mu_{20}\mu_{11}\mu_{22}\mu_{13}$
$2+4$	$2+0+2+2$	c_1	$\mu_{20}^2\mu_{22}\mu_{04}$
	$1+1+2+2$	c_2	$\mu_{20}^2\mu_{13}^2$

$$D(\mu_{22}) = 2c_1\mu_{20}^2\mu_{13}\mu_{04} + 2c_4\mu_{20}\mu_{11}\mu_{13}^2 + 4c_7\mu_{20}\mu_{02}\mu_{22}\mu_{13} + 4c_{10}\mu_{11}^2\mu_{22}\mu_{13}$$
$$+ 2c_1\mu_{02}^2\mu_{40}\mu_{13} + 2c_4\mu_{11}\mu_{02}\mu_{31}\mu_{13}$$

$$D(\mu_{31}) = 3c_3\mu_{20}\mu_{11}\mu_{22}\mu_{04} + 3c_6\mu_{20}\mu_{02}\mu_{22}\mu_{13} + 3c_9\mu_{11}^2\mu_{22}\mu_{13} + 6c_2\mu_{02}^2\mu_{31}\mu_{22}$$
$$+ 3c_4\mu_{11}\mu_{02}\mu_{22}^2$$

$$D(\mu_{40}) = 4c_5\mu_{20}\mu_{02}\mu_{31}\mu_{04} + 4c_8\mu_{11}^2\mu_{31}\mu_{04} + 4c_1\mu_{02}^2\mu_{31}\mu_{22} + 4c_3\mu_{11}\mu_{02}\mu_{31}\mu_{13}$$

$$D(\mu_{11}) = c_3\mu_{20}\mu_{02}\mu_{31}\mu_{04} + c_4\mu_{20}\mu_{02}\mu_{22}\mu_{13} + 2c_8\mu_{11}\mu_{02}\mu_{40}\mu_{04} + 2c_9\mu_{11}\mu_{02}\mu_{31}\mu_{13}$$
$$+ 2c_{10}\mu_{11}\mu_{02}\mu_{22}^2 + c_3\mu_{02}^2\mu_{40}\mu_{13} + c_4\mu_{02}^2\mu_{31}\mu_{22}$$

$$D(\mu_{20}) = 4c_1\mu_{20}\mu_{11}\mu_{22}\mu_{04} + 4c_2\mu_{20}\mu_{11}\mu_{13}^2 + 2c_3\mu_{11}^2\mu_{31}\mu_{04} + 2c_4\mu_{11}^2\mu_{22}\mu_{13}$$
$$+ 2c_5\mu_{11}\mu_{02}\mu_{40}\mu_{04} + 2c_6\mu_{11}\mu_{02}\mu_{31}\mu_{13} + 2c_7\mu_{11}\mu_{02}\mu_{22}^2.$$

The equation

$$D(\mu_{13}) + D(\mu_{22}) + D(\mu_{31}) + D(\mu_{40}) + D(\mu_{11}) + D(\mu_{20}) = 0 \qquad (3.79)$$

should hold for arbitrary values of the moments. From this constraint we can put together the matrix of the system of linear equations for the coefficients as can be seen in Table 3.3.

After the SVD, the diagonal of the $\mathbf{W}$ matrix contains the values 9.25607, 7.34914, 6.16761, $3.03278 \cdot 10^{-16}$, 2.41505, 3.51931, 4.02041, 4.65777, 5.11852 and $4.49424 \cdot 10^{-16}$. We can see two values, $3.03278 \cdot 10^{-16}$ and $4.49424 \cdot 10^{-16}$, are less than the threshold, so the system of equations has two independent solutions. Integer coefficients of

Table 3.3 The matrix of the system of linear equations for the coefficients. The empty elements are zero. The solution is in the last two rows.

Term	c_1	c_2	c_3	c_4	c_5	c_6	c_7	c_8	c_9	c_{10}
$\mu_{20}^2\mu_{13}\mu_{04}$	2	2								
$\mu_{20}\mu_{11}\mu_{22}\mu_{04}$	4		3	1						
$\mu_{20}\mu_{02}\mu_{31}\mu_{04}$			1		4	1				
$\mu_{11}^2\mu_{31}\mu_{04}$			2					4	1	
$\mu_{11}\mu_{02}\mu_{40}\mu_{04}$			1		2			2		
$\mu_{20}\mu_{11}\mu_{13}^2$		4		2						
$\mu_{20}\mu_{02}\mu_{22}\mu_{13}$			1			3	4			
$\mu_{11}^2\mu_{22}\mu_{13}$			2						3	4
$\mu_{11}\mu_{02}\mu_{31}\mu_{13}$			4	2		2			2	
$\mu_{02}^2\mu_{40}\mu_{13}$	2		1							
$\mu_{11}\mu_{02}\mu_{22}^2$				3			2			2
$\mu_{02}^2\mu_{31}\mu_{22}$	4	6	1							
I_{S1}	1	−1	−2	2			2	−2	1	−1
I_{S2}					1	−4	3	−1	4	−3

the solutions are in the last two rows of Table 3.3. We obtain two solutions:

$$I_a = (\mu_{20}^2\mu_{22}\mu_{04} - \mu_{20}^2\mu_{13}^2 - 2\mu_{20}\mu_{11}\mu_{31}\mu_{04} + 2\mu_{20}\mu_{11}\mu_{22}\mu_{13} + 2\mu_{20}\mu_{02}\mu_{31}\mu_{13}$$
$$- 2\mu_{20}\mu_{02}\mu_{22}^2 + \mu_{11}^2\mu_{40}\mu_{04} - \mu_{11}^2\mu_{22}^2 - 2\mu_{11}\mu_{02}\mu_{40}\mu_{13} + 2\mu_{11}\mu_{02}\mu_{31}\mu_{22}$$
$$+ \mu_{02}^2\mu_{40}\mu_{22} - \mu_{02}^2\mu_{31}^2)/\mu_{00}^{10}$$

and

$$I_b = (\mu_{20}\mu_{02}\mu_{40}\mu_{04} - 4\mu_{20}\mu_{02}\mu_{31}\mu_{13} + 3\mu_{20}\mu_{02}\mu_{22}^2 - \mu_{11}^2\mu_{40}\mu_{04} + 4\mu_{11}^2\mu_{31}\mu_{13}$$
$$- 3\mu_{11}^2\mu_{22}^2)/\mu_{00}^{10}.$$

We can easily recognize the link to the AMIs yielded by the graph method:

$$I_a = I_9, \quad I_b = I_1 I_6.$$

3.6 Numerical experiments

In this section, we demonstrate the performance of both the AMIs and the normalized moments on artificial data in simulated recognition experiments and in a controlled experiment on real data, where the known ground truth allows the success rate to be evaluated. Practical experiments on real data can be found in Chapter 8.

3.6.1 Digit recognition

This experiment illustrates recognition of digits 1, 2, . . . , 9, 0 (see Figure 3.14) deformed by various affine transforms. We deliberately choose a font where the digit 9 is not exactly a rotated version of 6.

The original binary image of each digit of the size 48×32 was deformed by 100 affine transforms, whose parameters were generated as random values with Gaussian distribution. The "mean value" was identity transform, the standard deviation of the translation was 8 pixels and the standard deviation σ of the other parameters was set to 0.25. Examples of the deformed digits can be seen in Figure 3.15.

1 2 3 4 5 6 7 8 9 0

Figure 3.14 The original digits used in the recognition experiment.

1 2 3 4 5 6 7 8 9 0

Figure 3.15 The examples of the deformed digits.

All deformed digits were classified independently by a minimum-distance classifier acting in the space of 30 AMIs. The values of the invariants were normalized by their standard deviations to have a comparable dynamic range. The results are summarized in Table 3.4(a). It is clearly visible that the AMIs yielded an excellent 100% success rate. Part of the results is graphically visualized in Figure 3.16, where one can see the distribution of digits in the space of two AMIs I_1 and I_6. All digits form compact clusters, well separated from each other (the clusters of "6" and "9" are close to one another but still separable). The same situation can be observed in other feature subspaces.

To test the robustness with respect to noise, we ran this experiment again. Gaussian white noise inducing SNR 26 dB was added to each deformed digit. (To eliminate the role of the background, the noise was added to the smallest rectangle circumscribed to the digits only.) The results are summarized in Table 3.4(b). We may still observe an excellent recognition rate with the exception of certain "9" misclassified as "6".

When repeating this experiment with a heavier noise of 14 dB, the success rate dropped (see Table 3.4(c)).

In the next part of this experiment, we demonstrate the importance of choosing independent invariants. We classified the noisy digits (SNR $=$ 14 dB) again but only by eight invariants ($I_1, \ldots I_4, I_6, \ldots I_9,$). As seen in Table 3.5(a), the results are for obvious reasons slightly worse than those achieved by 30 AMIs.

Then, an irreducible but dependent invariant I_{10} was added to the feature set and the digits were reclassified. The results are presented in Table 3.5(b). As can be seen, the overall percentage of correct results increased only slightly. Then, the invariant I_{10} was replaced by the independent invariant I_{25}. As seen in Table 3.5(c), the percentage of the correct recognitions increased significantly to the level achieved by 30 invariants.

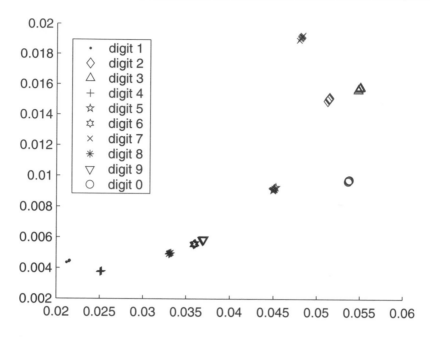

Figure 3.16 The digits in the feature space of AMIs I_1 and I_6 (ten samples of each digit).

Table 3.4 The recognition rate (in %) of the AMIs.

	1	2	3	4	5	6	7	8	9	0	Overall
(a) Noise-free case	100	100	100	100	100	100	100	100	100	100	100
(b) Additive noise, SNR = 26 dB	100	100	100	100	100	100	100	100	71	100	97.1
(c) Additive noise, SNR = 14 dB	99	82	78	36	100	52	95	100	27	70	73.9

Table 3.5 The percentage of correctly recognized figures by the limited number of affine moment invariants.

	1	2	3	4	5	6	7	8	9	0	Overall
(a) Independent invariant set I_1, I_2, I_3, I_4, I_6, I_7, I_8, I_9	76	80	31	34	100	46	76	100	68	51	66.2
(b) Dependent invariant set I_1, I_2, I_3, I_4, I_6, I_7, I_8, I_9, I_{10}	51	100	46	73	53	56	74	100	72	70	69.5
(c) Independent invariant set I_1, I_2, I_3, I_4, I_6, I_7, I_8, I_9, I_{25}	89	77	73	63	94	66	79	100	54	50	74.5

3.6.2   Recognition of symmetric patterns

This experiment demonstrates the power of the features to recognize affinely distorted symmetric objects. The simple binary patterns used in the test are shown in Figure 3.17(a). They have N-FRS with $N = 1, 2, 3, 4, 5$ and ∞, respectively. Each pattern was deformed successively by ten randomly generated affine transforms (for one instance, see Figure 3.17(b)). The affine deformations were constructed as a matrix product

$$\begin{pmatrix} a_1 & a_2 \\ b_1 & b_2 \end{pmatrix} = \begin{pmatrix} 1 & 0 \\ 0 & z \end{pmatrix} \begin{pmatrix} \cos \varrho & -\sin \varrho \\ \sin \varrho & \cos \varrho \end{pmatrix} \begin{pmatrix} \omega & 0 \\ 0 & \omega \end{pmatrix} \begin{pmatrix} \delta & 0 \\ 0 & 1/\delta \end{pmatrix} \begin{pmatrix} \cos \alpha & -\sin \alpha \\ \sin \alpha & \cos \alpha \end{pmatrix}, \quad (3.80)$$

where α and ϱ were uniformly distributed between 0 and 2π, ω and δ were normally distributed with mean value 1 and standard deviation 0.1. The value of z was set randomly to ± 1. We added zero-mean Gaussian noise of standard deviation 0.25 to the deformed objects (the original image values were from $\{0, 1\}$).

Figure 3.17   Test patterns: (a) originals, (b) examples of distorted patterns, and (c) the standard positions.

First, we used AMIs up to the fifth order. All patterns in all instances were recognized correctly. An example of the feature space of the AMIs I_6 and I_{47} is in Figure 3.18.

Then, the normalized moments were used as features for recognition. In the bottom row of Figure 3.17 the normalized positions of the test patterns are shown. Recall that this is for illustration only; transforming the objects is not required for calculation of the normalized moments. In the experiment with noisy patterns (as well as in reality), the choice of threshold used for the decision whether or not a moment is zero is very important and may greatly affect the results. In this case, the threshold 0.175 provided an error-free result, the threshold 0.17 as well as the threshold 0.18 caused one error. This is a consequence of the heavy noise. For an example of the feature space of the normalized moments, see Figure 3.19.

The clusters of both types of feature are a little spread, but because they do not intersect each other, the recognition power of the moments is still preserved.

3.6.3   The children's mosaic

To show the behavior of the invariants on real images, we have chosen an experiment with a mosaic carpet (Figure 3.20) consisting of various tiles. Each tile contains a simple pattern – geometric figure (a polygon or a conic section) or an animal silhouette. Each tile was snapped

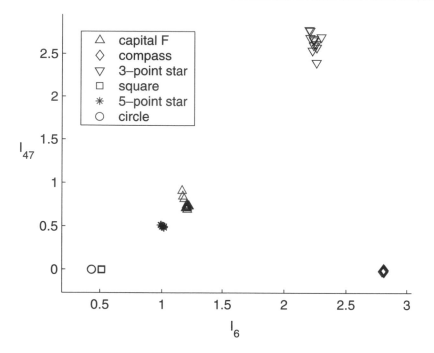

Figure 3.18 The space of two AMIs I_6 and I_{47}.

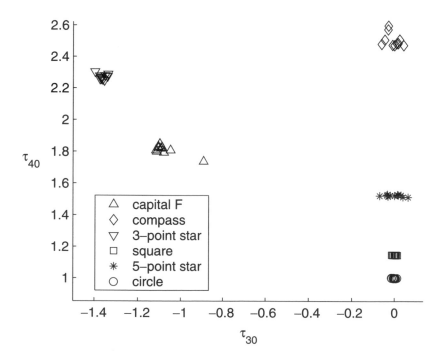

Figure 3.19 The space of two normalized moments τ_{30} and τ_{40}.

by a camera perpendicularly and then with a slight projective distortion (3.1) from eight sides, so we have nine photographs of each tile. The slant angle of the camera was about 20° from the perpendicular direction. For an example of a tile picture, see Figure 3.21. We can see a noticeable trapezoidal distortion of the tile. The tiles are colored but in this experiment we used the color information for the segmentation only; the moments were then computed from binary images. The perpendicular snaps were used as class representatives (templates) and the other snaps were classified by a minimum distance classifier.

Figure 3.20 The part of the mosaic used in the experiment.

Table 3.6 The numbers of errors in recognition of 200 tile snaps.

Maximum order of invariants	6	7	8	9
AMIs		6	3	0	7
Normalized moments		2	4	1	4

The tiles were first classified by the AMIs and then by the normalized moments. In both cases, the magnitudes of the individual features were scaled to the same range. The recognition rate depends significantly on the order (and the number) of the invariants used. In Table 3.6, we can see the total number of errors that occurred for various maximum orders of the invariants. Clearly, the maximum order 8 is an optimal choice both for the AMIs as well as for the normalized moments. For the detailed contingency table for the AMIs, see Table 3.7. Note that the geometric patterns include objects that can be affinely transformed

Figure 3.21 The image of a tile with slight projective distortion.

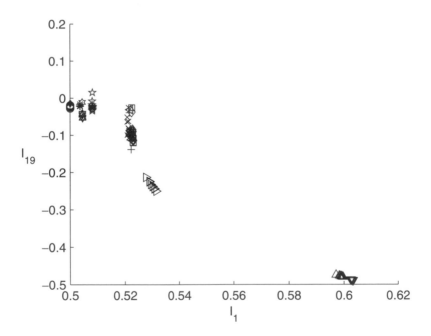

Figure 3.22 The geometric patterns. The feature space of I_1 and I_{19}: $\triangledown$ – equilateral triangle, $\triangle$ – isosceles triangle, $\square$ – square, $\Diamond$ – rhombus, $+$ – big rectangle, $\times$ – small rectangle, $\triangleleft$ – rhomboid, $\triangleright$ – trapezoid, $\star$ – pentagon, $*$ – regular hexagon, $\maltese$ – irregular hexagon, $\bigcirc$ – circle, $\cdot$ – ellipse.

one to another. The first such group consists of an equilateral and an isosceles triangle, the second group includes a square, a rhombus, small and big rectangles and a rhomboid, the third group includes a regular and an irregular hexagon, and the last group of equivalent objects

Table 3.7 The contingency table of the AMIs up to the eighth order.

Equilateral triangle	4 4 0 0 0 0 0 0 0 0 0 0 0 0 0 0 0 0 0 0 0 0 0 0 0 0
Isosceles triangle	4 4 0 0 0 0 0 0 0 0 0 0 0 0 0 0 0 0 0 0 0 0 0 0 0 0
Square	0 0 0 0 0 0 0 0 0 0 0 0 0 0 0 0 0 0 0 0 0 0 0 0 0 0
Rhombus	0 0 1 4 0 0 1 0 0 0 0 0 0 0 0 0 0 0 0 0 0 0 0 0 0 0
Big rectangle	0 0 7 4 8 7 7 0 0 0 0 0 0 0 0 0 0 0 0 0 0 0 0 0 0 0
Small rectangle	0 0 0 0 0 1 0 0 0 0 0 0 0 0 0 0 0 0 0 0 0 0 0 0 0 0
Rhomboid	0 0 0 0 0 0 0 0 0 0 0 0 0 0 0 0 0 0 0 0 0 0 0 0 0 0
Trapezoid	0 0 0 0 0 0 0 8 0 0 0 0 0 0 0 0 0 0 0 0 0 0 0 0 0 0
Pentagon	0 0 0 0 0 0 0 0 8 0 0 0 0 0 0 0 0 0 0 0 0 0 0 0 0 0
Regular hexagon	0 0 0 0 0 0 0 0 0 2 0 0 0 0 0 0 0 0 0 0 0 0 0 0 0 0
Irregular hexagon	0 0 0 0 0 0 0 0 0 8 6 0 0 0 0 0 0 0 0 0 0 0 0 0 0 0
Circle	0 0 0 0 0 0 0 0 0 0 0 8 6 0 0 0 0 0 0 0 0 0 0 0 0 0
Ellipse	0 0 0 0 0 0 0 0 0 0 0 0 2 0 0 0 0 0 0 0 0 0 0 0 0 0
Bear	0 0 0 0 0 0 0 0 0 0 0 0 0 8 0 0 0 0 0 0 0 0 0 0 0 0
Squirrel	0 0 0 0 0 0 0 0 0 0 0 0 0 0 8 0 0 0 0 0 0 0 0 0 0 0
Pig	0 0 0 0 0 0 0 0 0 0 0 0 0 0 0 8 0 0 0 0 0 0 0 0 0 0
Cat	0 0 0 0 0 0 0 0 0 0 0 0 0 0 0 0 8 0 0 0 0 0 0 0 0 0
Bird	0 0 0 0 0 0 0 0 0 0 0 0 0 0 0 0 0 8 0 0 0 0 0 0 0 0
Dog	0 0 0 0 0 0 0 0 0 0 0 0 0 0 0 0 0 0 8 0 0 0 0 0 0 0
Cock	0 0 0 0 0 0 0 0 0 0 0 0 0 0 0 0 0 0 0 8 0 0 0 0 0 0
Hedgehog	0 0 0 0 0 0 0 0 0 0 0 0 0 0 0 0 0 0 0 0 8 0 0 0 0 0
Rabbit	0 0 0 0 0 0 0 0 0 0 0 0 0 0 0 0 0 0 0 0 0 8 0 0 0 0
Duck	0 0 0 0 0 0 0 0 0 0 0 0 0 0 0 0 0 0 0 0 0 0 8 0 0 0
Dolphin	0 0 0 0 0 0 0 0 0 0 0 0 0 0 0 0 0 0 0 0 0 0 0 8 0 0
Cow	0 0 0 0 0 0 0 0 0 0 0 0 0 0 0 0 0 0 0 0 0 0 0 0 0 8

contains a circle and an ellipse. If a pattern in a group is classified as another pattern within the same group, it cannot be considered a mistake.

For an example of the feature subspace of the AMIs, see Figures 3.22 and 3.23. The patterns were divided into two graphs, for lucidity. In the first graph, the affinely equivalent geometric patterns actually create one mixed cluster. In the second graph, the clusters of animal silhouettes are partially overlapping, the correct recognition by means of two features is not possible. Fortunately, by using 39 features up to the eighth order, a correct recognition can be achieved. For analogous graphs for the normalized moments, see Figures 3.24 and 3.25.

For a comparison, we tried to recognize the tiles also by the basic TRS invariants. The best achievable result was 18 errors out of 200 trials, which clearly illustrates the necessity of having affine-invariant features in this case.

To investigate the influence of the perspective projection, we repeated this experiment with a different setup. The tiles were captured by a camera with the slant angle of approximately 45°. As seen in Figure 3.26, the trapezoidal distortion is now significant. The best possible result of the AMIs up to the eighth order yielded 56 errors, the optimal result of the normalized moments produced 53 errors. It illustrates that these features cannot be successfully used in the case of heavy projective distortion.

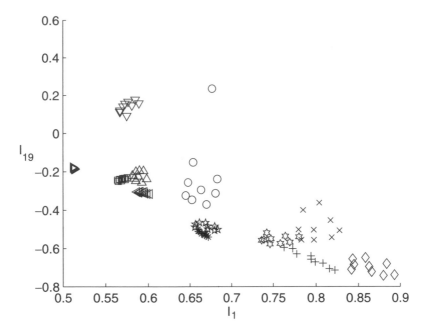

Figure 3.23 The animal silhouettes. The feature space of I_1 and I_{19}: $\triangledown$ – bear, $\triangle$ – squirrel, $\square$ – pig, $\diamond$ – cat, + – bird, × – dog, $\triangleleft$ – cock, $\triangleright$ – hedgehog, ☆ – rabbit, ✱ – duck, ✿ – dolphin, $\bigcirc$ – cow.

3.7   Affine invariants of color images

All invariants that have been presented in this chapter (and also in Chapter 2) are defined for graylevel images. In most applications, this definition is restricted to binary images. On the other hand, in this section we show how the definition of moment invariants can be extended to *color images* and in general to *multichannel images*.

Color images can be understood as three RGB channels, each of them viewable as a graylevel image. Without introducing any new theory, we could calculate moment invariants from each channel independently, and use them as a three-component vector. In that way we should obtain $3(m - 6)$ affine invariants, where m is the number of moments used. However, the RGB channels are not independent and the affine transform is supposed to be the same for each channel, so we should have $3m - 6$ independent invariants.

To exploit the between-channel relations, Mindru *et al.* [40] used moments computed from certain products and powers of the channels. They defined the *generalized color moment* of degree $d = \alpha + \beta + \gamma$ as

$$M_{pq}^{\alpha\beta\gamma} = \iint_D x^p y^q (R(x, y))^\alpha (G(x, y))^\beta (B(x, y))^\gamma \, dx \, dy, \qquad (3.81)$$

where R, G and B are image functions of three channels of a color image and α, β, γ are non-negative integers. They use these moments for the construction of affine invariants and also of combined invariants to the affine transform and brightness/contrast linear changes.

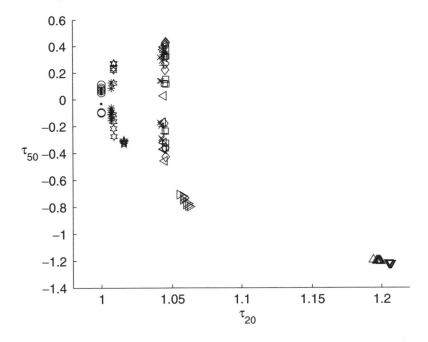

Figure 3.24 The geometric patterns. The feature space of τ_{20} and τ_{50}: $\triangledown$ – equilateral triangle, $\triangle$ – isosceles triangle, $\square$ – square, $\diamondsuit$ – rhombus, $+$ – big rectangle, $\times$ – small rectangle, $\triangleleft$ – rhomboid, $\triangleright$ – trapezoid, $\star$ – pentagon, $*$ – regular hexagon, $\maltese$ – irregular hexagon, $\bigcirc$ – circle, $\cdot$ – ellipse.

These features unfortunately exhibit very high redundancy. In the case of an infinite set of moments of all orders, only the moments where $d = 1$ are independent. This redundancy decreases as the maximum order of moments decreases; for low-order moments this method may yield meaningful results. However, even for low orders, using higher powers of brightness in individual channels is more sensitive to nonlinearity of the contrast changes and may lead to misclassification.

The theory of affine invariants offers another approach using simultaneous affine invariants. By a *simultaneous invariant* we understand an AMI containing moments of at least two different orders.[11] This method producing *joint invariants* does not yield dependent descriptors.

We can choose two channels (let us label them a and b), take an arbitrary simultaneous invariant, e.g. I_3, and calculate the second-order moments from one channel and the third-order moments from the other channel. We obtain a joint invariant

$$J_{2,3} = (\mu_{20}^{(a)}\mu_{21}^{(b)}\mu_{03}^{(b)} - \mu_{20}^{(a)}(\mu_{12}^{(b)})^2 - \mu_{11}^{(a)}\mu_{30}^{(b)}\mu_{03}^{(b)} + \mu_{11}^{(a)}\mu_{21}^{(b)}\mu_{12}^{(b)} + \mu_{02}^{(a)}\mu_{30}^{(b)}\mu_{12}^{(b)}$$
$$- \mu_{02}^{(a)}(\mu_{21}^{(b)})^2)/\mu_{00}^7$$

(the normalizing moment μ_{00} is computed from the whole color image, $\mu_{00} = \mu_{00}^{(a)} + \mu_{00}^{(b)} + \mu_{00}^{(c)}$).

---

[11] An invariant containing the moments of one order only is called homogeneous.

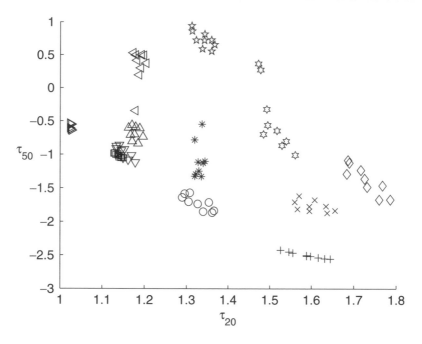

Figure 3.25 The animal silhouettes. The feature space of τ_{20} and τ_{50}: $\triangledown$ – bear, $\triangle$ – squirrel, $\square$ – pig, $\diamondsuit$ – cat, $+$ – bird, $\times$ – dog, $\triangleleft$ – cock, $\triangleright$ – hedgehog, $\star$ – rabbit, $\ast$ – duck, $\star$ – dolphin, $\bigcirc$ – cow.

Figure 3.26 The image of a tile with heavy projective distortion.

We can also utilize algebraic invariants of two or more binary forms of the same order. An example of such an invariant is of the second order

$$J_2 = (\mu_{20}^{(a)} \mu_{02}^{(b)} + \mu_{20}^{(b)} \mu_{02}^{(a)} - 2\mu_{11}^{(a)} \mu_{11}^{(b)})/\mu_{00}^4.$$

If we use moments of one channel only (i.e. $a = b$), we obtain essentially I_1. Another such invariant is of the third order

$$J_3 = (\mu_{30}^{(a)} \mu_{03}^{(b)} - 3\mu_{21}^{(a)} \mu_{12}^{(b)} + 3\mu_{21}^{(b)} \mu_{12}^{(a)} - \mu_{30}^{(b)} \mu_{03}^{(a)})/\mu_{00}^5.$$

If we use moments of one channel only, then $J_3 = 0$. A third-order invariant of degree two from one channel does not exist, while that from two channels does.

In color invariants, we can also employ the zero and the first-order moments. The central moments are defined with respect to the main centroid of the whole image

$$m_{00} = m_{00}^{(a)} + m_{00}^{(b)} + m_{00}^{(c)},$$

$$x_c = (m_{10}^{(a)} + m_{10}^{(b)} + m_{10}^{(c)})/m_{00}, \qquad y_c = (m_{01}^{(a)} + m_{01}^{(b)} + m_{01}^{(c)})/m_{00}, \tag{3.82}$$

$$\mu_{pq}^{(a)} = \int_{-\infty}^{\infty} \int_{-\infty}^{\infty} (x - x_c)^p (y - y_c)^q a(x, y) \, dx \, dy \quad p, q = 0, 1, 2, \dots. \tag{3.83}$$

Since the centroids of the individual channels are generally different from the main centroid, the first-order moments need not be zero. We can use them for construction of additional joint invariants, e.g.

$$J_1 = (\mu_{10}^{(a)} \mu_{01}^{(b)} - \mu_{10}^{(b)} \mu_{01}^{(a)})/\mu_{00}^3.$$

An example of a simultaneous invariant of the first and second orders is

$$J_{1,2} = (\mu_{20}^{(a)} (\mu_{01}^{(b)})^2 + \mu_{02}^{(a)} (\mu_{10}^{(b)})^2 - 2\mu_{11}^{(a)} \mu_{10}^{(b)} \mu_{01}^{(b)})/\mu_{00}^5.$$

There exists even a zero-order joint affine invariant

$$J_0 = \mu_{00}^{(a)}/\mu_{00}^{(b)}.$$

Using joint invariants, we obtain 12 additional independent invariants of color images. This idea can be further extended to multichannel images consisting of an arbitrary number of channels. This may be particularly useful in satellite and aerial image recognition because contemporary multi- and hyper-spectral sensors produce up to several hundreds of channels (spectral bands).

3.8   Generalization to three dimensions

As already mentioned in Chapter 2, 3D images have become common, namely in medical imaging, in the last two decades. Although the task of recognizing 3D objects appears less frequently than that in 2D, generalization of the AMIs to 3D may not only come in useful but is also of theoretical interest. Any method of deriving affine invariants described in this chapter can be extended to 3D. However, their extension difficulty is different.

A 3D affine transform of the coordinates (x, y, z) to (x', y', z') can be expressed in a matrix form as

$$\mathbf{x}' = \mathbf{A}\mathbf{x} + \mathbf{b},$$

$$\text{where } \mathbf{x}' = \begin{pmatrix} x' \\ y' \\ z' \end{pmatrix}, \ \mathbf{A} = \begin{pmatrix} a_1 & a_2 & a_3 \\ b_1 & b_2 & b_3 \\ c_1 & c_2 & c_3 \end{pmatrix}, \ \mathbf{x} = \begin{pmatrix} x \\ y \\ z \end{pmatrix}, \ \mathbf{b} = \begin{pmatrix} a_0 \\ b_0 \\ c_0 \end{pmatrix}. \tag{3.84}$$

In this section, we present a derivation of 3D AMIs using the method of geometric primitives, which is a generalization of the graph method well known from the 2D case, and by affine normalization. The tensor method and the method using the Cayley–Aronhold equation are mentioned only briefly.

3.8.1   Method of geometric primitives

The method of geometric primitives was introduced by Xu and Li [41] and performs a 3D version of the graph method. It uses tetrahedrons with one vertex in the centroid of the object and three vertices moving through the object. The volume of such a body is

$$V(O, 1, 2, 3) = \frac{1}{6} \begin{vmatrix} x_1 & x_2 & x_3 \\ y_1 & y_2 & y_3 \\ z_1 & z_2 & z_3 \end{vmatrix}. \tag{3.85}$$

When we denote

$$C_{123} = 6V(O, 1, 2, 3),$$

then, having $(r \geq 3)$ points, an invariant can be obtained by integration of the product

$$I(f) = \int_{-\infty}^{\infty} \prod_{j,k,\ell=1}^{r} C_{jk\ell}^{n_{jk\ell}} \cdot \prod_{i=1}^{r} f(x_i, y_i, z_i) \, dx_i \, dy_i \, dz_i, \tag{3.86}$$

where $n_{jk\ell}$ are non-negative integers,

$$w = \sum_{j,k,\ell} n_{jk\ell}$$

is the weight of the invariant and r is its degree. When undergoing an affine transform, $I(f)$ changes as

$$I(f)' = J^w |J|^r I(f).$$

The normalization of $I(f)$ by μ_{000}^{w+r} yields an absolute affine invariant.
   For example, if $r = 3$ and $n_{123} = 2$, then we obtain

$$I_1^{3D} = \frac{1}{6} \int_{-\infty}^{\infty} C_{123}^2 f(x_1, y_1, z_1) f(x_2, y_2, z_2)$$

$$\times f(x_3, y_3, z_3) \, dx_1 \, dy_1 \, dz_1 \, dx_2 \, dy_2 \, dz_2 \, dx_3 \, dy_3 \, dz_3 / \mu_{000}^5$$

$$= (\mu_{200}\mu_{020}\mu_{002} + 2\mu_{110}\mu_{101}\mu_{011} - \mu_{200}\mu_{011}^2 - \mu_{020}\mu_{101}^2 - \mu_{002}\mu_{110}^2)/\mu_{000}^5. \tag{3.87}$$

Another example we obtain for $r = 4$, $n_{123} = 1$, $n_{124} = 1$, $n_{134} = 1$ and $n_{234} = 1$:

$$I_2^{3D} = \frac{1}{36} \int_{-\infty}^{\infty} C_{123}C_{124}C_{134}C_{234} f(x_1, y_1, z_1) f(x_2, y_2, z_2) f(x_3, y_3, z_3) f(x_4, y_4, z_4)$$

$$dx_1\, dy_1\, dz_1\, dx_2\, dy_2\, dz_2\, dx_3\, dy_3\, dz_3\, dx_4\, dy_4\, dz_4 / \mu_{000}^8$$

$$= (\mu_{300}\mu_{003}\mu_{120}\mu_{021} + \mu_{300}\mu_{030}\mu_{102}\mu_{012} + \mu_{030}\mu_{003}\mu_{210}\mu_{201} - \mu_{300}\mu_{120}\mu_{012}^2$$

$$- \mu_{300}\mu_{102}\mu_{021}^2 - \mu_{030}\mu_{210}\mu_{102}^2 - \mu_{030}\mu_{201}^2\mu_{012} - \mu_{003}\mu_{210}^2\mu_{021} - \mu_{003}\mu_{201}\mu_{120}^2$$

$$- \mu_{300}\mu_{030}\mu_{003}\mu_{111} + \mu_{300}\mu_{021}\mu_{012}\mu_{111} + \mu_{030}\mu_{201}\mu_{102}\mu_{111}$$

$$+ \mu_{003}\mu_{210}\mu_{120}\mu_{111} + \mu_{210}^2\mu_{012}^2 + \mu_{201}^2\mu_{021}^2 + \mu_{120}^2\mu_{102}^2 - \mu_{210}\mu_{120}\mu_{102}\mu_{012}$$

$$- \mu_{210}\mu_{201}\mu_{021}\mu_{012} - \mu_{201}\mu_{120}\mu_{102}\mu_{021} - 2\mu_{210}\mu_{012}\mu_{111}^2 - 2\mu_{201}\mu_{021}\mu_{111}^2$$

$$- 2\mu_{120}\mu_{102}\mu_{111}^2 + 3\mu_{210}\mu_{102}\mu_{021}\mu_{111} + 3\mu_{201}\mu_{120}\mu_{012}\mu_{111} + \mu_{111}^4) / \mu_{000}^8.$$

$$(3.88)$$

The 3D invariants cannot be described by ordinary multigraphs, where each edge connects two nodes. In 3D, we would need some *hyperedges* that connect three nodes. This concept is known in the literature as a *hypergraph*. The hypergraphs were used, e.g. in reference [42], for estimation of parameters of an affine transform.

The *k-uniform hypergraph* is a pair $G = (V, E)$, where $V = \{1, 2, \ldots, n\}$ is a finite set of nodes and $E \subseteq \binom{V}{k}$ is a set of hyperedges. Each hyperedge connects up to k nodes. In this context, ordinary graphs are two-uniform hypergraphs. We need three-uniform hypergraphs for description of the 3D AMIs, i.e. each hyperedge connects three nodes. The hypergraphs corresponding to I_1^{3D} and I_2^{3D} are shown in Figure 3.27, where the hyperedges are drawn as sets containing the connected nodes.

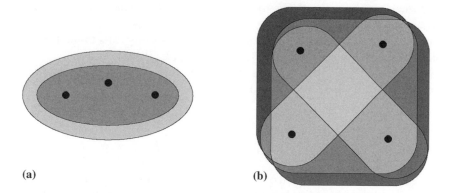

(a)   (b)

Figure 3.27 The hypergraphs corresponding to invariants (a) I_1^{3D}, and (b) I_2^{3D}.

In tensor notation, the 3D moment tensor is

$$M^{i_1 i_2 \cdots i_k} = \int_{-\infty}^{\infty} \int_{-\infty}^{\infty} \int_{-\infty}^{\infty} x^{i_1} x^{i_2} \cdots x^{i_k} f(x^1, x^2, x^3)\, dx^1\, dx^2\, dx^3, \qquad (3.89)$$

where $x^1 = x, x^2 = y$ and $x^3 = z$. $M^{i_1 i_2 \cdots i_k} = m_{pqr}$ if p indices equal 1, q indices equal 2 and r indices equal 3. The unit polyvector in 3D: $\varepsilon_{123} = \varepsilon_{312} = \varepsilon_{231} = 1$, $\varepsilon_{321} = \varepsilon_{132} = \varepsilon_{213} = -1$ and the remaining 21 elements are zero. In this notation, I_1^{3D} can be obtained from

$$M^{ij} M^{kl} M^{mn} \varepsilon_{ikm} \varepsilon_{jln}$$
$$= 6(m_{200}m_{020}m_{002} + 2m_{110}m_{101}m_{011} - m_{200}m_{011}^2 - m_{020}m_{101}^2 - m_{002}m_{110}^2). \tag{3.90}$$

3.8.2   Normalized moments in 3D

An extension of the normalization method into 3D is more difficult than the method of geometric primitives. The affine matrix $\mathbf{A}$ can be decomposed into two rotations and nonuniform scaling between them

$$\mathbf{A} = \mathbf{R}_1 \, \mathbf{S} \, \mathbf{R}_2$$

$$= \begin{pmatrix} \cos \alpha_{xx} & \cos \alpha_{xy} & \cos \alpha_{xz} \\ \cos \alpha_{yx} & \cos \alpha_{yy} & \cos \alpha_{yz} \\ \cos \alpha_{zx} & \cos \alpha_{zy} & \cos \alpha_{zz} \end{pmatrix} \begin{pmatrix} \delta_x & 0 & 0 \\ 0 & \delta_y & 0 \\ 0 & 0 & \delta_z \end{pmatrix} \begin{pmatrix} \cos \beta_{xx} & \cos \beta_{xy} & \cos \beta_{xz} \\ \cos \beta_{yx} & \cos \beta_{yy} & \cos \beta_{yz} \\ \cos \beta_{zx} & \cos \beta_{zy} & \cos \beta_{zz} \end{pmatrix},$$
$$\tag{3.91}$$

where $\mathbf{R}_1$ and $\mathbf{R}_2$ are rotation matrices.

The rotations in 3D create a group that is often called SO(3). Any rotation from this group is described by its *rotation matrix* $\mathbf{R}$:

$$\mathbf{x}' = \mathbf{R}\mathbf{x}. \tag{3.92}$$

The axis of the rotation must pass through the origin of the coordinates. The rotation matrix is the matrix of direction cosines, e.g. α_{zy} is the angle between the original position of the z-axis and the new position of the y-axis. We can choose three of these angles; they must not be in one row or in one column. The rotation matrix is *orthonormal*, i.e. $\mathbf{R}^{-1} = \mathbf{R}^{\mathrm{T}}$, and its determinant equals one.

We use the central moments (2.25) for normalization of moments to the translation and zero-order moment for normalization to scaling (2.26), (2.27).

Similarly as in 2D, the principal axes can be used for normalization to the first rotation and nonuniform scaling [43]. We diagonalize the moment matrix

$$\mathbf{M} = \begin{pmatrix} \mu_{200} & \mu_{110} & \mu_{101} \\ \mu_{110} & \mu_{020} & \mu_{011} \\ \mu_{101} & \mu_{011} & \mu_{002} \end{pmatrix}. \tag{3.93}$$

If we represent a 3D object centered on the origin of the coordinate system by moments up to second order only, then the object is equivalent to an ellipsoid of definite size and orientation. Its axes are the principal axes of the object, eigenvalues of the matrix (3.93) define length of axes of the ellipsoid and its eigenvectors define its rotation. Therefore, if we rotate the object by the inverse (transposed) matrix of the eigenvectors and divide the result by the eigenvalues, then the object is normalized to the first rotation and nonuniform scaling.

In 2D, we do not normalize to stretching directly by the eigenvalues λ_1 and λ_2 of the matrix (2.21), but we use the coefficient δ (3.45) computed from them. Similarly in 3D, we use m_{000} for normalization to uniform scaling and we have an additional constraint $\delta_x \delta_y \delta_z = 1$ on the normalizing coefficients δ_x, δ_y and δ_z from that. It implies that

$$\delta_x = \sqrt{\frac{\sqrt[3]{\lambda_1 \lambda_2 \lambda_3}}{\lambda_1}}, \quad \delta_y = \sqrt{\frac{\sqrt[3]{\lambda_1 \lambda_2 \lambda_3}}{\lambda_2}}, \quad \delta_z = \sqrt{\frac{\sqrt[3]{\lambda_1 \lambda_2 \lambda_3}}{\lambda_3}}, \tag{3.94}$$

where λ_1, λ_2 and λ_3 are eigenvalues of the matrix $\mathbf{M}$ from (3.93).

Now, the normalization to the second rotation has to be solved. We propose a 3D analogy of complex moments for this task.

3D complex moments are based on a concept of *spherical harmonics*. Their detailed description can be found in references [44] to [46]. The *spherical harmonic function* of degree ℓ and order m is defined as

$$Y_\ell^m(\theta, \varphi) = \sqrt{\frac{(2\ell + 1)}{4\pi} \frac{(\ell - m)!}{(\ell + m)!}} P_\ell^m(\cos \theta) e^{im\varphi}, \tag{3.95}$$

where $\ell = 0, 1, 2, \ldots$, $m = -\ell, -\ell + 1, \ldots, \ell$, and P_ℓ^m is an *associated Legendre function*

$$P_\ell^m(x) = (-1)^m (1 - x^2)^{(m/2)} \frac{d^m}{dx^m} P_\ell(x). \tag{3.96}$$

$P_\ell(x)$ is a *Legendre polynomial*

$$P_\ell(x) = \frac{1}{2^\ell \ell!} \frac{d^\ell}{dx^\ell} (x^2 - 1)^\ell. \tag{3.97}$$

The relation

$$P_\ell^{-m}(x) = (-1)^m \frac{(\ell - m)!}{(\ell + m)!} P_\ell^m(x) \tag{3.98}$$

is used for definition of the associated Legendre function of the negative order.

The first several spherical harmonics have the forms

$$Y_0^0(\theta, \varphi) = \frac{1}{2}\sqrt{\frac{1}{\pi}},$$

$$Y_1^0(\theta, \varphi) = \frac{1}{2}\sqrt{\frac{3}{\pi}} \cos\theta, \qquad\qquad Y_1^1(\theta, \varphi) = -\frac{1}{2}\sqrt{\frac{3}{2\pi}} \sin\theta e^{i\varphi},$$

$$Y_2^0(\theta, \varphi) = \frac{1}{4}\sqrt{\frac{5}{\pi}}(3\cos^2\theta - 1), \qquad Y_2^1(\theta, \varphi) = -\frac{1}{2}\sqrt{\frac{15}{2\pi}} \sin\theta \cos\theta e^{i\varphi},$$

$$Y_2^2(\theta, \varphi) = \frac{1}{4}\sqrt{\frac{15}{2\pi}} \sin^2\theta e^{2i\varphi},$$

$$Y_3^0(\theta, \varphi) = \frac{1}{4}\sqrt{\frac{7}{\pi}}(5\cos^3\theta - 3\cos\theta), \qquad Y_3^1(\theta, \varphi) = -\frac{1}{8}\sqrt{\frac{21}{\pi}} \sin\theta(5\cos^2\theta - 1)e^{i\varphi},$$

$$Y_3^2(\theta, \varphi) = \frac{1}{4}\sqrt{\frac{105}{2\pi}} \sin^2\theta \cos\theta e^{2i\varphi}, \qquad Y_3^3(\theta, \varphi) = -\frac{1}{8}\sqrt{\frac{35}{\pi}} \sin^3\theta e^{3i\varphi}.$$

$$\tag{3.99}$$

The spherical harmonics of a negative order can be computed as $Y_\ell^{-m} = (-1)^m (Y_\ell^m)^*$. The first few spherical harmonics can be seen in Figure 3.28, where the absolute real value is

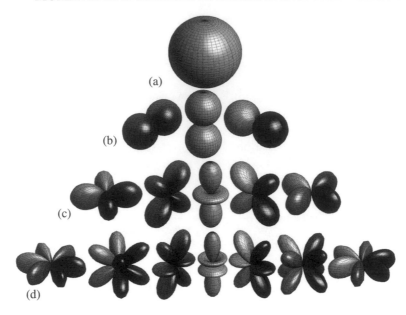

Figure 3.28 The spherical harmonics. (a) $\ell = 0$, $m = 0$, (b) $\ell = 1$, $m = -1, 0, 1$, (c) $\ell = 2$, $m = -2, -1, 0, 1, 2$, (d) $\ell = 3$, $m = -3, -2, -1, 0, 1, 2, 3$. Imaginary parts are displayed for $m < 0$ and real parts for $m \geq 0$.

expressed by the distance of the surface from the origin of the coordinates. Since the spherical harmonics with negative order are (negative) complex conjugates, the imaginary parts are shown instead. The gray level describes the complex phase, Y_ℓ^0 are real, with zero phase.

If we substitute Cartesian coordinates x, y and z for the spherical coordinates ϱ, θ and φ

$$x = \varrho \sin \theta \cos \varphi \qquad\qquad \varrho = \sqrt{x^2 + y^2 + z^2}$$

$$y = \varrho \sin \theta \sin \varphi \qquad\qquad \sin \theta e^{i\varphi} = (x + iy)/\varrho \qquad (3.100)$$

$$z = \varrho \cos \theta \qquad\qquad \cos \theta = z/\varrho$$

in (3.99), we obtain the spherical harmonics in Cartesian coordinates, e.g.

$$Y_1^1(x, y, z) = -\frac{1}{2}\sqrt{\frac{3}{2\pi}} \frac{x + iy}{\varrho}. \qquad (3.101)$$

A generalization of complex moments in 3D is obtained by substituting the spherical harmonics for the angular factor $e^{i(p-q)\theta}$

$$c_{s\ell}^m = \int_0^{2\pi} \int_0^\pi \int_0^\infty \varrho^{s+1} Y_\ell^m(\theta, \varphi) f(r, \theta, \varphi) \, d\varrho \, d\theta \, d\varphi, \qquad (3.102)$$

where s is the order of the moment (it stands for $p + q$ in 2D), ℓ can be called *latitudinal repetition* (it stands for the index difference $p - q$ in 2D) and m is a new index called *longitudinal repetition*.

We can obtain complex moment in Cartesian coordinates

$$c_{s\ell}^m = \int\limits_{-\infty}^{\infty} \varrho^s Y_\ell^m(x, y, z) f(x, y, z) \, dx \, dy \, dz. \tag{3.103}$$

If $s = 0, 1, 2, \ldots$, then $\ell = 0, 2, 4, \ldots, s - 2, s$ for even s and $\ell = 1, 3, 5, \ldots, s - 2, s$ for odd s, $m = -\ell, -\ell + 1, \ldots, \ell$. The number of complex moments corresponds to the number of geometric moments.

A notion of composition of rotations around different axes is not easy. We can imagine so-called *Euler angles* φ, θ and ψ. φ is the angle of rotation around the z-axis, the rotation matrix is

$$\Phi = \begin{pmatrix} \cos\varphi & -\sin\varphi & 0 \\ \sin\varphi & \cos\varphi & 0 \\ 0 & 0 & 1 \end{pmatrix} \tag{3.104}$$

θ is the angle of rotation around the x-axis, the rotation matrix is

$$\Theta = \begin{pmatrix} 1 & 0 & 0 \\ 0 & \cos\theta & -\sin\theta \\ 0 & \sin\theta & \cos\theta \end{pmatrix} \tag{3.105}$$

and ψ is the angle of rotation around the z-axis again, the rotation matrix is

$$\Psi = \begin{pmatrix} \cos\psi & -\sin\psi & 0 \\ \sin\psi & \cos\psi & 0 \\ 0 & 0 & 1 \end{pmatrix} \tag{3.106}$$

$\varphi, \psi \in \langle 0, 2\pi \rangle$, while $\theta \in \langle 0, \pi \rangle$ (or $\langle -\pi, \pi \rangle$ and $\langle -\pi/2, \pi/2 \rangle$ respectively). The matrices can be multiplied to the resultant rotation matrix

$$\mathbf{A} = \Psi\Theta\Phi. \tag{3.107}$$

The system zxz of the axes can be changed; it is generally $a_1 a_2 a_1$, where a_1 is some axis and a_2 is an axis perpendicular to a_1. We can imagine a sphere, where longitude (azimuth φ) and latitude (elevation θ) are defined. We have chosen some significant point with orientation on the surface of the sphere. First, we rotate the sphere such that the significant point has zero longitude, it is on zero meridian after the rotation. The axis of rotation passes through the "north pole" and the "south pole". Now, we rotate the sphere such that the significant point moves along the zero meridian to the north pole, the axis of rotation connects the points with $\varphi = \pm 90°$, $\theta = 0$. Finally, we rotate the sphere with the significant point on the north pole so the significant point is oriented in some standard direction (ψ). The axis of rotation is again the north pole – south pole.

The angle φ can be obtained as a phase of some complex moment $c_{s\ell}^m$

$$\varphi = \frac{1}{m} \arctan\left(\frac{\mathcal{Im}(c_{s\ell}^m)}{\mathcal{Re}(c_{s\ell}^m)}\right) \tag{3.108}$$

typically $c_{s\ell}^m = c_{31}^1$ for a nonsymmetric object. The angle θ can be obtained by comparing the absolute value of a complex moment, e.g. c_{31}^0 with the value of m_{000} (or c_{00}^0). A volume of a sphere equals $\frac{4}{3}\pi\varrho^3$, so we can consider

$$\varrho = \sqrt[3]{\frac{3}{4\pi}m_{000}} \tag{3.109}$$

as a radius of a characteristic sphere of the object. If we use moments normalized to scaling, then $m_{000} = 1$. Then we obtain for the characteristic value of θ

$$\cos\theta = 2\sqrt{\frac{\pi}{3}\frac{c_{31}^0}{\varrho^3}}, \qquad \sin\theta = \sqrt{1 - \cos^2\theta} \tag{3.110}$$

and we can construct the matrix Θ (3.105).

It is better to rotate the values of moments by $\Theta\Phi$. The general formula for the conversion of the moment values in an affine transform is derived from the polynomial theorem

$$\mu'_{pqr} = \sum_{i_1=0}^{p}\sum_{j_1=0}^{p-i_1}\sum_{i_2=0}^{q}\sum_{j_2=0}^{q-i_2}\sum_{i_3=0}^{r}\sum_{j_3=0}^{r-i_3}$$

$$\times \frac{p!}{i_1!j_1!(p-i_1-j_1)!}\frac{q!}{i_2!j_2!(q-i_2-j_2)!}\frac{r!}{i_3!j_3!(r-i_3-j_3)!}$$

$$\times a_{11}^{i_1}a_{12}^{j_1}a_{13}^{p-i_1-j_1}a_{21}^{i_2}a_{22}^{j_2}a_{23}^{q-i_2-j_2}a_{31}^{i_3}a_{32}^{j_3}a_{33}^{r-i_3-j_3}$$

$$\times \mu_{i_1+i_2+i_3,\,j_1+j_2+j_3,\,p+q+r-i_1-j_1-i_2-j_2-i_3-j_3} \tag{3.111}$$

where

$$A = \begin{pmatrix} a_{11} & a_{12} & a_{13} \\ a_{21} & a_{22} & a_{23} \\ a_{31} & a_{32} & a_{33} \end{pmatrix}$$

is the matrix of the affine transform; in our case it is the rotation matrix $\Theta\Phi$. The angle ψ can be more simply computed from these converted values of the moments. The angle ψ is then computed by (3.108), the rotation matrix Ψ (3.106) performs the final normalization of the object to the second rotation.

If the magnitude of c_{31}^1 is close to zero, we must search for some higher-order normalizing moment. The index m plays the part of index difference from 2D in this case. If the object has N-FRS with respect to the current axis, we should use a moment with $m = N$ or generally with $m = kN$, where k is a positive integer as small as possible.

3.8.3   Half normalization in 3D

The method of the half normalization can also be generalized to 3D, but not in a straight-forward way. We can normalize the image up to the nonuniform scaling as in the previous section and then use 3D rotation invariants. However, we can use only one second-order invariant, because the second-order moments were already used for the normalization to the first rotation and the nonuniform scaling. Lo and Don [43] published the third-order rotation invariants, but the detailed description of their method is beyond the scope of this book. The advantage of their method is that it can be used for derivation of higher-order invariants. Here, we present only the basics.

The spherical harmonics are transformed under rotation by the formula [47]

$$Y_\ell^m(\mathbf{R}^{-1}\mathbf{x}) = \sum_{n=-\ell}^{\ell} D_{nm}^\ell(\mathbf{R})Y_\ell^n(\mathbf{x}), \tag{3.112}$$

i.e. the spherical harmonics after rotation by the rotation matrix $\mathbf{R}$ are linear combinations of the spherical harmonics with the same degree, where coefficients of these linear combinations are so-called *Wigner D-functions* $D_{nm}^\ell(\mathbf{R})$. We need to find such a polynomial of the complex moments that would provide phase cancelation. The coefficients of the polynomial can be found as a solution of corresponding equations and the result can be expressed in terms of so-called *Clebsch–Gordan coefficients*. They are labeled as $\langle \ell, m, \ell', m'|\ell, \ell', j, k\rangle$.

Then, we need to construct *composite complex moment forms*

$$v_s(\ell, \ell')_j^k = \sum_{m=-\ell}^{\ell} \langle \ell, m, \ell', k - m|\ell, \ell', j, k\rangle\, c_{s\ell}^m\, c_{s\ell'}^{k-m} \tag{3.113}$$

$v_s(\ell, \ell')_0^0$ is the rotational invariant itself; other invariants can be obtained as a tensor product of the composite complex moment forms and complex moments, where the sum of the upper indices equals zero. An example of the product of two forms is

$$v_s(\ell, \ell')_j v_{s'}(\ell'', \ell''')_j = \frac{1}{\sqrt{2j+1}} \sum_{k=-j}^{j} (-1)^{j-k} v_s(\ell, \ell')_j^k v_{s'}(\ell'', \ell''')_j^{-k} \tag{3.114}$$

which is an invariant. If both forms are identical, we obtain the "norm" of the form $v_s^2(\ell, \ell')_j = v_s(\ell, \ell')_j v_s(\ell, \ell')_j$ from (3.114). An example of the product of a form and a complex moment is

$$v_s(\ell, \ell')_j v_{s'} = \frac{1}{\sqrt{2j+1}} \sum_{k=-j}^{j} (-1)^{j-k} v_s(\ell, \ell')_j^k c_{s'j}^{-k}. \tag{3.115}$$

By this approach, the rotation invariants of the second order can be obtained in the form

$$c_{20}^0 = \frac{1}{2}\sqrt{\frac{1}{\pi}}(\mu_{200} + \mu_{200} + \mu_{200})$$

$$v_2(2, 2)_0^0 = \frac{1}{\sqrt{5}}(2c_{22}^2 c_{22}^{-2} - 2c_{22}^1 c_{22}^{-1} + (c_{22}^0)^2)$$

$$= \frac{5}{4\pi}(\mu_{200}^2 + \mu_{020}^2 + \mu_{002}^2 - \mu_{200}\mu_{020} - \mu_{200}\mu_{002} - \mu_{020}\mu_{002}$$

$$+ 3\mu_{110}^2 + 3\mu_{101}^2 + 3\mu_{011}^2)$$

$$v_2(2, 2)_2 v_2 = \frac{1}{\sqrt{35}}(6\sqrt{2}c_{22}^2 c_{22}^0 c_{22}^{-2} - 3\sqrt{3}(c_{22}^1)^2 c_{22}^{-2} - 3\sqrt{3}c_{22}^2(c_{22}^{-1})^2$$

$$+ 3\sqrt{2}c_{22}^1 c_{22}^0 c_{22}^{-1} - \sqrt{2}(c_{22}^0)^3)$$

$$= \frac{5}{32\pi}\sqrt{\frac{5}{2\pi}}(-2\mu_{200}^3 - 2\mu_{020}^3 - 2\mu_{002}^3 + 3\mu_{200}^2\mu_{020} + 3\mu_{200}^2\mu_{002}$$

$$+ 3\mu_{020}^2\mu_{200} + 3\mu_{020}^2\mu_{002} + 3\mu_{002}^2\mu_{200} + 3\mu_{002}^2\mu_{020}$$

$$- 12\mu_{200}\mu_{020}\mu_{002} + 18\mu_{200}\mu_{011}^2 + 18\mu_{020}\mu_{101}^2 + 18\mu_{002}\mu_{110}^2$$

$$- 9\mu_{200}\mu_{110}^2 - 9\mu_{200}\mu_{101}^2 - 9\mu_{020}\mu_{110}^2 - 9\mu_{020}\mu_{011}^2$$

$$- 9\mu_{002}\mu_{101}^2 - 9\mu_{002}\mu_{011}^2 - 54\mu_{110}\mu_{101}\mu_{011}). \qquad (3.116)$$

They are readily related to the invariants (2.28) from Chapter 2 as

$$c_{20}^0 = \frac{1}{2}\sqrt{\frac{1}{\pi}}\psi_1$$

$$v_2(2,2)_0^0 = \frac{5}{4\pi}(\psi_1^2 - 3\psi_2) \qquad (3.117)$$

$$v_2(2,2)_2 v_2 = \frac{5}{32\pi}\sqrt{\frac{5}{2\pi}}(-2\psi_1^3 + 9\psi_1\psi_2 - 27\psi_3).$$

Third-order invariants are listed in Appendix 3.C.

3.8.4 Direct solution of the Cayley–Aronhold equation

The extension of this method to 3D is relatively straightforward; however, the number of terms of the invariants is higher than in 2D.

A 3D affine transform can be decomposed into three nonuniform scalings, six skews and three translations. The solution is analogous to the 2D case. There are six Cayley–Aronhold differential equations, e.g.

$$\sum_p \sum_q \sum_r p\mu_{p-1,q+1,r} \frac{\partial I}{\partial \mu_{pqr}} = 0. \qquad (3.118)$$

Solving the other five equations can be substituted by using the symmetry condition. Since we have six permutations of three indices, we can obtain up to six symmetric terms with the same coefficient.

In the computation, we must first enumerate all possible terms of an invariant. We can imagine the problem as a table that must be filled in such a way that all row sums equal the weight of the invariants and column sums equal orders of the moments. Creating and solving the system of linear equations based on (3.118) is then analogous to the 2D case.

3.9 Conclusion

In this chapter, we presented three different methods of automatic generation of the affine moment invariants. Each of them has its positive and negative properties.

The graph method is transparent and easy to implement but its drawback is an exponential computing complexity of generating invariants that builds a wall preventing it from being used for higher weights. Until now, the invariants of weight 12 and less have been found explicitly. This is not, however, a limitation for practical use – the explicit forms of the invariants are generated only once and the weight 12 is sufficient for all applications we have encountered.

The second method was based on image normalization. Its main advantage is the fact that it directly yields an independent set of invariants. In this way we are rid of the time-consuming elimination of reducible and dependent invariants. On the other hand, the need to search nonzero moments (and the need for the choice of proper thresholds), less numerical stability, propagation of potential errors into all invariants, possible discontinuity in rotation normalization, and also more complicated explicit forms of the invariants are clear drawbacks of this approach. Half normalization provides a compromise between the normalized moments and the AMIs. This approach retains the advantage of producing independent invariants and removes the problems with normalization to the second rotation, but still suffers from lower numerical stability than the AMIs and from the possibility of error propagation.

The last method we presented is based on an automatic direct solution of the Cayley–Aronhold differential equation. Its advantage is the relatively low computing complexity of invariant generation. The disadvantage of the method is a complicated implementation and, similarly to the graph method, producing reducible and dependent invariants that should be identified and removed afterwards.

Although the background, the complexity and the computational aspects of the above methods are different, we proved that they are theoretically equivalent because they end up with the sets of invariants between which a one-to-one mapping exists.

Finally, we briefly explained the generalization of the AMIs to color images and to 3D images.

Appendix 3.A

Proof of the Fundamental theorem (Theorem 3.1)

Fourier transform $F(u, v)$ of image $f(x, y)$ is given as

$$F(u, v) = \int_{-\infty}^{\infty} \int_{-\infty}^{\infty} e^{-2\pi i (ux+vy)} f(x, y) \, dx \, dy. \tag{3.119}$$

It can be expanded into a power series

$$F(u, v) = \sum_{j=0}^{\infty} \sum_{k=0}^{\infty} \frac{(-2\pi i)^{j+k}}{j!k!} m_{jk} u^j v^k \tag{3.120}$$

where m_{jk} are geometric moments of the image $f(x, y)$. If the spatial coordinates x, y are transformed by an affine transform

$$\mathbf{x}' = \mathbf{A}\mathbf{x} \tag{3.121}$$

and the frequencies u, v are transformed by (3.9), then the Fourier transform is affinely transformed as

$$\int_{-\infty}^{\infty} \int_{-\infty}^{\infty} e^{-2\pi i (ux+vy)} f(x, y) \, dx \, dy = \frac{1}{|J|} \int_{-\infty}^{\infty} \int_{-\infty}^{\infty} e^{-2\pi i (u'x'+v'y')} f(x, y) \, dx' \, dy'. \tag{3.122}$$

Applying the expansion (3.120) on both sides, we obtain

$$\sum_{j=0}^{\infty} \sum_{k=0}^{\infty} \frac{(-2\pi i)^{j+k}}{j!k!} m_{jk} u^j v^k = \frac{1}{|J|} \sum_{j=0}^{\infty} \sum_{k=0}^{\infty} \frac{(-2\pi i)^{j+k}}{j!k!} m'_{jk} u'^j v'^k. \tag{3.123}$$

The monomials $u^j v^k$ of the same order are under affine transformation (3.9), transformed among themselves (e.g. from a quadratic monomial we again obtain a linear combination of quadratic monomials). Therefore, it is possible to decompose the equality (3.123) and compare the terms with the same monomial order. For one particular order $p = j + k$ we thus obtain

$$\sum_{j=0}^{\infty} \frac{p!}{j!(p-j)!} m_{j,p-j} u^j v^{p-j} = \frac{1}{|J|} \sum_{j=0}^{\infty} \frac{p!}{j!(p-j)!} m'_{j,p-j} u'^j v'^{p-j} \qquad (3.124)$$

which is an expression of the transformation of binary algebraic form of order p; see (3.8) and (3.10). Thus, if we substitute the coefficients of the binary forms for moments

$$a_j = m_{j,p-j} \qquad (3.125)$$

and

$$a'_j = \frac{1}{|J|} m'_{j,p-j}, \qquad (3.126)$$

the Fundamental theorem is proven. □

Appendix 3.B

In this appendix, we present explicit forms of higher-order invariants that were used in the experiments.

$$
\begin{aligned}
I_{11} = (&\mu_{30}^2\mu_{12}^2\mu_{04} - 2\mu_{30}^2\mu_{12}\mu_{03}\mu_{13} + \mu_{30}^2\mu_{03}^2\mu_{22} - 2\mu_{30}\mu_{21}^2\mu_{12}\mu_{04} \\
&+ 2\mu_{30}\mu_{21}^2\mu_{03}\mu_{13} + 2\mu_{30}\mu_{21}\mu_{12}^2\mu_{13} - 2\mu_{30}\mu_{21}\mu_{03}^2\mu_{31} - 2\mu_{30}\mu_{12}^3\mu_{22} \\
&+ 2\mu_{30}\mu_{12}^2\mu_{03}\mu_{31} + \mu_{21}^4\mu_{04} - 2\mu_{21}^3\mu_{12}\mu_{13} - 2\mu_{21}^3\mu_{03}\mu_{22} + 3\mu_{21}^2\mu_{12}^2\mu_{22} \\
&+ 2\mu_{21}^2\mu_{12}\mu_{03}\mu_{31} + \mu_{21}^2\mu_{03}^2\mu_{40} - 2\mu_{21}\mu_{12}^3\mu_{31} - 2\mu_{21}\mu_{12}^2\mu_{03}\mu_{40} + \mu_{12}^4\mu_{40})/\mu_{00}^{13}
\end{aligned}
$$

$$
\begin{aligned}
I_{19} = (&\mu_{20}\mu_{30}\mu_{12}\mu_{04} - \mu_{20}\mu_{30}\mu_{03}\mu_{13} - \mu_{20}\mu_{21}^2\mu_{04} + \mu_{20}\mu_{21}\mu_{12}\mu_{13} \\
&+ \mu_{20}\mu_{21}\mu_{03}\mu_{22} - \mu_{20}\mu_{12}^2\mu_{22} - 2\mu_{11}\mu_{30}\mu_{12}\mu_{13} + 2\mu_{11}\mu_{30}\mu_{03}\mu_{22} \\
&+ 2\mu_{11}\mu_{21}^2\mu_{13} - 2\mu_{11}\mu_{21}\mu_{12}\mu_{22} - 2\mu_{11}\mu_{21}\mu_{03}\mu_{31} + 2\mu_{11}\mu_{12}^2\mu_{31} \\
&+ \mu_{02}\mu_{30}\mu_{12}\mu_{22} - \mu_{02}\mu_{30}\mu_{03}\mu_{31} - \mu_{02}\mu_{21}^2\mu_{22} + \mu_{02}\mu_{21}\mu_{12}\mu_{31} \\
&+ \mu_{02}\mu_{21}\mu_{03}\mu_{40} - \mu_{02}\mu_{12}^2\mu_{40})/\mu_{00}^{10}
\end{aligned}
$$

$$
\begin{aligned}
I_{25} = (&\mu_{20}^3\mu_{12}^2\mu_{04} - 2\mu_{20}^3\mu_{12}\mu_{03}\mu_{13} + \mu_{20}^3\mu_{03}^2\mu_{22} - 4\mu_{20}^2\mu_{11}\mu_{21}\mu_{12}\mu_{04} \\
&+ 4\mu_{20}^2\mu_{11}\mu_{21}\mu_{03}\mu_{13} + 2\mu_{20}^2\mu_{11}\mu_{12}^2\mu_{13} - 2\mu_{20}^2\mu_{11}\mu_{03}^2\mu_{31} \\
&+ 2\mu_{20}^2\mu_{02}\mu_{30}\mu_{12}\mu_{04} - 2\mu_{20}^2\mu_{02}\mu_{30}\mu_{03}\mu_{13} - 2\mu_{20}^2\mu_{02}\mu_{21}\mu_{12}\mu_{13} \\
&+ 2\mu_{20}^2\mu_{02}\mu_{21}\mu_{03}\mu_{22} + \mu_{20}^2\mu_{02}\mu_{12}^2\mu_{22} - 2\mu_{20}^2\mu_{02}\mu_{12}\mu_{03}\mu_{31} \\
&+ \mu_{20}^2\mu_{02}\mu_{03}^2\mu_{40} + 4\mu_{20}\mu_{11}^2\mu_{21}^2\mu_{04} - 8\mu_{20}\mu_{11}^2\mu_{21}\mu_{03}\mu_{22} \\
&- 4\mu_{20}\mu_{11}^2\mu_{12}^2\mu_{22} + 8\mu_{20}\mu_{11}^2\mu_{12}\mu_{03}\mu_{31} - 4\mu_{20}\mu_{11}\mu_{02}\mu_{30}\mu_{21}\mu_{04}
\end{aligned}
$$

$$+ 4\mu_{20}\mu_{11}\mu_{02}\mu_{30}\mu_{03}\mu_{22} + 4\mu_{20}\mu_{11}\mu_{02}\mu_{21}^2\mu_{13}$$

$$- 4\mu_{20}\mu_{11}\mu_{02}\mu_{21}\mu_{12}\mu_{22} + 4\mu_{20}\mu_{11}\mu_{02}\mu_{12}^2\mu_{31}$$

$$- 4\mu_{20}\mu_{11}\mu_{02}\mu_{12}\mu_{03}\mu_{40} + \mu_{20}\mu_{02}^2\mu_{30}^2\mu_{04} - 2\mu_{20}\mu_{02}^2\mu_{30}\mu_{21}\mu_{13}$$

$$+ 2\mu_{20}\mu_{02}^2\mu_{30}\mu_{12}\mu_{22} - 2\mu_{20}\mu_{02}^2\mu_{30}\mu_{03}\mu_{31} + \mu_{20}\mu_{02}^2\mu_{21}^2\mu_{22}$$

$$- 2\mu_{20}\mu_{02}^2\mu_{21}\mu_{12}\mu_{31} + 2\mu_{20}\mu_{02}^2\mu_{21}\mu_{03}\mu_{40} - 8\mu_{11}^3\mu_{21}^2\mu_{13}$$

$$+ 16\mu_{11}^3\mu_{21}\mu_{12}\mu_{22} - 8\mu_{11}^3\mu_{12}^2\mu_{31} + 8\mu_{11}^2\mu_{02}\mu_{30}\mu_{21}\mu_{13}$$

$$- 8\mu_{11}^2\mu_{02}\mu_{30}\mu_{12}\mu_{22} - 4\mu_{11}^2\mu_{02}\mu_{21}^2\mu_{22} + 4\mu_{11}^2\mu_{02}\mu_{12}^2\mu_{40}$$

$$- 2\mu_{11}\mu_{02}^2\mu_{30}^2\mu_{13} + 4\mu_{11}\mu_{02}^2\mu_{30}\mu_{12}\mu_{31} + 2\mu_{11}\mu_{02}^2\mu_{21}^2\mu_{31}$$

$$- 4\mu_{11}\mu_{02}^2\mu_{21}\mu_{12}\mu_{40} + \mu_{02}^3\mu_{30}^2\mu_{22} - 2\mu_{02}^3\mu_{30}\mu_{21}\mu_{31} + \mu_{02}^3\mu_{21}^2\mu_{40})/\mu_{00}^{14}$$

$$I_{47} = (-\mu_{50}^2\mu_{05}^2 + 10\mu_{50}\mu_{41}\mu_{14}\mu_{05} - 4\mu_{50}\mu_{32}\mu_{23}\mu_{05} - 16\mu_{50}\mu_{32}\mu_{14}^2 + 12\mu_{50}\mu_{23}^2\mu_{14}$$

$$- 16\mu_{41}^2\mu_{23}\mu_{05} - 9\mu_{41}^2\mu_{14}^2 + 12\mu_{41}\mu_{32}^2\mu_{05} + 76\mu_{41}\mu_{32}\mu_{23}\mu_{14} - 48\mu_{41}\mu_{23}^3$$

$$- 48\mu_{32}^3\mu_{14} + 32\mu_{32}^2\mu_{23}^2)/\mu_{00}^{14}.$$

The set $\{I_1, I_2, I_3, I_4, I_6, I_7, I_8, I_9, I_{19}\}$ is a complete and independent set of features up to the fourth order. Alternatively, I_{11} or I_{25} can substitute I_{19}. I_{47} is the simplest homogeneous invariant of the fifth order.

Appendix 3.C

This is a list of third-order 3D rotation invariants. The simplest five of them are listed explicitly:

$$\psi_4 = v_3(1, 1)_0^0 = \frac{1}{\sqrt{3}}(2c_{31}^1 c_{31}^{-1} - (c_{31}^0)^2)$$

$$\psi_5 = v_3(3, 3)_0^0 = \frac{1}{\sqrt{7}}(2c_{33}^3 c_{33}^{-3} - 2c_{33}^2 c_{33}^{-2} + 2c_{33}^1 c_{33}^{-1} - (c_{33}^0)^2)$$

$$\psi_6 = v_3(1, 1)_2 v_2 = \frac{1}{\sqrt{15}}(\sqrt{3}(c_{31}^1)^2 c_{22}^{-2} - \sqrt{6}c_{31}^0 c_{31}^1 c_{22}^{-1} + \sqrt{2}((c_{31}^0)^2 + c_{31}^1 c_{31}^{-1})c_{22}^0$$

$$- \sqrt{6}c_{31}^0 c_{31}^{-1} c_{22}^1 c_{22}^0 + \sqrt{3}(c_{31}^{-1})^2 c_{22}^2)$$

$$\psi_7 = v_3(3, 1)_2 v_2 = \frac{1}{\sqrt{105}}((-\sqrt{5}c_{33}^2 c_{31}^0 + \sqrt{15}c_{33}^3 c_{31}^{-1} + c_{33}^1 c_{31}^1)c_{22}^{-2}$$

$$- (\sqrt{10}c_{33}^2 c_{31}^{-1} + \sqrt{3}c_{33}^0 c_{31}^1 - 2\sqrt{2}c_{33}^1 c_{31}^0)c_{22}^{-1}$$

$$+ (-3c_{33}^0 c_{31}^0 + \sqrt{6}c_{33}^1 c_{31}^{-1} + \sqrt{6}c_{33}^{-1} c_{31}^1)c_{22}^0$$

$$- (\sqrt{10}c_{33}^{-2} c_{31}^1 + \sqrt{3}c_{33}^0 c_{31}^{-1} - 2\sqrt{10}c_{33}^{-1} c_{31}^0)c_{22}^1$$

$$+ (-\sqrt{5}c_{33}^{-2} c_{31}^0 + \sqrt{15}c_{33}^{-3} c_{31}^1 + c_{33}^{-1} c_{31}^{-1})c_{22}^2)$$

$$\psi_8 = v_3(3, 3)_2 v_2 = \frac{1}{\sqrt{210}}((10c_{33}^3 c_{33}^{-1} - 10\sqrt{2}c_{33}^2 c_{33}^0 + 2\sqrt{15}(c_{33}^1)^2)c_{22}^{-2}$$

$$- (5\sqrt{10}c_{33}^3 c_{33}^{-2} - 5\sqrt{6}c_{33}^2 c_{33}^{-1} + 2\sqrt{5}c_{33}^1 c_{33}^0)c_{22}^{-1}$$

$$+ (5\sqrt{10}c_{33}^3 c_{33}^{-3} - 3\sqrt{10}c_{33}^1 c_{33}^{-1} + 2\sqrt{10}(c_{33}^0)^2)c_{22}^0$$

$$- (5\sqrt{10}c_{33}^{-3} c_{33}^2 - 5\sqrt{6}c_{33}^{-2} c_{33}^1 + 2\sqrt{5}c_{33}^{-1} c_{33}^0)c_{22}^1$$

$$+ (10c_{33}^{-3} c_{33}^1 - 10\sqrt{2}c_{33}^{-2} c_{33}^0 + 2\sqrt{15}(c_{33}^{-1})^2)c_{22}^2). \qquad (3.127)$$

Other four third-order invariants can be obtained as products (3.114) of two composite moment forms

$$\psi_9 = v_3^2(3, 1)_2,$$

$$\psi_{10} = v_3^2(3, 3)_2,$$

$$\psi_{11} = v_3(3, 1)_2 v_3(1, 1)_2,$$

$$\psi_{12} = v_3(3, 3)_2 v_3(3, 1)_2,$$

$$(3.128)$$

where the composite complex moment forms are

$$v_3(3, 3)_2^2 = 2\sqrt{\frac{5}{42}}c_{33}^3 c_{33}^{-1} - 2\sqrt{\frac{5}{21}}c_{33}^2 c_{33}^0 + \sqrt{\frac{2}{7}}(c_{33}^1)^2,$$

$$v_3(3, 3)_2^1 = \frac{5}{\sqrt{21}}c_{33}^3 c_{33}^{-2} - \sqrt{\frac{5}{7}}c_{33}^2 c_{33}^{-1} + \frac{2}{\sqrt{42}}c_{33}^1 c_{33}^0,$$

$$v_3(3, 3)_2^0 = \frac{5}{\sqrt{21}}c_{33}^3 c_{33}^{-3} - \sqrt{\frac{3}{7}}c_{33}^1 c_{33}^{-1} + \frac{2}{\sqrt{21}}(c_{33}^0)^2,$$

$$v_3(3, 3)_2^{-1} = \frac{5}{\sqrt{21}}c_{33}^2 c_{33}^{-3} - \sqrt{\frac{5}{7}}c_{33}^1 c_{33}^{-2} + \frac{2}{\sqrt{42}}c_{33}^0 c_{33}^{-1},$$

$$v_3(3, 3)_2^{-2} = 2\sqrt{\frac{5}{42}}c_{33}^1 c_{33}^{-3} - 2\sqrt{\frac{5}{21}}c_{33}^0 c_{33}^{-2} + \sqrt{\frac{2}{7}}(c_{33}^{-1})^2,$$

$$v_3(3, 1)_2^2 = \sqrt{\frac{5}{7}}c_{33}^3 c_{31}^{-1} - \sqrt{\frac{5}{21}}c_{33}^2 c_{31}^0 + \frac{1}{\sqrt{21}}c_{33}^1 c_{31}^1,$$

$$v_3(3, 1)_2^1 = \sqrt{\frac{10}{21}}c_{33}^2 c_{31}^{-1} - 2\sqrt{\frac{2}{21}}c_{33}^1 c_{31}^0 + \frac{1}{\sqrt{7}}c_{33}^0 c_{31}^1,$$

$$v_3(3, 1)_2^0 = \sqrt{\frac{2}{7}}c_{33}^1 c_{31}^{-1} - \sqrt{\frac{3}{7}}c_{33}^0 c_{31}^0 + \sqrt{\frac{2}{7}}c_{33}^{-1} c_{31}^1,$$

$$v_3(3, 1)_2^{-1} = \sqrt{\frac{10}{21}}c_{33}^{-2} c_{31}^1 - 2\sqrt{\frac{2}{21}}c_{33}^{-1} c_{31}^0 + \frac{1}{\sqrt{7}}c_{33}^0 c_{31}^{-1},$$

$$v_3(3, 1)_2^{-2} = \sqrt{\frac{5}{7}}c_{33}^{-3} c_{31}^1 - \sqrt{\frac{5}{21}}c_{33}^{-2} c_{31}^0 + \frac{1}{\sqrt{21}}c_{33}^{-1} c_{31}^{-1},$$

$$v_3(1, 1)_2^2 = (c_{31}^1)^2,$$

$$v_3(1, 1)^1_{\frac{1}{2}} = \sqrt{2}c^0_{31}c^1_{31},$$

$$v_3(1, 1)^0_{\frac{1}{2}} = \sqrt{\frac{2}{3}}((c^0_{31})^2 + c^1_{31}c^{-1}_{31}),$$

$$v_3(1, 1)^{-1}_{\frac{1}{2}} = \sqrt{2}c^0_{31}c^{-1}_{31},$$

$$v_3(1, 1)^{-2}_{\frac{1}{2}} = (c^{-1}_{31})^2. \tag{3.129}$$

References

[1] Weiss, I. (1988) "Projective invariants of shapes," in *Proceedings of Computer Vision and Pattern Recognition CVPR'88* (Ann Arbor, Michigan), pp. 1125–34, IEEE Computer Society.

[2] Rothwell, C. A., Zisserman, A., Forsyth, D. A. and Mundy, J. L. (1992) "Canonical frames for planar object recognition," in *Proceedings of the Second European Conference on Computer Vision ECCV'92* (St. Margherita, Italy), LNCS vol. 588, pp. 757–72, Springer.

[3] Weiss, I. (1992) "Differential invariants without derivatives," in *Proceedings of the Eleventh International Conference on Pattern Recognition ICPR'92* (Hague, The Netherlands), pp. 394–8, IEEE Computer Society.

[4] Suk, T. and Flusser, J. (1996) "Vertex-based features for recognition of projectively deformed polygons," *Pattern Recognition*, vol. 29, no. 3, pp. 361–7.

[5] Lenz, R. and Meer, P. (1994) "Point configuration invariants under simultaneous projective and permutation transformations," *Pattern Recognition*, vol. 27, no. 11, pp. 1523–32.

[6] Rao, N. S. V., Wu, W. and Glover, C. W. (1992) "Algorithms for recognizing planar polygonal configurations using perspective images," *IEEE Transactions on Robotics and Automation*, vol. 8, no. 4, pp. 480–86.

[7] Mundy, J. L. and Zisserman, A. (1992) *Geometric Invariance in Computer Vision.* Cambridge, Massachusetts: MIT Press.

[8] Rothwell, C. A., Forsyth, D. A., Zisserman, A. and Mundy, J. L. (1993) "Extracting projective structure from single perspective views of 3D point sets," in *Proceedings of the Fourth International Conference on Computer Vision ICCV'94*, pp. 573–82, IEEE Computer Society.

[9] Van Gool, E. P. L., Moons, T. and Oosterlinck, A. (1995) "Vision and Lie's approach to invariance," *Image and Vision Computing*, vol. 13, no. 4, pp. 259–77.

[10] Voss, K. and Süße, H. (1995) *Adaptive Modelle und Invarianten für zweidimensionale Bilder.* Aachen: Shaker (in German).

[11] Yuanbin, W., Bin, Z. and Tianshun, Y. (2008) "Moment invariants of restricted projective transformations," in *Proceedings of the International Symposium on Information Science and Engineering ISISE'08* (Shanghai, China), vol. 1, pp. 249–53, IEEE Computer Society.

[12] Suk, T. and Flusser, J. (2004) "Projective moment invariants," *IEEE Transactions on Pattern Analysis and Machine Intelligence*, vol. 26, no. 10, pp. 1364–7.

[13] Hu, M.-K. (1962) "Visual pattern recognition by moment invariants," *IRE Transactions on Information Theory*, vol. 8, no. 2, pp. 179–87.

[14] Reiss, T. H. (1991) "The revised fundamental theorem of moment invariants," *IEEE Transactions on Pattern Analysis and Machine Intelligence*, vol. 13, no. 8, pp. 830–4.

[15] Flusser, J. and Suk, T. (1993) "Pattern recognition by affine moment invariants," *Pattern Recognition*, vol. 26, no. 1, pp. 167–74.

[16] Flusser, J. and Suk, T. (1991) "Pattern recognition by means of affine moment invariants," Research Report 1726, Institute of Information Theory and Automation (in Czech).

[17] Sylvester, J. J., assisted by Franklin, F. (1879) "Tables of the generating functions and groundforms for the binary quantics of the first ten orders," *American Journal of Mathematics*, vol. 2, pp. 223–51.

[18] Sylvester, J. J., assisted by Franklin, F. (1879) "Tables of the generating functions and groundforms for simultaneous binary quantics of the first four orders taken two and two together," *American Journal of Mathematics*, vol. 2, pp. 293–306, 324–9.

[19] Schur, I. (1968) *Vorlesungen über Invariantentheorie*. Berlin: Springer (in German).

[20] Gurevich, G. B. (1964) *Foundations of the Theory of Algebraic Invariants*. Groningen, The Netherlands: Nordhoff.

[21] Hilbert, D. (1993) *Theory of Algebraic Invariants*. Cambridge: Cambridge University Press.

[22] Mamistvalov, A. G. (1998) "n-dimensional moment invariants and conceptual mathematical theory of recognition n-dimensional solids," *IEEE Transactions on Pattern Analysis and Machine Intelligence*, vol. 20, no. 8, pp. 819–31.

[23] Mamistvalov, A. G. (1970) "On the fundamental theorem of moment invariants," *Bulletin of the Academy of Sciences of the Georgian SSR*, vol. 59, pp. 297–300 (in Russian).

[24] Suk, T. and Flusser, J. (2004) "Graph method for generating affine moment invariants," in *Proceedings of the 17th International Conference on Pattern Recognition ICPR'04* (Cambridge, UK), pp. 192–5, IEEE Computer Society.

[25] Reiss, T. H. (1993) *Recognizing Planar Objects using Invariant Image Features*, LNCS vol. 676. Berlin: Springer.

[26] Suk, T. and Flusser, J. (2005) "Tables of affine moment invariants generated by the graph method," Research Report 2156, Institute of Information Theory and Automation.

[27] Gurevich, G. B. (1937) *Osnovy teorii algebraicheskikh invariantov*. Moskva, The Union of Soviet Socialist Republics: OGIZ (in Russian).

[28] Cyganski, D. and Orr, J. A. (1985) "Applications of tensor theory to object recognition and orientation determination," *IEEE Transactions Pattern Analysis and Machine Intelligence*, vol. 7, no. 6, pp. 662–73.

[29] Rothe, I., Süsse, H. and Voss, K. (1996) "The method of normalization to determine invariants," *IEEE Transactions on Pattern Analysis and Machine Intelligence*, vol. 18, no. 4, pp. 366–76.

[30] Zhang, Y., Wen, C., Zhang, Y. and Soh, Y. C. (2003) "On the choice of consistent canonical form during moment normalization," *Pattern Recognition Letters*, vol. 24, no. 16, pp. 3205–15.

[31] Pei, S.-C. and Lin, C.-N. (1995) "Image normalization for pattern recognition," *Image and Vision Computing*, vol. 13, no. 10, pp. 711–23.

[32] Suk, T. and Flusser, J. (2005) "Affine normalization of symmetric objects," in *Proceedings of the Seventh International Conference on Advanced Concepts for Intelligent Vision Systems Acivs'05*, pp. 100–7, Springer.

[33] Shen, D. and Ip, H. H. S. (1997) "Generalized affine invariant image normalization," *IEEE Transactions on Pattern Analysis and Machine Intelligence*, vol. 19, no. 5, pp. 431–40.

[34] Sprinzak, J. and Werman, M. (1994) "Affine point matching," *Pattern Recognition Letters*, vol. 15, no. 4, pp. 337–9.

[35] Heikkilä, J. (2004) "Pattern matching with affine moment descriptors," *Pattern Recognition*, vol. 37, pp. 1825–34.

[36] Abu-Mostafa, Y. S. and Psaltis, D. (1985) "Image normalization by complex moments," *IEEE Transactions Pattern Analysis and Machine Intelligence*, vol. 7, no. 1, pp. 46–55.

[37] Suk, T. and Flusser, J. (2006) "The independence of the affine moment invariants," in *Proceedings of the Fifth International Workshop on Information Optics WIO'06* (Toledo, Spain), pp. 387–96, American Institute of Physics.

[38] Flusser, J. and Suk, T. (1994) "A moment-based approach to registration of images with affine geometric distortion," *IEEE Transactions on Geoscience and Remote Sensing*, vol. 32, no. 2, pp. 382–7.

[39] Suk, T. and Flusser, J. (2008) "Affine moment invariants generated by automated solution of the equations," in *Proceedings of the 19th International Conference on Pattern Recognition ICPR'08* (Tampa, Florida), IEEE Computer Society.

[40] Mindru, F., Tuytelaars, T., Gool, L. V. and Moons, T. (2004) "Moment invariants for recognition under changing viewpoint and illumination," *Computer Vision and Image Understanding*, vol. 94, no. 1–3, pp. 3–27.

[41] Xu, D. and Li, H. (2006) "3-D affine moment invariants generated by geometric primitives," in *Proceedings of the 18th International Conference on Pattern Recognition ICPR'06* (Hong Kong), pp. 544–7, IEEE Computer Society.

[42] Rota Bulò, S., Albarelli, A., Torsello, A. and Pelillo, M. (2008) "A hypergraph-based approach to affine parameters estimation," in *Proceedings of the 19th International Conference on Pattern Recognition ICPR'08* (Tampa, Florida), IEEE Computer Society.

[43] Lo, C.-H. and Don, H.-S. (1989) "3-D moment forms: Their construction and application to object identification and positioning," *IEEE Transactions on Pattern Analysis and Machine Intelligence*, vol. 11, no. 10, pp. 1053–64.

[44] Byerly, W. E. (1893) *An Elementary Treatise on Fourier's Series and Spherical, Cylindrical, and Ellipsoidal Harmonics with Applications to Problems in Mathematical Physics*. New York: Dover Publications.

[45] Hobson, E. W. (1931) *The Theory of Spherical and Ellipsoidal Harmonics*. Cambridge: University Press.

[46] Ferrers, N. M. (1877) *An Elementary Treatise on Spherical Harmonics and Subjects Connected with Them*. London: Macmillan.

[47] Yen, Y.-F. (2004) "3D rotation matching using spherical harmonic transformation of k-space navigator," in *Proceedings of ISMRM 12th Scientific Meeting*, p. 2154, International Society for Magnetic Resonance in Medicine.

4

Implicit invariants to elastic transformations[1]

4.1 Introduction

A common denominator of the moment invariants introduced in the previous chapters is that they did not expand beyond the framework of linear spatial deformations of the image. The linear model is, however, insufficient in numerous practical cases, where elastic and/or local deformations may appear. Let us imagine, for instance, recognition of objects printed on the surface of a ball or a bottle (Figure 4.1) and recognition of images taken by a fish-eye lens camera (Figure 4.2). In such cases it would be possible, at least in principle – if we knew all the parameters of the scene and camera, to estimate the exact deformation model, re-project the image into "flat" coordinates and then recognize the objects by means of any standard technique. In other cases this is impossible, especially when photographing a flexible time-variant surface whose model is unknown (see Figure 4.3).

It is not possible to find invariants under general elastic transform (more precisely, it is but they would have almost no discrimination power). In this chapter, we restrict ourselves to polynomial or close-to-polynomial transformations. This is not a serious limitation because polynomials are dense in the set of continuous functions. We show that moment invariants in the traditional sense cannot exist, the reason being that polynomial transformations of any finite degree (except linear ones) do not preserve the order of the moments and do not form a group. A composition of two transforms of degree n is a transform of degree n^2; an inverse transform to a polynomial is not a polynomial transform at all. This nongroup property seemingly makes any construction of invariants impossible. Invariants to a polynomial transform of finite degree $n > 1$ cannot exist. If they existed, it would also be invariant to any composition of such transforms, implying that it must be invariant to any polynomial transform of an arbitrary high degree.

To overcome this, we extend the notion of invariants from Chapter 1 and introduce so-called *implicit invariants*, which act as a distance or similarity measure between two objects independently of their deformations.

[1]Our colleagues Jaroslav Kautsky, Flinders University of South Australia, Adelaide, Australia, and Filip Šroubek, Institute of Information Theory and Automation of the ASCR, Prague, Czech Republic, contributed significantly to this chapter.

Figure 4.1 Nonlinear deformations of characters printed on curved surfaces.

Figure 4.2 Nonlinear deformation of the text captured by a fish-eye lens camera.

All moment invariants studied so far in this book may be called *explicit* invariants. For an explicit invariant E must hold $E(f) = E(\mathcal{D}(f))$ for any image f and any admissible deformation $\mathcal{D}$. On the other hand, implicit invariant I is a functional defined on image pairs such that $I(f, \mathcal{D}(f)) = I(f, f) = 0$ for any image f and deformation $\mathcal{D}$ and $I(f, g) > 0$ for

Figure 4.3 Deformations due to the print on a nonrigid time-variant surface.

any f, g such that g is not a $\mathcal{D}$-deformed version of f. According to this definition, explicit invariants are only particular cases of implicit invariants. Clearly, if an explicit invariant E exists, we can set $I(f, g) = |E(f) - E(g)|$.

Unlike explicit invariants, implicit invariants do not provide a description of a single image because they are always defined for a pair of images. This is why they cannot be used for shape encoding and reconstruction. Implicit invariants were designed as a tool for object recognition and matching. We consider $I(f, g)$ to be a "distance measure" (even if it does not exhibit all properties of a metric) between f and g factorized by $\mathcal{D}$. We can, for each database template g_i, calculate the value of $I(f, g_i)$ and then classify f according to the minimum.

It should be emphasized that the idea of implicit invariants is much more general; it is not restricted to invariants from moments and to spatial transformations. Implicit invariants are general tools that can help us in solving many classification tasks where explicit invariants do not exist. The word "implicit" here has nothing to do with implicit curves and implicit polynomials. It simply expresses the fact that there is no explicit formula enabling this invariant to be evaluated for the given image.

A question arises, of course, if some techniques other than moments can be used for recognition of objects under polynomial transformation. There have been many papers on "elastic matching" and "deformable templates" (see reference [1] for an example and further references), which are powerful in recognition but usually slow. Most of them essentially perform an exhaustive search in the parameter space of functions approximating the deformation and look for an extremum of some similarity or dissimilarity measure. Popular choices include various splines as approximating functions and cross-correlation and mutual information as a similarity measure. Even if the authors speedup the search by pyramidal image representation and/or sophisticated optimization algorithms, these methods still perform relatively slowly because they do not use any invariant representation.

In the literature, one can find many papers on "nonlinear" invariants (see reference [2] for a survey and other references). Most of them are limited to projective transformation, which is the simplest nonlinear transform, and their main goal is to find local invariant descriptors allowing recognition of partially occluded objects (see, for instance, references [3] and [4]). Since this problem differs from that with which we deal in this chapter, those methods cannot easily be adopted for our purpose.

4.2   General moments under a polynomial transform

Basic definitions of general and geometric moments in 2D have already been introduced in Chapter 1. For the sake of notation simplicity and transparency, let us now consider one-dimensional (1D) images. If $p_0, p_1, \ldots, p_{n-1}$ are polynomial basis functions defined on a bounded $D \subset \mathbb{R}$, then a general moment m_j of f is given as

$$m_j = \int_D f(x) p_j(x) \, dx.$$

Using a matrix notation, we can write

$$\mathbf{p}(x) = \begin{pmatrix} p_0(x) \\ p_1(x) \\ \vdots \\ p_{n-1}(x) \end{pmatrix} \quad \text{and} \quad \mathbf{m} = \begin{pmatrix} m_0 \\ m_1 \\ \vdots \\ m_{n-1} \end{pmatrix}. \tag{4.1}$$

Let $r : D \to \tilde{D}$ be a polynomial transformation of the domain D into $\tilde{D}$ and let $\tilde{f} : \tilde{D} \to \mathbb{R}$ be a spatially deformed version of f, i.e.

$$\tilde{f}(r(x)) = f(x) \tag{4.2}$$

for $x \in D$. Since in general $r(D) \subset \tilde{D}$ ($\tilde{D}$ is usually a minimum rectangle circumscribed around $r(D)$) we define $\tilde{f}(\tilde{x}) = 0$ for $\tilde{x} \in \tilde{D} \div r(D)$. This means that image $\tilde{f}$ is a spatially deformed version of f, zero-padded outside $r(D)$.

We are interested in the relation between the moments $\mathbf{m}$ and the moments

$$\tilde{\mathbf{m}} = \int_{\tilde{D}} \tilde{f}(\tilde{x}) \tilde{\mathbf{p}}(\tilde{x}) \, d\tilde{x}$$

of the transformed image with respect to some other $\tilde{n}$ basis polynomials

$$\tilde{\mathbf{p}}(\tilde{x}) = \begin{pmatrix} \tilde{p}_0(\tilde{x}) \\ \tilde{p}_1(\tilde{x}) \\ \vdots \\ \tilde{p}_{n-1}(\tilde{x}) \end{pmatrix}$$

defined on $\tilde{D}$. By substituting $\tilde{x} = r(x)$ into the definition of $\tilde{\mathbf{m}}$ and by using the composite function integration, we obtain the following result:

Theorem 4.1 *Denote by $J_r(x)$ the Jacobian of the transform function r. If*

$$\tilde{\mathbf{p}}(r(x)) |J_r(x)| = \mathbf{A}\mathbf{p}(x) \tag{4.3}$$

for some $\tilde{n} \times n$ matrix $\mathbf{A}$, then

$$\tilde{\mathbf{m}} = \mathbf{A}\mathbf{m}. \tag{4.4}$$

The power of this theorem depends on our ability to choose the basis functions so that we can, for a given transform r, express the left-hand side of (4.3) in terms of the basis functions $\mathbf{p}$ and thus construct the matrix $\mathbf{A}$. This is always possible for any polynomial transform r.

4.3 Explicit and implicit invariants

Let us assume that the transformation r depends on a finite number, say m, $m < \tilde{n}$, parameters $\mathbf{a} = (a_1, \ldots, a_m)$. Traditional explicit moment invariants with respect to r can be obtained in two steps.

1. Eliminate $\mathbf{a} = (a_1, \ldots, a_m)$ from the system (4.4). This leaves us $\tilde{n} - m$ equations that depend only on the two sets of general moments. We call it a *reduced system*.

2. Rewrite these equations equivalently in the form:

$$q_j(\tilde{\mathbf{m}}) = q_j(\mathbf{m}), \quad j = 1, \ldots, \tilde{n} - m \tag{4.5}$$

for some functions q_j. Then, the explicit moment invariants are $E(f) = q_j(\mathbf{m}^{(f)})$.

To demonstrate deriving explicit invariants in a simple case, consider a 1D affine transform $r(x) = ax + b$, $a > 0$, and choose the standard power basis $p_j(x) = \tilde{p}_j(x) = x^j$, $j = 0, 1, 2, 3$ (here, $n = \tilde{n} = 4$ suffices). We have $J_r(x) = a$ and

$$\mathbf{A} = a \begin{pmatrix} 1 & 0 & 0 & 0 \\ b & a & 0 & 0 \\ b^2 & 2ba & a^2 & 0 \\ b^3 & 3b^2a & 3ba^2 & a^3 \end{pmatrix}.$$

Solving the first two equations (4.4) with this matrix $\mathbf{A}$ for a and b gives

$$a = \frac{\tilde{m}_0}{m_0} \quad \text{and} \quad b = \frac{\tilde{m}_1 m_0^2 - \tilde{m}_0 m_1^2}{\tilde{m}_0 m_0^2}$$

and, after substituting these into the remaining two equations and some manipulation, we obtain two equations in the form (4.5) with

$$q_1(\mathbf{m}) = \frac{m_2 m_0 - m_1^2}{m_0^4} \quad \text{and} \quad q_2(\mathbf{m}) = \frac{m_3 m_0^2 - 3m_2 m_1 m_0 + 2m_1^3}{m_0^6}.$$

As another example, consider, again for 1D standard powers, the transform $r(x) = ax^2$. We now have $J_r(x) = 2ax$ and (for $\tilde{n} = 2$, we need $n = 4$)

$$\mathbf{A} = 2a \begin{pmatrix} 0 & 1 & 0 & 0 \\ 0 & 0 & 0 & a \end{pmatrix}.$$

The first equation (4.4) gives

$$a = \frac{\tilde{m}_0}{2m_1}$$

while the second equation rewrites, after substitution, as

$$\frac{\tilde{m}_1}{\tilde{m}_0^2} = \frac{m_3}{2m_1^2}.$$

that cannot be rewritten in the required form (4.5). This shows that explicit invariants do not exist for this type of transform (and, similarly, for all other transformations that do not preserve the order of moments and/or do not form a group).

The previous example shows that, for some transforms, we may be able to eliminate the transform parameters in the quest for finding moment invariants, while the second step – finding the explicit forms q_j – may not be possible. Introducing implicit invariants can overcome this drawback.

Consider the equations obtained from the system (4.4) by eliminating the parameters specifying the transformation r. This reduced system is independent of the particular transformation. To classify an object, we traditionally compare values of its descriptors (explicit moment invariants) with those of the database images, i.e. we look for such database images, as satisfying the equations (4.5). However, it is equivalent to checking for that database image the above reduced system is satisfied.

Therefore, in case we are unable to find explicit moment invariants in the form (4.5), we can use this system as a set of implicit invariants. More precisely, we arrange each equation of the system so as to have zero on its right-hand side. The magnitude of the left-hand side then represents one implicit invariant; in case of more than one equation the classification is performed according to the ℓ_2 norm of the vector of implicit invariants. In other words, the images are classified according to the minimum error with which the reduced system is satisfied.

We will demonstrate the above idea of implicit moment invariants on a 1D quadratic transform

$$r(x) = x + ax^2,$$

where $a \in (0, 1/2)$, which maps interval $D = \langle-1, 1\rangle$ on interval $\tilde{D} = \langle a-1, a+1\rangle$. As $m = 1$ and we want to find two implicit invariants, we need $\tilde{n} = 3$ and $n = 6$. The Jacobian is $J_r(x) = 1 + 2ax$ and we would obtain

$$\mathbf{A} = \begin{pmatrix} 1 & 2a & 0 & 0 & 0 & 0 \\ 0 & 1 & 3a & 2a^2 & 0 & 0 \\ 0 & 0 & 1 & 4a & 5a^2 & 2a^3 \end{pmatrix}.$$

However, now we have to evaluate the moments of the transformed signal over the domain $\tilde{D}$ that depends on the unknown parameter a. To resolve this problem, imagine a shifted power basis

$$\tilde{p}_j(\tilde{x}) = p_j(\tilde{x} - a), \qquad j = 0, 1, \ldots, \tilde{n} - 1.$$

Then we have, after the shift of variable $\tilde{x} = \hat{x} + a$,

$$\tilde{m}_j = \int_{a-1}^{a+1} \tilde{f}(\tilde{x})\tilde{p}_j(\tilde{x})\, d\tilde{x} = \int_{-1}^{1} \tilde{f}(\hat{x} + a)p_j(\hat{x})\, d\hat{x}$$

which is now independent of a as $\hat{f}(\hat{x}) = \tilde{f}(\hat{x} + a)$ has a domain $\langle-1, 1\rangle$. Note that in order to calculate moments of transformed image $\tilde{f}$, we do not have to consider the shifted basis $\tilde{p}$. For $\mathbf{p}$ as standard powers and $\tilde{\mathbf{p}}$ as defined above, we obtain a different transform matrix

$$\mathbf{A} = \begin{pmatrix} 1 & 2a & 0 & 0 & 0 & 0 \\ -a & 1-2a^2 & 3a & 2a^2 & 0 & 0 \\ a^2 & 2a(a^2-1) & 1-6a^2 & 4a(1-a^2) & 5a^2 & 2a^3 \end{pmatrix}.$$

Another way to resolve this problem without imagining a different power basis is by assuming such $r(x)$, which maps the interval D on itself. Both approaches are equivalent and they do not restrict the applicability of implicit invariants. To understand this issue better, consider classification tasks in practice. We usually segment objects we want to classify from images. Without loss of generality, assume that the segmented objects are defined on the same domain, which implies that the transform inscribes $\tilde{D}$ in D.

The first of equations (4.4) gives

$$a = \frac{\tilde{m}_0 - m_0}{2m_1}$$

while the two reduced equations rewrite, after substitution, as

$$2m_1^2(\tilde{m}_1 - m_1) = m_1(3m_2 - \tilde{m}_0)(\tilde{m}_0 - m_0) + m_3(\tilde{m}_0 - m_0)^2$$

$$4m_1^3(\tilde{m}_2 - m_2) = 4m_1^2(2m_3 - m_1)(\tilde{m}_0 - m_0)$$

$$+ m_1(5m_4 + \tilde{m}_0 - 6m_2)(\tilde{m}_0 - m_0)^2$$

$$+ (m_5 - 2m_3)(\tilde{m}_0 - m_0)^3. \tag{4.6}$$

Note that parameter a was eliminated and is not present in both implicit invariants. Thanks to this, the method does not require its knowledge.

In the examples above, the derivation of the transform matrix $\mathbf{A}$ for simple transformations r and small number of moments was straightforward. Due to programming complexity and numerical stability, this intuitive approach cannot be used for higher-order polynomial transform r and/or for more invariants. To obtain a numerically stable method it is important to use suitable polynomial bases, such as orthogonal polynomials, without using their expansions into standard (monomial) powers (see Chapter 6 for a review of OG moments). A proper implementation is based on the representation of polynomial bases by matrices of a special structure [5, 6]. Such a representation allows the polynomials to be evaluated efficiently by means of recurrent relations. For constructing the matrix $\mathbf{A}$ and for other implementation details, see reference [7].

4.4 Implicit invariants as a minimization task

Depending on r, the elimination of the m parameters of the transformation function from the $\tilde{n}$ equations of (4.4) to obtain a parameter-free reduced system may require numerical solving of nonlinear equations. This may be undesirable or impossible. Even the simple transform r used in the experimental section would lead to cubic equations in terms of its parameters. Obtaining a neat reduced system may be very difficult. Furthermore, even if successful, we would create an unbalanced method – we have demanded some of the equations in (4.4) to hold exactly and use the accuracy in the resulting system as a matching criterion to find the transformed image. We therefore recommend another implementation of the implicit invariants. Instead of eliminating the parameters, we calculate the "uniform best fit" from all equations in (4.4). For a given set of values of the moments $\mathbf{m}$ and $\tilde{\mathbf{m}}$, we find values of the m parameters to satisfy (4.4) as well as possible in the ℓ_2 norm; the error of this fit then becomes the value of the respective implicit invariant.

Implementation of the recognition by implicit invariants as a minimization task can be described as follows.

1. Given is a library (database) of images $g_j(x, y)$, $j = 1, \ldots, L$, and a deformed image $\tilde{f}(\tilde{x}, \tilde{y})$ that is assumed to have been obtained by a transform of a known polynomial form $r(x, y, \mathbf{a})$ with unknown values of m parameters $\mathbf{a}$.

2. Choose the appropriate domains and polynomial bases $\mathbf{p}$ and $\tilde{\mathbf{p}}$.

3. Derive a program to evaluate the matrix $\mathbf{A}(\mathbf{a})$. This critical error-prone step is performed by a symbolic algorithmic procedure that produces the program used then in numerical calculations. This step is performed only once for the given task. (It has to be repeated only if we change the polynomial bases or the form of transform $r(x, y, \mathbf{a})$, which basically means only if we move to another application).

4. Calculate the moments $\mathbf{m}^{(g_j)}$ of all library images $g_j(x, y)$.

5. Calculate the moments $\tilde{\mathbf{m}}^{(\tilde{f})}$ of the deformed image $\tilde{f}(\tilde{x}, \tilde{y})$.

6. For all $j = 1, \ldots, L$ calculate, using an optimizer, the values of the implicit invariant

$$I(\tilde{f}, g_j) = \min_{\mathbf{a}} \| \tilde{\mathbf{m}}^{(\tilde{f})} - \mathbf{A}(\mathbf{a})\mathbf{m}^{(g_j)} \| \tag{4.7}$$

and denote

$$M = \min_j I(\tilde{f}, g_j).$$

The norm used here should be weighted, for example relatively to the components corresponding to the same degree.

7. The identified image is such g_k for which $I(\tilde{f}, g_k) = M$. The ratio

$$\gamma = \frac{I(\tilde{f}, g_\ell)}{I(\tilde{f}, g_k)},$$

where $I(\tilde{f}, g_\ell)$ is the second minimum may be used as a confidence measure of the identification. Note that $1 \leq \gamma < \infty$, the higher γ the more reliable classification. We may accept only decisions the confidence of which exceeds a certain threshold.

4.5   Numerical experiments

As shown earlier, the implicit moment invariants can be constructed for a very broad class of image transforms including all polynomial transforms. Here, we will demonstrate the performance of the method on images transformed by the following function

$$\begin{pmatrix} \tilde{x} \\ \tilde{y} \end{pmatrix} = \mathbf{r}(x, y) = \begin{pmatrix} ax + by + c(ax + by)^2 \\ -bx + ay \end{pmatrix}, \tag{4.8}$$

that is, a rotation with scaling (parameters a and b) followed by a quadratic deformation in the $\tilde{x}$ direction (parameter c). We have chosen this particular transform for our tests because it is general enough to approximate many real-life situations such as deformations caused by the fact that the photographed object was drawn/printed on a cylindrical or conical surface. As discussed earlier, explicit invariants to this kind of transform cannot exist because they do not preserve moment orders.

The transformation maps $D = \langle -2, 2 \rangle^2$ into $\tilde{D} = \langle -\sigma(1 - c\sigma), \sigma(1 + c\sigma) \rangle \times \langle -\sigma, \sigma \rangle$ where we denoted $\sigma = 2(|a| + |b|)$. We need to restrict $|c| \leq 1/(2\sigma)$ to have a one-to-one transform.

To illustrate the performance of the method, we carried out the following three sets of experiments. In the first set, we evaluated the robustness of the implicit invariants with respect to noise and proved experimentally their invariance to image rotation and quadratic warping. This experiment was carried out on artificial deformations of the test image. The second experimental set was intended to demonstrate the use of implicit invariants as a shape similarity measure between real distorted objects. This test was performed on a standard benchmark database of the Amsterdam Library of Object Images (ALOI) [8]. The last experiment was carried out on real images taken in the lab, and illustrates a good performance and high recognition power of the implicit invariants even where theoretical assumptions about the degradation are not met.

4.5.1 Invariance and robustness test

In order to demonstrate the numerical behavior of the proposed method, we conducted an experiment on the test image already used in Chapter 2, that was artificially deformed and corrupted by additive noise.

First, we applied the spatial transform (4.8). We did not introduce any scaling, so the transform then became a two-parameter one, fully determined by the rotation angle and quadratic deformation. For the sake of simplicity, we used a normalized parameter $q = 2\sigma c$, that lies always in the interval $\langle -1, 1 \rangle$, in the following discussion. The range of parameter values we used in the test was from $-40°$ to $40°$ with a $4°$ step for rotation angle and from -1 to 1 with a 0.1 step for q (see Figure 4.4 for an example of a deformed image). For each deformed image and the original, we calculated the values of six implicit invariants according to (4.7). This was implemented as a numeric optimization in the parameter space. In Figure 4.5(a), we plotted the ℓ_2 norm of the vector of implicit invariants. Observe that it is more or less constant and very close to zero, irrespective of the degree of geometrical deformation. (The norm is slightly bigger when the rotation angle approaches $\pm 40°$ because of resampling errors.) This illustrates a perfect invariance with respect to both rotation and quadratic deformation.

Then we performed a similar experiment. We fixed the rotation angle at $30°$ and added Gaussian white noise to the transformed images $\tilde{f}$ (see Figure 4.4 for an example). For each deformed image and each noise level, ten realizations of noise were generated. Figure 4.5(b) shows again the ℓ_2 norm of the vector of implicit invariants for quadratic deformation q between ± 1 and SNR from 50 dB (low noise) to 0 dB (high noise). As in the previous case, the graph is very flat. Up to 10 dB, no disturbing slope is visible. Only for severely corrupted data with SNR around 0 dB (the noise variance is equal to the image variance) dues the norm start to increase. This experiment proved the high noise robustness of the implicit invariants, a direct consequence of the robustness of moments. Note, however, that the vulnerability of the moments increases slowly as the moment order increases and the same is also true for implicit invariants.

Figure 4.4 An example of the deformed and noisy image used in the experiment.

4.5.2   ALOI classification experiment

The aim of the second experiment was to test the discriminative power of the implicit invariants. We took 100 images from the commonly used benchmark database ALOI and deformed each of them by the warping model (4.8) (see Figure 4.6 for some examples). The coefficients of the deformations were generated randomly; q from the interval $\langle -1, 1 \rangle$ and the rotation angle from $\langle -40°, 40° \rangle$, both with uniform distribution. Each deformed image was then classified against the undistorted database by two different methods: by six implicit invariants according to minimal norm and by the AMIs I_1, I_2, I_3, I_4, I_6, I_7 using the minimum-distance rule.

We ran the whole experiment several times with different deformation parameters. In each run the recognition rate we achieved was 99% or 100% for the implicit invariants and from 34% to 40% for the AMIs. These results illustrate two important facts. First, the implicit invariants can serve as an efficient tool for object recognition when the object deformation corresponds to the assumed model. Second, in the case of nonlinear distortions the implicit invariants significantly outperform the AMIs, corresponding to our theoretical expectation – if q is close to zero, the nonlinear term is negligible, the distortion can be approximated by affine transform and the AMIs have a reasonable chance of recognizing the object correctly. However, for larger q, such an approximation is very rough and the AMIs naturally fail.

When only rotation is present, both methods should be equivalent. To prove this, we ran the experiment once again with fixed $q = 0$. Then the recognition rate of both methods was 100%, as expected.

4.5.3   Character recognition on a bottle

The last experiment illustrates the applicability of the implicit invariants to classification of real images in situations where the image deformation is not known and, moreover, is not exactly polynomial. Such a situation occurs in practice when the intention is to recognize, for instance, letters or digits which are painted/printed on curved surfaces, such as balls, cans, bottles, etc.

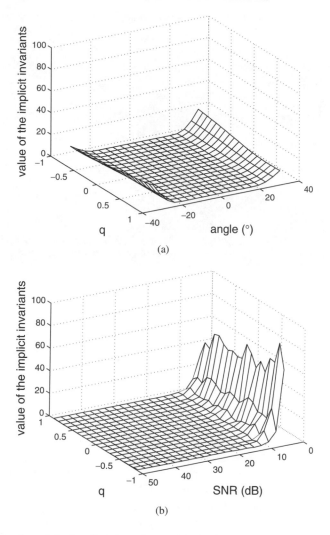

Figure 4.5 Properties of the implicit invariants: (a) the dependence of the implicit invariants on rotation angle and quadratic warping coefficient q; (b) the dependence of the implicit invariants on the noise level (SNR) and quadratic warping coefficient q. Note that both graphs are almost constant and close to zero.

With a standard digital camera (Olympus C-5050), we took six photos of capital letters "M", "N", "E" and "K" printed on a label and glued to a bottle. The label was intentionally tilted by a random degree of rotation; see Figure 4.7. This resulted in six differently deformed instances of each letter. The task was to recognize (classify) these letters against a database containing the full English alphabet of the identical font.

One can see that, apart from the deformation induced by the cylindrical shape of the bottle, rotation and slight perspective is present. Scaling must also be considered, since the size of the captured letters and database templates may not match.

Figure 4.6 Sample original images from the ALOI database used in the experiment.

In an "ideal" case, when the camera is in infinity and the letters are not rotated, the image distortion can be described by an orthogonal projection of the cylinder onto a plane, i.e.

$$\begin{pmatrix} \tilde{x} \\ \tilde{y} \end{pmatrix} = \begin{pmatrix} b \sin \left(\dfrac{x}{b} \right) \\ y \end{pmatrix},$$

where b is the bottle radius, x, y are the coordinates on the bottle surface and $\tilde{x}$, $\tilde{y}$ are the coordinates on the acquired images. In our case the object-to-camera distance was finite, so a small perspective effect also appears. Although in this case we would be able to measure the camera-to-object distance and the radius b, we did not do that. We intentionally approached this experiment as a blind one, without employing any model and/or parameter knowledge.

We assume that the deformation can be approximated by linear polynomials in the horizontal and vertical directions which model the similarity transform, and by a quadratic polynomial in the horizontal direction that approximates cylinder-to-plane projection. This leads to a class of geometric transforms of the shape (4.8) for which we have built the implicit invariants.

The deformed letters were segmented and binarized first. After the segmentation, we used six implicit invariants to classify them against the undistorted alphabet of 26 letters. The recognition was quite successful – only the most deformed letter "N" was misclassified as "H", all other letters were recognized correctly.

For comparison, we also tried to recognize the letters using two other methods. First, we employed the AMIs similarly as in the experiment with the ALOI images with a similar result – the AMIs failed in many cases because they are not designed to handle nonlinear deformations. Then we employed the spline-based elastic matching proposed by Kybic and Unser in reference [1]. This method searches the space of parameters of cubic B-spline deformation of the templates, calculates the mean square error as a measure of dissimilarity between the images and finds the minimum of this error function using

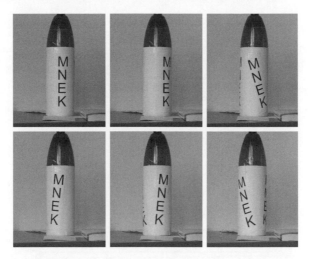

Figure 4.7 The six bottle images used in the experiment.

hierarchical optimization. This method was designed in medical imaging for recognition of templates undergoing elastic deformations and was proved to be very powerful [1]. We used the author's original implementation of the method.[2] The recognition rate was 100%, which is not surprising – the method in fact performs an exhaustive search in the space of deformations while working with full uncompressed representation of each template.

It is interesting to compare the methods in terms of their speed and confidence. The computational cost of both methods is of vastly different magnitude. On a computer with the Intel Pentium Core Duo T2400 processor, the time required to classify all 24 deformed letters against 26 letters in the database was, in the case of elastic matching, over one hour. On the other hand, it took less then one minute when using implicit invariants. Such a significant difference was caused mainly by the fact that elastic matching works with full image representation, while implicit invariants use a large compression in a few moments. The efficiency of the implicit invariants could be further improved by using fast algorithms for moment calculation, particularly in the case of binary images.

Surprisingly, the use of different object representation (i.e. different amount of data) did not lead to different confidences. The confidences γ of both methods are not very high but similar; see Table 4.1. It is interesting to notice that extremely low confidence (close to 1) is frequent for the letter "M" often having the second closest match "N". The explanation is that the quadratic deformation of "M" can merge the right diagonal line with the right vertical line, thus creating an impression of the letter "N".

4.6 Conclusion

In this chapter we introduced a new method for recognition of objects undergoing unknown elastic deformation. The method is based on so-called implicit invariants that are quantities that do not change under a certain class of spatial transformation. The implicit invariants

---

[2]We thank Jan Kybic, Czech Technical University, Prague, for kindly providing us with his code.

Table 4.1   Classification confidence of four letters (each having six different degrees of distortion) using the implicit invariants (top) and the spline-based elastic matching (bottom). MC means a misclassified letter.

Invariants' confidence					
M	9.4	1.3	2.3	6.4	1.1	1.6
N	28.5	11.9	1.3	9.7	10.9	(MC)
E	5.8	2.2	10.4	5.5	3.3	3.0
K	12.1	10.0	2.1	5.4	6.4	2.6

Elastic matching confidence					
M	11.5	1.7	1.9	9.8	10.5	1.4
N	8.3	3.9	2.5	9.3	6.8	2.5
E	6.4	3.1	2.2	5.3	3.8	2.0
K	10.8	3.0	1.9	5.0	3.4	2.2

described in this chapter are functions of image moments (preferably but not necessarily of OG moments). We demonstrated how to derive implicit moment invariants with respect to general polynomial transformation of the spatial coordinates. It should be emphasized that this idea is much more general; it is not restricted to invariants from moments and to spatial transformations. Implicit invariants are general tools that can help solve many classification tasks where explicit invariants do not exist. They are particularly useful when the transformations in question do not form a group.

We explained that any implicit invariant can be viewed as a distance measure between two images factorized by an admissible transformation. However, implicit invariants need not be a metric in a strict sense. For polynomial transform r and for moment invariants defined in this chapter, only some properties of a metric are fulfilled. For any f and g we have $I(f, g) \geq 0$ and $I(f, f) = I(f, \tilde{f}) = 0$. If $I(f, g) = 0$ for invariants of all moments, then $g = \tilde{f}$. This property follows from the uniqueness theorem of moments of compactly supported functions and holds in an ideal infinite case only. It is not guaranteed in practice, when always working with a (small) finite set of moments. Since nonlinear polynomials are not invertible on the set of polynomials, I is not symmetric, i.e. $I(f, g) \neq I(g, f)$. This is why in practical applications we have to consider carefully the "direction" of the transform and construct the invariants accordingly. Triangular inequality $I(f, g) \leq I(f, h) + I(h, g)$ is also not guaranteed.

There is a wide variety of potential applications – apart from the above-mentioned images printed on curved surfaces, we envisage applications in omnidirectional vision (using this technique, objects might be classified directly without any re-projection onto a plane) and in recognition of images taken by a fish-eye-lens camera.

References

[1] Kybic, J. and Unser, M. (2003) "Fast parametric elastic image registration," *IEEE Transactions on Image Processing*, vol. 12, no. 11, pp. 1427–42.

[2] Mundy, J. L. and Zisserman, A. (1992) *Geometric Invariance in Computer Vision.* Cambridge, Massachusetts: MIT Press.

[3] Weiss, I. (1988) "Projective invariants of shapes," in *Proceedings of Computer Vision and Pattern Recognition CVPR'88* (Ann Arbor, Michigan), pp. 1125–34, IEEE Computer Society.

[4] Pizlo, Z. and Rosenfeld, A. (1992) "Recognition of planar shapes from perspective images using contour-based invariants," *CVGIP: Image Understanding*, vol. 56, no. 3, pp. 330–50.

[5] Golub, G. H. and Kautsky, J. (1983) "Calculation of Gauss quadratures with multiple free and fixed knots," *Numerische Mathematik*, vol. 41, no. 2, pp. 147–63.

[6] Golub, G. H. and Kautsky, J. (1983) "On the calculation of Jacobi matrices," *Linear Algebra and Its Applications*, vol. 52/3, pp. 439–55, July.

[7] Flusser, J., Kautsky, J. and Šroubek, F. (2009) "Implicit moment invariants," *International Journal of Computer Vision*, online June 2009, doi:10.1007/s11263-009-0259-4.

[8] Geusebroek, J. M., Burghouts, G. J. and Smeulders, A. W. M. (2005) "The Amsterdam library of object images," *International Journal of Computer Vision*, vol. 61, no. 1, pp. 103–12.

5

Invariants to convolution

5.1 Introduction

In the previous chapters we described moment invariants with respect to various transformations of the spatial coordinates – translation, rotation, scaling, affine transform and finely to elastic transforms. According to the terminology introduced in Chapter 1, all these deformations fall into the category of *geometric degradations* of an image. However, geometric degradations are not the only source of image imperfection. As pointed out in Chapter 1, heavy image corruption may be caused by *radiometric degradations* that influence the intensity values. The linear contrast changes discussed in Chapter 2 may be considered very simple and easy-to-handle radiometric degradations but most of the degradations of this category are much more complicated and deriving invariants with respect to them is a challenging problem.

Radiometric degradations which we consider in this chapter are caused by such factors as a wrong focus, camera and/or scene motion, camera vibrations and by taking images through a turbulent medium such as atmosphere or water. Along with various second-order factors such as lens aberrations and finite-size sampling pulse, they result in the phenomenon called *image blurring* (see Figures 5.1, 5.2 and 5.3).

Assuming that the image acquisition time is so short that the blurring factors do not change during the image formation and also assuming that the blurring is of the same kind for all colors/graylevels, we can describe the observed blurred image $g(x, y)$ of a scene $f(x, y)$ by a *superposition integral*

$$g(x, y) = \int_{-\infty}^{\infty} \int_{-\infty}^{\infty} f(s, t) h(x, y, s, t) \, \mathrm{d}s \, \mathrm{d}t + n(x, y), \tag{5.1}$$

where $h(x, y, s, t)$ is the *space-variant point-spread function* (PSF) of the system and $n(x, y)$ is an additive random noise. If the PSF is *space-invariant*, i.e. if

$$h(x, y, s, t) = h(x - s, y - t),$$

then equation (5.1) is simplified to the form of a standard 2D convolution

$$g(x, y) = (f * h)(x, y) + n(x, y). \tag{5.2}$$

Moments and Moment Invariants in Pattern Recognition   Jan Flusser, Tomáš Suk and Barbara Zitová
© 2009 John Wiley & Sons, Ltd

Figure 5.1 An image of a 3D scene blurred by a space-variant PSF. The blurring is caused by the narrow depth of field of the camera.

Figure 5.2 Space-invariant out-of-focus blur of a flat scene.

Such an imaging system is called a *linear space-invariant system* and is fully determined by its PSF. However, the PSF is in most practical situations unknown. The model (5.2) is a frequently used compromise between universality and simplicity – it is general enough to describe sufficiently many practical situations such as out-of-focus blur of a flat scene, motion

Figure 5.3 Motion blur due to the camera shake. The PSF is approximately space-invariant.

blur of a flat scene in case of linear constant-velocity motion, and media turbulence blur. At the same time, it can be analyzed and numerically processed.

As in the previous chapters, it is desirable to find a representation of the image $g(x, y)$ that does not depend on the imaging system without any prior knowledge of system parameters including the PSF. As usual, there are two different approaches to this problem: image normalization or direct image description by invariants.

Image normalization in this case consists of *blind deconvolution* that removes or suppresses the blurring. Blind deconvolution of a single image has been extensively studied in the image processing literature, see reference [1] for a survey. It is an ill-posed problem of high computing complexity and the quality of the results is questionable.[1] This is why we do not follow the normalization approach in this chapter and rather, look for invariant descriptors of the image that do not depend on the PSF. An invariant approach is much more effective here than the normalization approach because we avoid a difficult and ambiguous inversion of equation (5.2). Unlike deconvolution, invariant descriptors do not restore the original image; this may be understood as a weakness in some applications. On the other hand, in many cases we do not need to know the whole original image; we only need, for instance, to localize or recognize some objects on it (typical examples are matching of a blurred template against a database and recognition of blurred characters; see Figure 5.4). In such situations, only the knowledge of a certain incomplete but robust representation of the image is sufficient. However, such a representation should be independent of the imaging system and should really describe the original image, not the degraded one. In other words, we are looking for a functional I that is invariant to the degradation operator, i.e. $I(g) = I(f)$. Particularly, if the

---

[1]Blind deconvolution of multiple images and blind deconvolution of a single image supported by another underexposed image were proven to be solvable and numerically stable but these methods are not applicable here.

degradation is described by (5.2), then the equality

$$I(f) = I(f * h) \tag{5.3}$$

must hold for any $h(x, y)$. The invariants satisfying condition (5.3) are called *blur invariants* or *convolution invariants*. They were first introduced by Flusser *et al.* [2] and Flusser and Suk in [3].

(a)

(b)

Figure 5.4 (a) Blurred image to be matched against (b) a database: a typical situation where the convolution invariants may be employed.

Although the PSF is supposed to be unknown, we have to accept certain assumptions about it because, for an arbitrary PSF, no blur invariants exist. The first natural assumption is that the imaging system does not change the overall brightness of the image, i.e.

$$\int\limits_{-\infty}^{\infty} \int\limits_{-\infty}^{\infty} h(x, y) \, dx \, dy = 1.$$

Other assumptions we will accept in this chapter concern the symmetry of the PSF. We will see that the higher the symmetry, the more invariants exist.

In this chapter, we introduce several systems of blur invariants based on image moments. They differ from each other by the assumptions imposed on the PSF and, consequently, by the choice of moments from which the invariants are created.

5.2   Blur invariants for centrosymmetric PSFs

First, we show how the geometric moments change under convolution.

Lemma 5.1 *Let $f(x, y)$ and $h(x, y)$ be two arbitrary image functions and let $g(x, y) = (f * h)(x, y)$. Then $g(x, y)$ is also an image function and it holds for its moments*

$$m_{pq}^{(g)} = \sum_{k=0}^{p} \sum_{j=0}^{q} \binom{p}{k}\binom{q}{j} m_{kj}^{(h)} m_{p-k,q-j}^{(f)},$$

$$\mu_{pq}^{(g)} = \sum_{k=0}^{p} \sum_{j=0}^{q} \binom{p}{k}\binom{q}{j} \mu_{kj}^{(h)} \mu_{p-k,q-j}^{(f)}$$

and

$$c_{pq}^{(g)} = \sum_{k=0}^{p} \sum_{j=0}^{q} \binom{p}{k}\binom{q}{j} c_{kj}^{(h)} c_{p-k,q-j}^{(f)}$$

for any p and q.

For the proof of Lemma 5.1, see Appendix 5.A. We can immediately see from this Lemma that the simplest convolution invariant is μ_{00}. This is also the only existing invariant for nonsymmetric PSFs.

The simplest symmetry of the PSF we assume is a twofold rotation symmetry, also called a *centrosymmetry* or *central symmetry*, i.e. $h(x, y) = h(-x, -y)$ (to be precise, it is sufficient to assume centrosymmetry with respect to the centroid of $h(x, y)$ that may be generally different from $(0, 0)$). Then, all its odd-order central geometric moments and complex moments are identically zero, i.e. $\mu_{pq}^{(h)} = 0$ and $c_{pq}^{(h)} = 0$ if $p + q$ is odd. This property of the PSFs allows the fundamental theorem of blur invariants to be formulated.

Theorem 5.2 *Let $f(x, y)$ be an image function and p, q be non-negative integers. Let us define the following functionals $C(p, q)^{(f)}$:*
If $(p + q)$ is even, then

$$C(p, q)^{(f)} = 0.$$

If $(p + q)$ is odd, then

$$C(p, q)^{(f)} = \mu_{pq}^{(f)} - \frac{1}{\mu_{00}^{(f)}} \sum_{n=0}^{p} \sum_{\substack{m=0 \\ 0<n+m<p+q}}^{q} \binom{p}{n}\binom{q}{m} C(p-n, q-m)^{(f)} \cdot \mu_{nm}^{(f)}. \quad (5.4)$$

Then $C(p, q)$ is a centrosymmetric blur invariant for any p and q, i.e. $C(p, q)^{(f)} = C(p, q)^{(f*h)}$.

For the proof of Theorem 5.2, see Appendix 5.B. Theorem 5.2 also holds true when substituting c_{pq} for μ_{pq}. Since the functionals $C(p, q)$ are based on central moments, they are naturally shift invariant. Invariance to uniform scaling can be obtain by using normalized central moments

$$\nu_{pq} = \frac{\mu_{pq}}{\mu_{00}^{(p+q)/2+1}}.$$

Note that only odd-order nontrivial convolution invariants exist for centrosymmetric convolution kernels. It is not a shortcoming of Theorem 5.2, there are deep theoretical reasons for this, being the fact that the even-order moments of centrosymmetric functions are nonzero and cannot be eliminated to achieve invariance.[2] As will be seen later, this set of convolution invariants is complete and independent.

Applying the definition formula (5.4), we can construct invariants of any odd order and express them in an explicit form. A complete set of invariants of the third, fifth and seventh orders is listed below:

- third order:

$$C(3, 0) = \mu_{30},$$

$$C(2, 1) = \mu_{21},$$

$$C(1, 2) = \mu_{12},$$

$$C(0, 3) = \mu_{03}$$

- fifth order:

$$C(5, 0) = \mu_{50} - \frac{10\mu_{30}\mu_{20}}{\mu_{00}},$$

$$C(4, 1) = \mu_{41} - \frac{2}{\mu_{00}}(3\mu_{21}\mu_{20} + 2\mu_{30}\mu_{11}),$$

$$C(3, 2) = \mu_{32} - \frac{1}{\mu_{00}}(3\mu_{12}\mu_{20} + \mu_{30}\mu_{02} + 6\mu_{21}\mu_{11}),$$

$$C(2, 3) = \mu_{23} - \frac{1}{\mu_{00}}(3\mu_{21}\mu_{02} + \mu_{03}\mu_{20} + 6\mu_{12}\mu_{11}),$$

$$C(1, 4) = \mu_{14} - \frac{2}{\mu_{00}}(3\mu_{12}\mu_{02} + 2\mu_{03}\mu_{11}),$$

$$C(0, 5) = \mu_{05} - \frac{10\mu_{03}\mu_{02}}{\mu_{00}}$$

---

[2]There are a few papers presenting some even-order centrosymmetric convolution invariants, but their results can be disproved easily.

- seventh order:

$$C(7, 0) = \mu_{70} - \frac{7}{\mu_{00}}(3\mu_{50}\mu_{20} + 5\mu_{30}\mu_{40}) + \frac{210\mu_{30}\mu_{20}^2}{\mu_{00}^2},$$

$$C(6, 1) = \mu_{61} - \frac{1}{\mu_{00}}(6\mu_{50}\mu_{11} + 15\mu_{41}\mu_{20} + 15\mu_{40}\mu_{21} + 20\mu_{31}\mu_{30})$$

$$+ \frac{30}{\mu_{00}^2}(3\mu_{21}\mu_{20}^2 + 4\mu_{30}\mu_{20}\mu_{11}),$$

$$C(5, 2) = \mu_{52} - \frac{1}{\mu_{00}}(\mu_{50}\mu_{02} + 10\mu_{30}\mu_{22} + 10\mu_{32}\mu_{20} + 20\mu_{31}\mu_{21} + 10\mu_{41}\mu_{11}$$

$$+ 5\mu_{40}\mu_{12}) + \frac{10}{\mu_{00}^2}(3\mu_{12}\mu_{20}^2 + 2\mu_{30}\mu_{20}\mu_{02} + 4\mu_{30}\mu_{11}^2 + 12\mu_{21}\mu_{20}\mu_{11}),$$

$$C(4, 3) = \mu_{43} - -\frac{1}{\mu_{00}}(\mu_{40}\mu_{03} + 18\mu_{21}\mu_{22} + 12\mu_{31}\mu_{12} + 4\mu_{30}\mu_{13} + 3\mu_{41}\mu_{02}$$

$$+ 12\mu_{32}\mu_{11} + 6\mu_{23}\mu_{20}) + \frac{6}{\mu_{00}^2}(\mu_{03}\mu_{20}^2 + 4\mu_{30}\mu_{11}\mu_{02} + 12\mu_{21}\mu_{11}^2$$

$$+ 12\mu_{12}\mu_{20}\mu_{11} + 6\mu_{21}\mu_{02}\mu_{20}),$$

$$C(3, 4) = \mu_{34} - \frac{1}{\mu_{00}}(\mu_{04}\mu_{30} + 18\mu_{12}\mu_{22} + 12\mu_{13}\mu_{21} + 4\mu_{03}\mu_{31} + 3\mu_{14}\mu_{20}$$

$$+ 12\mu_{23}\mu_{11} + 6\mu_{32}\mu_{02}) + \frac{6}{\mu_{00}^2}(\mu_{30}\mu_{02}^2 + 4\mu_{03}\mu_{11}\mu_{20} + 12\mu_{12}\mu_{11}^2$$

$$+ 12\mu_{21}\mu_{02}\mu_{11} + 6\mu_{12}\mu_{20}\mu_{02}),$$

$$C(2, 5) = \mu_{25} - \frac{1}{\mu_{00}}(\mu_{05}\mu_{20} + 10\mu_{03}\mu_{22} + 10\mu_{23}\mu_{02} + 20\mu_{13}\mu_{12} + 10\mu_{14}\mu_{11}$$

$$+ 5\mu_{04}\mu_{21}) + \frac{10}{\mu_{00}^2}(3\mu_{21}\mu_{02}^2 + 2\mu_{03}\mu_{02}\mu_{20} + 4\mu_{03}\mu_{11}^2 + 12\mu_{12}\mu_{02}\mu_{11}),$$

$$C(1, 6) = \mu_{16} - \frac{1}{\mu_{00}}(6\mu_{05}\mu_{11} + 15\mu_{14}\mu_{02} + 15\mu_{04}\mu_{12} + 20\mu_{13}\mu_{03})$$

$$+ \frac{30}{\mu_{00}^2}(3\mu_{12}\mu_{02}^2 + 4\mu_{03}\mu_{02}\mu_{11}),$$

$$C(0, 7) = \mu_{07} - \frac{7}{\mu_{00}}(3\mu_{05}\mu_{02} + 5\mu_{03}\mu_{04}) + \frac{210\mu_{03}\mu_{02}^2}{\mu_{00}^2}.$$

The blur invariants introduced in Theorem 5.2 have their counterparts in the Fourier domain. First we show that, seemingly independently of moments, blur invariants can be derived also in the Fourier domain.

Due to the well-known convolution theorem, the corresponding relation to equation (5.2) in the spectral domain (when neglecting the noise) has the form

$$G(u, v) = F(u, v) \cdot H(u, v), \tag{5.5}$$

where $G(u, v)$, $F(u, v)$ and $H(u, v)$ are the Fourier transforms of the functions $g(x, y)$, $f(x, y)$ and $h(x, y)$, respectively. Considering the amplitude and phase separately, we obtain

$$|G(u, v)| = |F(u, v)| \cdot |H(u, v)| \tag{5.6}$$

and

$$\text{ph}G(u, v) = \text{ph}F(u, v) + \text{ph}H(u, v) \tag{5.7}$$

(note that the last equation is correct only for those points where $G(u, v) \neq 0$; $\text{ph}G(u, v)$ is not defined otherwise).

Due to the central symmetry of $h(x, y)$, its Fourier transform $H(u, v)$ is real (that means the phase of $H(u, v)$ is only a two-valued function):

$$\text{ph}H(u, v) \in \{0; \pi\}.$$

It follows immediately from the periodicity of tangent that

$$\tan(\text{ph}G(u, v)) = \tan(\text{ph}F(u, v) + \text{ph}H(u, v)) = \tan(\text{ph}F(u, v)). \tag{5.8}$$

Thus, a tangent of the phase of the image is an invariant with respect to convolution with any centrosymmetric PSF. The phase itself is not an invariant. Using a tangent is the simplest but not the only way to make it invariant to blur. For instance, Ojansivu and Heikkilä [4] removed this π-ambiguity by using a double phase.

Now we can formulate a theorem showing the close connection between the moment-based and Fourier-based convolution invariants.

Theorem 5.3 *The tangent of the Fourier transform phase of any image $f(x, y)$ can be expanded as a power series (except for the points in which $F(u, v) = 0$ or $\text{ph}F(u, v) = \pm\pi/2$)*

$$\tan(\text{ph}F(u, v)) = \sum_{k=0}^{\infty} \sum_{j=0}^{\infty} a_{kj} u^k v^j, \tag{5.9}$$

where

$$a_{kj} = \frac{(-1)^{(k+j-1)/2} \cdot (-2\pi)^{k+j}}{k! \cdot j! \cdot \mu_{00}} M(k, j)^{(f)}. \tag{5.10}$$

Functional $M(k, j)$ is essentially the same as $C(k, j)$ with central moments replaced by geometric ones.

For the proof of Theorem 5.3, see Appendix 5.C.

Theorem 5.3 demonstrates that from the theoretical point of view, both groups of blur invariants are equivalent when using complete sets. In reality, when working in a discrete domain, we always use a limited set of moment invariants. A one-to-one correspondence is then violated because to calculate a tangent of the phase at *any* frequency (u, v), *all* moment invariants $C(p, q)$ are needed.

Unlike many geometric invariants, the invariants from Theorem 5.2 do not have a clear "physical" meaning. However, understanding what image properties they reflect is important for their practical application. Two mutually connected key questions pertain to the joint null-space of the invariants (i.e. what is the class of images having all invariants $C(p, q)$ zero) and to their discrimination power. It can be shown that for all centrosymmetric images the invariants equal zero. This is an intrinsic limitation – any shape descriptor invariant to a certain group of transformations cannot, in principle, distinguish objects that differ from each other only by transformations from this group. Thus, any invariant (even different from those presented here) to convolution with a centrosymmetric PSF cannot distinguish different centrosymmetric objects because it must give a constant response on all centrosymmetric images. This is because any centrosymmetric image can be considered as a blurring PSF acting on a delta function. On the other hand, there are no functions other than the centrosymmetric ones in the null-space of the invariants. If $C(p, q)^{(f)} = 0$ for any p, q, then all central moments $\mu_{pq}^{(f)}$ of odd orders are also zero. This implies that $F(u, v)$ is real and, consequently, $f(x, y)$ must be centrally symmetric.

Since we have learned that in the null-space of the invariants (5.4) there are centrosymmetric functions only, we can consider the invariants as a "measure of nonsymmetry" of the image.

The independence of the invariants (5.4) follows immediately from the independence of the moments themselves. We show that they are also complete in the following sense.

Theorem 5.4 *Let $f(x, y)$, $g(x, y)$ and $h(x, y)$ be arbitrary image functions such that*

$$g(x, y) = (f * h)(x, y).$$

Let $C(p, q)^{(f)} = C(p, q)^{(g)}$ for any $p, q \geq 0$. Then, $h(x, y)$ is centrally symmetric.

Proof. Since $C(p, q)^{(f)} = C(p, q)^{(g)}$ for any p, q and $\mu_{00}^{(f)} = \mu_{00}^{(g)}$, it follows from Theorem 5.3 that

$$\tan[\mathrm{ph}F(u, v)] = \tan[\mathrm{ph}G(u, v)].$$

Due to the periodicity of tangent,

$$\mathrm{ph}G(u, v) = \mathrm{ph}F(u, v) + \phi(u, v),$$

where $\phi(u, v) \in \{0; \pi\}$. Since $g(x, y) = (f * h)(x, y)$, it holds that $G(u, v) = F(u, v) \cdot H(u, v)$ and, particularly,

$$\mathrm{ph}G(u, v) = \mathrm{ph}F(u, v) + \mathrm{ph}H(u, v).$$

From comparison of the last two equations, it implies that $H(u, v)$ is a real function. Thus, $h(x, y)$ is centrally symmetric. □

Theorem 5.4 says that under certain conditions we are always able to distinguish between two different image functions (modulo convolution with a centrally symmetric PSF) when considering an infinite set of all invariants.

Theorem 5.4 is relatively weak since it supposes that there exist $h(x, y)$ such that

$$g(x, y) = (f * h)(x, y).$$

A much stronger hypothesis can be formulated as follows:
Let $f(x, y)$ and $g(x, y)$ be the image functions such that $C(p, q)^{(f)} = C(p, q)^{(g)}$ for any $p, q \geq 0$. Then there exist centrally symmetric image functions $h_1(x, y)$ and $h_2(x, y)$ such that

$$(g * h_1)(x, y) = (f * h_2)(x, y).$$

Unfortunately, such a hypothesis is not valid in a general case. The convolution kernels $h_1(x, y)$ and $h_2(x, y)$, if they existed, might not have a bounded support. When attempting to prove this hypothesis, $h_1(x, y)$ and $h_2(x, y)$ are constructed in the Fourier domain to fulfill the constraint $G(u, v)H_1(u, v) = F(u, v)H_2(u, v)$. It is formally always possible but the inverse Fourier transform of H_1 and/or H_2 may not exist. However, when working in a discrete domain, the Fourier transform and its inverse always exist and we may consider the above hypothesis valid with a precision given by the number of moments with which we are working.

5.2.1   Template matching experiment

To demonstrate the performance of the above-described invariants, we apply them to the problem of matching a template to a blurred and noisy scene. We were inspired by a remote sensing application area where the *template matching* is a frequent requirement when registering a new image into the reference coordinates.

The template-matching problem is usually formulated as follows: having the templates and a digital image of a large scene, one has to find locations of the given templates in the scene. By the *template* we understand a small digital picture usually containing some significant object that was extracted previously from another image of the same scene and is now being stored in a database.

Numerous image-matching techniques have been proposed in the literature (see the survey paper [5], for instance). A common denominator of all these methods is the assumption that the templates as well as the scene image have already been preprocessed and the degradations such as blur, additive noise, etc. have been removed.

By means of our blur invariants, we try to perform matching without any previous de-blurring. We ran this experiment on real satellite images but the blur and noise were introduced artificially, presenting us with the possibility to control their amount and quantitatively to evaluate the results.

For computer implementation, we have first to convert equation (5.4) into the discrete domain. We employed the moment discretization (7.2). We use the invariants normalized by μ_{00} because of their invariance to image contrast. Moreover, further normalization is necessary to ensure approximately the same range of values regardless of p and q. If we had enough data for training, we could set up the weighting factors according to inter-class and intra-class variances of the features as proposed by Cash and Hamatian in reference [6]. However, this is not the case, because each class is represented by one template only. Thus, we used the invariants

$$\frac{C(p, q)}{\mu_{00}^{(p+q+4)/4}}$$

that yielded a satisfactory normalization of the range of values.

The matching algorithm itself is straightforward. We search the blurred image $g(x, y)$ and for each possible position of the template calculate the Euclidean distance in the space of blur invariants between the template and the corresponding window in $g(x, y)$. The matching position of the template is determined by the minimum distance. We use neither a fast hierarchical implementation nor optimization algorithms, we only perform a "brute-force" full search. The only user-defined parameter in the algorithm is the maximum order r of the invariants used. In this experiment we used $r = 7$; that means we applied 18 invariants from third to seventh orders. In real experiments, if we had more data for training, we could apply sophisticated feature selection algorithms to ensure maximum separability between the templates.

Three templates 48×48 were extracted from the 512×512 high-resolution SPOT image (City of Plzeň, Czech Republic, see Figure 5.5) and contain significant local landmarks – the confluence of two rivers, the apartment block and the road crossing; see Figure 5.6.

To simulate an acquisition by another sensor with a lower spatial resolution, the image was blurred by Gaussian masks of size 3×3 to 21×21 and corrupted by a Gaussian white noise with standard deviation ranging from 0 to 40 each. In all blurred and noisy frames, the algorithm attempted to localize the templates. As expected, the success rate of the matching depends on the amount of blur and on the noise level. In Figure 5.7, we can see an example of a frame in which all three templates were localized correctly. The results are summarized in Figure 5.9. The standard noise deviation is plotted on the horizontal axis, the size of the blurring mask on the vertical axis. The ratio w between the size of the blurring mask and the size of the template is an important factor. Due to the blurring, the pixels lying near the boundary of the template inside the image are affected by those pixels lying outside the template. The higher the w, the larger the part of the template is involved in this *boundary effect* (see Figure 5.8 for an illustration of this phenomenon). Since the invariants are calculated from a bounded area of the image where the blurring is not exactly a convolution, they are no longer invariant which might lead to mismatch, especially when w is higher than 0.15. The curve is a least-square fit of the results by a parabola, the area below the curve corresponds to the domain in which the algorithm performs mostly successfully. We can clearly see the robustness as well as the limitations of the matching by invariants.

5.2.2   Invariants to linear motion blur

As we have seen, it is advantageous to keep the null-space of the invariants as small as possible since it determines the discrimination power of the convolution invariants. If possible, the reduction of the null-space can be done via imposing additional assumptions concerning the blurring PSF. The more specific prior information about the PSF is included; the smaller the null-space, the more invariants with higher discriminability exist.

A typical example is a linear motion blur. In the case of a linear horizontal motion, the PSF has the following form:

$$h(x, y) = \begin{cases} \dfrac{1}{vt}\delta(y) & \Longleftrightarrow 0 \le x \le vt \\ 0 & \text{otherwise,} \end{cases} \tag{5.11}$$

where v is the motion velocity, t is the exposure time and δ is a Dirac function. Such a PSF is symmetric with respect to its centroid and we can, of course, apply all invariants (5.4). However, we can also use certain specific properties of the PSF.

If

$$g(x, y) = (f * h)(x, y),$$

where $h(x, y)$ has the form (5.11), then $\mu_{pq}^{(h)}$ is zero not only for $p + q$ odd as for any centrosymmetric PSF but also for every $q \ne 0$. Consequently, the assertion of Lemma 5.1 reduces to the form

$$\mu_{pq}^{(g)} = \sum_{k=0}^{[p/2]} \binom{p}{2k} \mu_{p-2k,q}^{(f)} \mu_{2k,0}^{(h)}.$$

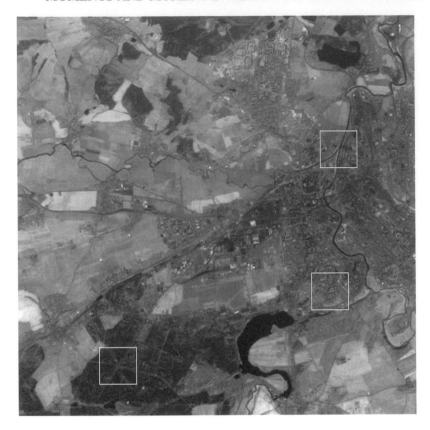

Figure 5.5 Original high-resolution satellite photograph, the City of Plzeň, Czech Republic, with three selected templates.

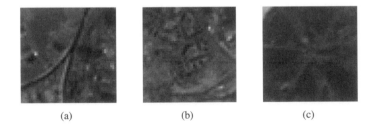

(a)   (b)   (c)

Figure 5.6 The templates: (a) the confluence of the rivers Mže and Radbuza, (b) the apartment block, and (c) the road crossing.

Since more moments of the PSF are zero, we can derive more convolution invariants. In addition to the odd-order invariants (5.4), we also have even-order invariants (we list only the first few of them in explicit forms)

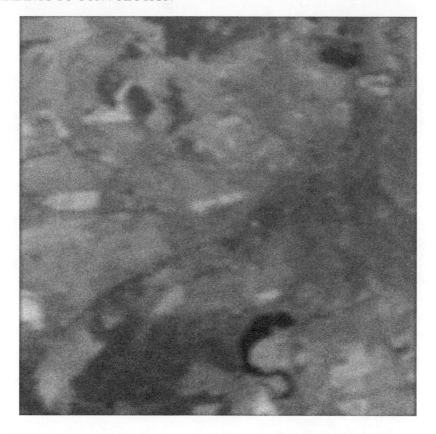

Figure 5.7 Blurred and noisy image. An example of a frame in which all three templates were localized correctly.

- second order:

$$M_1 = \mu_{11},$$
$$M_2 = \mu_{02}$$

- fourth order:

$$M_3 = \mu_{04},$$
$$M_4 = \mu_{13},$$
$$M_5 = \mu_{22} - \frac{\mu_{20}\mu_{02}}{\mu_{00}},$$
$$M_6 = \mu_{31} - \frac{3\mu_{20}\mu_{11}}{\mu_{00}}.$$

More details on the motion blur invariants can be found in reference [7] and also in reference [8], where a slightly different approach is presented.

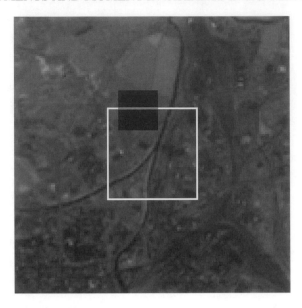

Figure 5.8 Boundary effect. Under a discrete convolution on a bounded support, the pixels near the template boundary (white square) are affected by the pixels lying outside the template. An impact of this effect depends on the size of the blurring mask (black square).

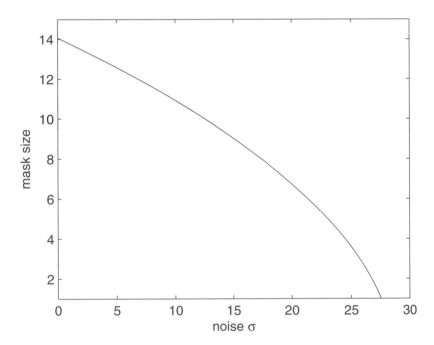

Figure 5.9 The graph summarizing the "mean" results of the experiment. The area below the curve denotes the domain in a "noise-blur space" in which the algorithm works mostly successfully.

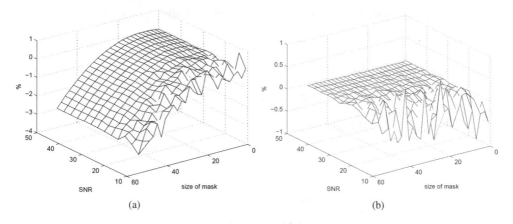

(a)   (b)

Figure 5.10 The influence of the boundary effect and noise on the numerical properties of the centrosymmetric convolution invariant $C(5, 0)$. Horizontal axes: the blurring mask size and the SNR in dB, respectively. Vertical axis: the relative error in %. (a) The invariant corrupted by a boundary effect, and (b) the same invariant calculated from a zero-padded template where no boundary effect appears.

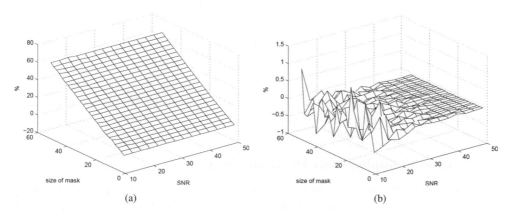

(a)   (b)

Figure 5.11 The influence of the boundary effect and noise on the numerical properties of the horizontal motion blur invariant M_6. Horizontal axes: the blurring mask size and the SNR in dB, respectively. Vertical axis: the relative error in %. (a) The invariant corrupted by a boundary effect, and (b) the same invariant calculated from a zero-padded template where no boundary effect appears.

5.2.3 Extension to n dimensions

The convolution invariants (5.4) can easily be extended to images of an arbitrary number of dimensions [9]. All formulas stay the same as in 2D, only the notation should be changed to reflect the number of dimensions.

Let $n > 1$ be a dimension of the spatial domain. Let us denote the n-dimensional variable as

$$\mathbf{x} \equiv (x_1, \ldots, x_n),$$

and the n-dimensional multi-index as

$$\mathbf{p} \equiv (p_1, \ldots, p_n).$$

Let us define the order of the multi-index as

$$|\mathbf{p}| \equiv \sum_{i=1}^{n} p_i$$

and the n-dimensional binomial coefficients as

$$\binom{\mathbf{p}}{\mathbf{k}} \equiv \prod_{i=1}^{n} \binom{p_i}{k_i}.$$

In n dimensions, the centrosymmetry of the PSF means that $h(\mathbf{x}) = h(-\mathbf{x})$ for any $\mathbf{x}$.

The n-dimensional centrosymmetric convolution invariants are then defined as

$$C(\mathbf{p})^{(f)} = \mu_{\mathbf{p}}^{(f)} - \frac{1}{\mu_0^{(f)}} \sum_{\substack{0 \le \mathbf{j} \le \mathbf{p} \\ 0 < |\mathbf{j}| < |\mathbf{p}|}} \binom{\mathbf{p}}{\mathbf{j}} C(\mathbf{p} - \mathbf{j})^{(f)} \cdot \mu_{\mathbf{j}}^{(f)} \qquad (5.12)$$

for $|\mathbf{p}|$ odd and as

$$C(\mathbf{p})^{(f)} = 0$$

for $|\mathbf{p}|$ even.

The proof of invariance is exactly the same as the proof of Theorem 5.2, so it is not repeated. The same is also true for the relationship between moment-based convolution invariants and those derived in the Fourier domain.

5.2.4   Possible applications and limitations

The convolution invariants (5.4) and their modifications have numerous applications, namely in image matching and registration in remote sensing [3, 4, 10] and medical imaging [11, 12]. Some authors reported their high robustness to noise and use these invariants even if no blur is present in the images. However, we cannot recommend that, because one would unnecessarily omit all even-order moments. The convolution invariants have also found successful applications in face recognition on out-of-focus photographs [2], in normalizing blurred images into the canonical forms [13, 14], in blurred digit and character recognition [15, 16], in robot control [17] and in focus/defocus quantitative measurements [18].

Besides the principal limitations originating from the globality of moments, which is common to all moment invariants, there are two other serious limitations to broader practical applications of the invariants (5.4). The first is its sensitivity to the boundary effect. If the field of view is limited (as it always is), the boundary effect inevitably appears. If the size of the blurring filter is small compared to the size of the image, the boundary effect is negligible; otherwise it may become substantial. The second limit is imposed by the fact that the invariants (5.4) consisting of geometric moments can easily be made scale-invariant but achieving the rotation invariance is complicated. Since in most applications at least a small rotation is present, this weakness motivated intensive research. In the next section we, among others, show how to overcome this limitation.

5.3   Blur invariants for N-fold symmetric PSFs

The above assumption about the centrosymmetry of the PSF is not very restrictive. The vast majority of real imaging systems have even higher symmetry, such as fourfold or circular symmetry. We already pointed out that the more we know about the PSF (which may mean the higher symmetry), the more invariants we can derive. At the same time we are aware of the discriminability limitation – the invariants cannot distinguish objects having the same symmetry as the PSF to which they are invariant. These two facts were a primary motivation in studying invariants with respect to convolution with an N-fold rotation symmetric PSF for a general $N = 2, 3, \ldots, \infty$ (see Chapter 2 for the definition of N-FRS) and using specific invariants to the given PSF in practice. These invariants are very difficult (although possible) to be expressed by means of geometric moments but can be easily derived in terms of complex moments. This is allowed because certain (depending on N) complex moments of N-fold symmetric functions are zero, as proved in Chapter 2. Employing favorable behavior of the complex moments under rotation, we can, in addition to the specific convolution invariance, derive so-called combined invariants to convolution and rotation.

Using complex moments, Theorem 5.2 can be generalized:

Theorem 5.5 *Let $f(x, y)$ be an image function, p, q be non-negative integers and $N > 1$ be a (finite) integer or $N = \infty$. Let us define the following functional $K_N(p, q)^{(f)}$:*

$$K_N(p, q)^{(f)} = c_{pq}^{(f)} - \frac{1}{c_{00}^{(f)}} \sum_{\substack{n=0 \\ 0<n+m \\ (n-m)/N \text{ is integer}}}^{p} \sum_{m=0}^{q} \binom{p}{n}\binom{q}{m} K_N(p - n, q - m)^{(f)} \cdot c_{nm}^{(f)}. \quad (5.13)$$

Then $K_N(p, q)$ is an N-fold blur invariant for any p and q, i.e.

$$K_N(p, q)^{(f)} = K_N(p, q)^{(f*h)}$$

for an arbitrary N-fold symmetric $h(x, y)$.

The recursive formula (5.13) is similar to (5.4) when substituting complex moments. The difference is that the summation goes over a different set of indexes which depends on N. Thanks to this, it is not necessary to define extra zero invariants $K_N(p, q)^{(f)} = 0$ for integer $(p - q)/N$ because these cases are also included in (5.13). The proof of this theorem is almost the same as the proof of Theorem 5.2 and thus we skip it.

Several consequences and particular cases of Theorem 5.5 are worth commenting on.

- Theorem 5.2 with complex moments is a particular case of Theorem 5.5 for $N = 2$, hence $C(p, q) = K_2(p, q)$.

- The invariants (5.13) are generally complex. If we prefer working with real-valued invariants, we should use real and imaginary parts separately.

- It holds for any invariant (5.13) $K_N(p, q) = K_N(q, p)^*$, so it is sufficient to consider the cases $p \geq q$ only. This assertion can be proved by induction for the order of the invariants.

- For any N (finite or infinite) it holds $K_N(0, 0) = c_{00}$ and $K_N(p, p) = 0$ for $p > 0$.

- Theorem 5.5 formally holds also for $N = 1$ (i.e. PSF without any symmetry) but all invariants except $K_1(0, 0)$ are identically zero, $K_1(p, q)^{(f)} = 0$ for any $f(x, y)$.

- The number of nontrivial invariants up to the given order increases as N increases, reaching its maximum for $N = \infty$. If we imagine the $K_N(p, q)$'s as elements of an $r \times r$ matrix, where $p \leq r$ and $q \leq r$, then zeros are always on the main diagonal regardless of N (except $p = q = 0$). Besides, zeros are also on all minor diagonals for which $(p - q)/N$ is an integer.

- The null space of the invariants decreases as N increases reaching its minimum for $N = \infty$.

- Let N be a product of L integers, $N = k_1 k_2 \cdots k_L$. Then the PSF has also k_n-fold symmetry for any $n = 1, 2, \ldots, L$. Thus, any $K_{k_n}(p, q)$ is also a blur invariant. However, using these invariants is useless since they are all dependent on $K_N(p, q)$'s. (Note that the dependence here does not mean equality; generally $K_N(p, q) \neq K_{k_n}(p, q)$.)

- The invariants (5.13) form an independent set for $p > q$ and they are complete in the sense similar to Theorem 5.4: let $f(x, y)$, $g(x, y)$ and $h(x, y)$ be arbitrary image functions such that

$$g(x, y) = (f * h)(x, y).$$

Let for a given N be $K_N(p, q)^{(f)} = K_N(p, q)^{(g)}$ for any $p, q \geq 0$. Then $h(x, y)$ has N-fold symmetry.

All the above assertions can be proved either directly or after simple modifications of the corresponding proofs from the previous section.

5.3.1   Blur invariants for circularly symmetric PSFs

Circularly symmetric PSFs (i.e. those having $N = \infty$, which means $h(x, y) = h(\sqrt{x^2 + y^2})$) appear quite often in imaging. The PSF of an out-of-focus camera is a cylinder-like function, the PSF describing longterm atmospheric turbulence blur can be approximated by a Gaussian, and the intrinsic PSF's of most optical sensors have this type of symmetry as well (let us recall, for instance, the Airy function describing light diffraction). Let us show what the invariants (5.13) look like in such a case.

In Chapter 2, we proved an important property of circularly symmetric functions: if $p \neq q$, then $c_{pq}^{(h)} = 0$. Consequently, Lemma 5.1 simplifies to the form

$$c_{pq}^{(g)} = \sum_{j=0}^{q} \binom{p}{j} \binom{q}{j} c_{jj}^{(h)} c_{p-j,q-j}^{(f)} \tag{5.14}$$

assuming that $p \geq q$.

Applying Theorem 5.5 we obtain a specific form of ∞-fold convolution invariants

$$K_\infty(p, q)^{(f)} = c_{pq}^{(f)} - \frac{1}{c_{00}^{(f)}} \sum_{n=1}^{q} \binom{p}{n}\binom{q}{n} K_\infty(p - n, q - n)^{(f)} \cdot c_{nn}^{(f)}.$$

Note that only $K_\infty(pp)$'s are identically zero (for $p > 0$) because the impact of the blurring on the c_{pp}'s cannot be eliminated. All other invariants are nontrivial. The invariants up to the sixth order are shown below in explicit forms:

$$K_\infty(1, 0) = c_{10}, \qquad K_\infty(2, 0) = c_{20}, \qquad K_\infty(3, 0) = c_{30},$$

$$K_\infty(4, 0) = c_{40}, \qquad K_\infty(5, 0) = c_{50}, \qquad K_\infty(6, 0) = c_{60},$$

$$K_\infty(2, 1) = c_{21} - 2\frac{c_{10}c_{11}}{c_{00}},$$

$$K_\infty(3, 1) = c_{31} - 3\frac{c_{20}c_{11}}{c_{00}},$$

$$K_\infty(4, 1) = c_{41} - 4\frac{c_{30}c_{11}}{c_{00}},$$

$$K_\infty(5, 1) = c_{51} - 5\frac{c_{40}c_{11}}{c_{00}},$$

$$K_\infty(3, 2) = c_{32} - 6\frac{K_\infty(2, 1)c_{11}}{c_{00}} - 3\frac{c_{10}c_{22}}{c_{00}},$$

$$K_\infty(4, 2) = c_{42} - 8\frac{K_\infty(3, 1)c_{11}}{c_{00}} - 6\frac{K_\infty(2, 1)c_{22}}{c_{00}}.$$

A detailed analysis of ∞-fold convolution invariants can be found in reference [19].

5.3.2 Blur invariants for Gaussian PSFs

We obtain even more specific invariants if we assume a Gaussian PSF. In this case, we benefit not only from its circular symmetry but also from the knowledge of its parametric form. However, Theorem 5.5 does not consider parametric forms of the PSFs at all, hence Gaussian blur invariants cannot be obtained directly as a special case of this theorem (more precisely, not all of them).

A 2D circularly symmetric Gaussian function $G_\sigma(x, y)$ is defined as

$$G_\sigma(x, y) = g_\sigma(x)g_\sigma(y),$$

where $\sigma > 0$ and $g_\sigma(x)$, $g_\sigma(y)$ are 1D Gaussian functions of a traditional shape

$$g_\sigma(x) = \frac{1}{\sqrt{2\pi}\sigma}e^{-(x^2)/(2\sigma^2)}.$$

As can be proved from definition, it holds for complex moments of $G_\sigma(x, y)$

$$c_{pq}^{(G_\sigma)} = \begin{cases} (2\sigma^2)^p p! & p = q \\ 0 & p \neq q. \end{cases}$$

Now, we can introduce Gaussian blur invariants.

Theorem 5.6 *Let $f(x, y)$ be an image function and $p \geq q$ be non-negative integers. Let us define the following functional*

$$K_G(p, q)^{(f)} = c_{pq}^{(f)} - \sum_{k=1}^{q} k! \binom{p}{k} \binom{q}{k} \left(\frac{c_{11}^{(f)}}{c_{00}^{(f)}} \right)^k K_G(p - k, q - k)^{(f)}. \tag{5.15}$$

Then $K_G(p, q)$ is a Gaussian blur invariant for any p and q.

For the proof of this theorem, see Appendix 5.D.[3]

Formally, equation (5.15) resembles the previous recursive invariants. Unlike them, the Gaussian blur invariants also have a nonrecursive form[4] that is more convenient for implementation

$$K_G(p, q)^{(f)} = \sum_{j=0}^{q} j! \binom{p}{j} \binom{q}{j} \left(-\frac{c_{11}^{(f)}}{c_{00}^{(f)}} \right)^j c_{p-j,q-j}^{(f)}. \tag{5.16}$$

The equivalence of equation (5.16) and equation (5.15) is proved in Appendix 5.E.

Similarly, as in the previous cases, only the invariants where $p \geq q$ are independent because $K_G(q, p) = K_G(p, q)^*$. Unlike the invariants $K_\infty(p, q)$, only one invariant $K_G(1, 1)$ equals zero regardless of the image; the other invariants are valid. This is because a Gaussian PSF has just one degree of freedom given by the parameter σ. To eliminate it, we "sacrificed" the moment c_{11}. As in the previous cases, these invariants form a complete and independent set in the sense explained above. The null space of the Gaussian blur invariants equals the set of all Gaussian functions; hence it is much smaller than the null space of the invariants $K_\infty(p, q)$ that indicates a better discrimination power of the Gaussian invariants.

Unfortunately, special blur invariants similar to equation (5.16) and equation (5.15) that would employ the parametric forms of the PSF do not exist for other common blurs such as out-of-focus blur, motion blur and diffraction (Airy) blur. If they existed, they would also have to stay invariant if the blurring is applied more than once (this is a basic property of all invariants). However, repeating these blurs goes beyond the considered degradation class. An out-of-focus blur applied twice is not an out-of-focus blur any more and the same is true for motion blur and diffraction. Convolutions with these PSFs do not form groups; hence respective invariants cannot exist in principle.

5.4   Combined invariants

In practice, image blurring is often coupled with spatial transformations of the image. To handle these situations, having *combined invariants* that are invariant simultaneously to convolution and to certain transforms of spatial coordinates is of great demand. Although a few attempts to construct combined invariants from geometric moments can be found in reference [3], only the introduction of complex moments into the blur invariants made this possible in a systematic way. Combined invariants to convolution with a centrosymmetric PSF and to rotation were introduced by Zitová and Flusser [21], who also reported their successful usage in satellite image registration [22] and in camera motion estimation [23]. Later on, they also introduced combined invariants to convolution with a circularly symmetric

---

[3] Some authors [20] realized without a deeper analysis that the complex moments c_{p0} are Gaussian blur invariants for any p. As we can see, this is a particular (and incomplete) case of Theorem 5.6.

[4] Gaussian invariants are probably the only reasonable class of blur invariants having a nonrecursive expression.

PSF and to rotation [19]. In the following section, we present the basics of a unified theory of the combined invariants to convolution and rotation.

5.4.1 Combined invariants to convolution and rotation

Theorem 5.5 introduced invariants to convolution with an arbitrary PSF having an N-FRS. Since an N-fold symmetry of the PSF is not violated by a rotation of the image (either before or after the blurring), it makes sense to look for combined invariants. Their construction is surprisingly easy and is very similar to the construction of pure rotation invariants from the complex moments (see Chapter 2).

Let us investigate the behavior of the invariants $K_N(p, q)$ under rotation of the image. Let us denote the invariants of the rotated image as $K'_N(p, q)$. We can observe that the invariants behave under rotation in the same way as the complex moments themselves, i.e.

$$K'_N(p, q) = e^{-i(p-q)\alpha} \cdot K_N(p, q), \tag{5.17}$$

where α is the angle of rotation. A formal proof of this statement can be easily accomplished by induction for the order.

Now we can repeat all the considerations performed in Chapter 2 when we constructed rotation invariants. Any product of the form

$$\prod_{j=1}^{n} K_N(p_j, q_j)^{k_j}, \tag{5.18}$$

where $n \geq 1$ and k_j, p_j and q_j; $j = 1, \ldots, n$, are non-negative integers such that

$$\sum_{j=1}^{n} k_j(p_j - q_j) = 0$$

is a combined invariant to convolution and rotation.

To create a basis of the combined invariants up to a certain order, we apply the same approach as in Chapter 2. We choose a low-order invariant as a basic element, $K_N(1, 2)$. (This invariant is nontrivial for any N but may be zero for certain images; if this is the case we have to search for the "nearest" nonzero invariant.) Then any nontrivial invariant $K_N(p, q)$, $p > q$, generates one element of the basis as

$$K_N(p, q)K_N(1, 2)^{p-q}.$$

The proof of independence and completeness (modulo rotation and convolution with an N-fold symmetric PSF) of this set is straightforward when using the propositions proved earlier. Additional invariance, for instance to scaling and/or to contrast changes, can be achieved by an obvious normalization (see reference [24] for more details and for the tests of numerical properties). An example of using the combined invariants for registration and fusion of differently blurred and rotated images will be presented in Chapter 8.

The rotation property (5.17) is preserved also for Gaussian invariants $K_G(p, q)$ (5.15), hence we can create the combined invariants to rotation and Gaussian blur (and their basis) in the same way. Compared to the above case, we obtain additional combined invariants $K_G(p, p)$ for any $p \neq 1$.

Extension of the convolution-rotation invariants into 3D is rather complicated and beyond the scope of this book. We refer to our recent paper [25] where a group-theoretic approach to extension of the combined invariants into an arbitrary number of dimensions was proposed.

5.4.2   Combined invariants to convolution and affine transform

After deriving the combined blur-rotation invariants in the previous section, the next logical step is an extension to affine transform. Since the only N-fold symmetry preserved under an affine transform is for $N = 2$, it is meaningful to consider centrosymmetric PSFs only (we allow commutative blur and affine transform of the image).

The first attempt to find the *combined blur-affine invariants* (CBAIs) was published by Zhang *et al.* [14]. They did not derive the combined invariants explicitly. They transformed the image into a normalized position (where the normalization constraints were invariant to blur) and then they calculated the pure blur invariants (5.4) of the normalized image.

Here, we present a direct approach consisting of substituting the blur invariants (5.4) for central moments in the definition formulas of the AMIs that were derived in Chapter 3. The following theorem not only guarantees the existence of the CBAIs but also provides an explicit algorithm of how to construct them.

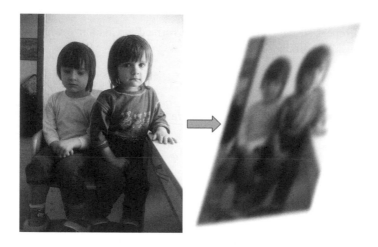

Figure 5.12 Original image and its blurred and affinely deformed version. The values of the CBAIs are the same for both of them.

Theorem 5.7 *Let $I(\mu_{00}, \ldots, \mu_{pq})$ be an AMI. Then $I(C(0, 0), \ldots, C(P, Q))$, where the $C(p, q)$'s are defined by equation (5.4), is a CBAI.*

For the proof of Theorem 5.7, see Appendix 5.F.

Certain care must be taken when applying this theorem because all even-order blur invariants are identically zero. Substituting them into the AMIs may lead to trivial combined invariants. In other words, not all AMIs generate nontrivial combined invariants. More details on the CBAIs and experimental studies can be found in reference [26]. The use of the CBAIs for aircraft silhouette recognition was reported in reference [27] and for the classification of winged insect [28]. A slightly different approach to the affine-blur invariant matching was presented in reference [29], where the combined invariants are constructed in Fourier domain. Most recently, Boldyš [30] generalized Theorem 5.7 into an arbitrary number of dimensions.

5.5   Conclusion

In this chapter, we presented a class of radiometric moment invariants that are invariant with respect to convolution (blurring) of an image with an unknown filter which exhibits a certain type of symmetry. The filter is often a distorting point-spread function of a nonideal imaging system. We recalled the traditional results for centrosymmetric filters and introduced a new unified theory for N-fold symmetric filters, showing at the same time the close connections between moment invariants and the Fourier phase of an image. The chapter is completed by describing combined invariants to convolution and rotation/affine transform.

Although various versions of the blur invariants have been successfully used in practical tasks, one should be aware of two intrinsic limitations. First, the invariants are sensitive to model errors. If the degradation is not an exact convolution, the invariance property may be violated. A boundary effect is a typical example of the model error. It can be removed by isolating and zero-padding the object before the blurring; however, this is mostly impossible. Second, the recognitive power is always restricted to "modulo convolution." As an immediate consequence, we are unable to distinguish objects having the same type of symmetry as the assumed filter.

Appendix 5.A

Proof of Lemma 5.1

Let us prove Lemma 5.1 for central moments, where the proof is more complicated than for geometric and complex moments.

First of all, let us prove that the centroid of $g(x, y)$ is a sum of the centroids of $f(x, y)$ and $h(x, y)$:

$$(x_c^{(g)}, y_c^{(g)}) = (x_c^{(f)} + x_c^{(h)}, y_c^{(f)} + y_c^{(h)}).$$

Using the definition of a centroid, we obtain

$$x_c^{(g)} = \frac{m_{10}^{(g)}}{m_{00}^{(g)}} = \frac{m_{10}^{(f)} m_{00}^{(h)} + m_{00}^{(f)} m_{10}^{(h)}}{m_{00}^{(f)} m_{00}^{(h)}} = \frac{m_{10}^{(f)}}{m_{00}^{(f)}} + \frac{m_{10}^{(h)}}{m_{00}^{(h)}} = x_c^{(f)} + x_c^{(h)}.$$

The proof for $y_c^{(g)}$ is similar.

Now we can prove the assertion of Lemma 5.1.

$$\mu_{pq}^{(g)} = \int\limits_{-\infty}^{\infty} \int\limits_{-\infty}^{\infty} (x - x_c^{(g)})^p (y - y_c^{(g)})^q \, g(x, y) \, dx \, dy$$

$$= \int\limits_{-\infty}^{\infty} \int\limits_{-\infty}^{\infty} (x - x_c^{(f)} - x_c^{(h)})^p (y - y_c^{(f)} - y_c^{(h)})^q (f * h)(x, y) \, dx \, dy$$

$$= \int\limits_{-\infty}^{\infty} \int\limits_{-\infty}^{\infty} (x - x_c^{(f)} - x_c^{(h)})^p (y - y_c^{(f)} - y_c^{(h)})^q$$

$$\cdot \left(\int_{-\infty}^{\infty} \int_{-\infty}^{\infty} h(a, b) f(x - a, y - b)\, da\, db \right) dx\, dy$$

$$= \int_{-\infty}^{\infty} \int_{-\infty}^{\infty} h(a, b) \left(\int_{-\infty}^{\infty} \int_{-\infty}^{\infty} (x - x_c^{(f)} - x_c^{(h)})^p (y - y_c^{(f)} - y_c^{(h)})^q \right.$$

$$\left. \cdot\, f(x - a, y - b)\, dx\, dy \right) da\, db$$

$$= \int_{-\infty}^{\infty} \int_{-\infty}^{\infty} h(a, b) \left(\int_{-\infty}^{\infty} \int_{-\infty}^{\infty} (x - x_c^{(f)} + a - x_c^{(h)})^p (y - y_c^{(f)} + b - y_c^{(h)})^q \right.$$

$$\left. \cdot\, f(x, y)\, dx\, dy \right) da\, db$$

$$= \int_{-\infty}^{\infty} \int_{-\infty}^{\infty} h(a, b) \left(\sum_{k=0}^{p} \sum_{j=0}^{q} \binom{p}{k} \binom{q}{j} (a - x_c^{(h)})^{p-k} (b - y_c^{(h)})^{q-j} \mu_{kj}^{(f)} \right) da\, db$$

$$= \sum_{k=0}^{p} \sum_{j=0}^{q} \binom{p}{k} \binom{q}{j} \mu_{kj}^{(f)} \mu_{p-k,q-j}^{(h)}. \qquad\qquad \square$$

Appendix 5.B

Proof of Theorem 5.2

The statement of the theorem is trivial for any even r. Let us prove the statement for odd r by induction.

- $r = 1$

$$C(0, 1)^{(g)} = C(0, 1)^{(f)} = 0,$$

$$C(1, 0)^{(g)} = C(1, 0)^{(f)} = 0$$

regardless of f and h.

- $r = 3$

There are four invariants of the third order: $C(1, 2)$, $C(2, 1)$, $C(0, 3)$ and $C(3, 0)$. Evaluating their recursive definition (5.4) we obtain them in the explicit forms:

$$C(1, 2) = \mu_{12},$$

$$C(2, 1) = \mu_{21},$$

$$C(0, 3) = \mu_{03},$$

$$C(3, 0) = \mu_{30}.$$

Let us prove the theorem for $C(1, 2)$; the proofs for the other invariants are similar.

$$C(1, 2)^{(g)} = \mu_{12}^{(g)} = \sum_{k=0}^{1}\sum_{j=0}^{2} \binom{1}{k}\binom{2}{j}\mu_{kj}^{(h)}\mu_{1-k,2-j}^{(f)} = \mu_{12}^{(f)} = C(1, 2)^{(f)}.$$

- Provided that the theorem is valid for all invariants of order $1, 3, \ldots, r - 2$. Then using Lemma 5.1, we obtain

$$C(p, q)^{(g)} = \mu_{pq}^{(g)} - \frac{1}{\mu_{00}^{(g)}} \sum_{\substack{n=0 \\ 0<n+m<p+q}}^{p}\sum_{m=0}^{q} \binom{p}{n}\binom{q}{m}C(p-n, q-m)^{(g)} \cdot \mu_{nm}^{(g)}$$

$$= \sum_{k=0}^{p}\sum_{j=0}^{q} \binom{p}{k}\binom{q}{j}\mu_{kj}^{(h)}\mu_{p-k,q-j}^{(f)}$$

$$- \frac{1}{\mu_{00}^{(f)}} \sum_{\substack{n=0 \\ 0<n+m<p+q}}^{p}\sum_{m=0}^{q} \binom{p}{n}\binom{q}{m}C(p-n, q-m)^{(f)}$$

$$\cdot \sum_{k=0}^{n}\sum_{j=0}^{m} \binom{n}{k}\binom{m}{j}\mu_{kj}^{(h)}\mu_{n-k,m-j}^{(f)}.$$

Using the identity

$$\binom{a}{b}\binom{b}{c} = \binom{a}{c}\binom{a-c}{b-c},$$

changing the order of the summation and shifting the indices, we obtain

$$C(p, q)^{(g)}$$

$$= C(p, q)^{(f)} + \sum_{\substack{k=0 \\ 0<k+j}}^{p}\sum_{j=0}^{q} \binom{p}{k}\binom{q}{j}\mu_{kj}^{(h)}\mu_{p-k,q-j}^{(f)}$$

$$- \frac{1}{\mu_{00}^{(f)}} \sum_{\substack{n=0 \\ 0<n+m<p+q}}^{p}\sum_{m=0}^{q} \sum_{\substack{k=0 \\ 0<k+j}}^{n}\sum_{j=0}^{m} \binom{p}{n}\binom{q}{m}\binom{n}{k}\binom{m}{j}$$

$$C(p-n, q-m)^{(f)}\mu_{kj}^{(h)}\mu_{n-k,m-j}^{(f)}$$

$$= C(p, q)^{(f)} + \sum_{\substack{k=0 \\ 0<k+j}}^{p}\sum_{j=0}^{q} \binom{p}{k}\binom{q}{j}\mu_{kj}^{(h)}\mu_{p-k,q-j}^{(f)}$$

$$- \frac{1}{\mu_{00}^{(f)}} \sum_{\substack{n=0 \\ 0<n+m<p+q}}^{p}\sum_{m=0}^{q} \sum_{\substack{k=0 \\ 0<k+j}}^{n}\sum_{j=0}^{m} \binom{p}{k}\binom{q}{j}\binom{p-k}{n-k}\binom{q-j}{m-j}$$

$$C(p-n, q-m)^{(f)}\mu_{kj}^{(h)}\mu_{n-k,m-j}^{(f)}$$

$$= C(p, q)^{(f)} + \sum_{\substack{k=0 \\ 0<k+j}}^{p} \sum_{j=0}^{q} \binom{p}{k}\binom{q}{j} \mu_{kj}^{(h)} \mu_{p-k,q-j}^{(f)}$$

$$- \frac{1}{\mu_{00}^{(f)}} \sum_{\substack{k=0 \\ 0<k+j}}^{p} \sum_{j=0}^{q} \sum_{\substack{n=k \\ n+m<p+q}}^{p} \sum_{m=j}^{q} \binom{p}{k}\binom{q}{j}\binom{p-k}{n-k}\binom{q-j}{m-j}$$

$$C(p-n, q-m)^{(f)} \mu_{kj}^{(h)} \mu_{n-k,m-j}^{(f)}$$

$$= C(p, q)^{(f)} + \sum_{\substack{k=0 \\ 0<k+j}}^{p} \sum_{j=0}^{q} \binom{p}{k}\binom{q}{j} \mu_{kj}^{(h)} \left(\mu_{p-k,q-j}^{(f)} \right.$$

$$- \frac{1}{\mu_{00}^{(f)}} \sum_{\substack{n=k \\ n+m<p+q}}^{p} \sum_{m=j}^{q} \binom{p-k}{n-k}\binom{q-j}{m-j}$$

$$\left. C(p-n, q-m)^{(f)} \mu_{n-k,m-j}^{(f)} \right)$$

$$= C(p, q)^{(f)} + \sum_{\substack{k=0 \\ 0<k+j}}^{p} \sum_{j=0}^{q} \binom{p}{k}\binom{q}{j} \mu_{kj}^{(h)} \left(\mu_{p-k,q-j}^{(f)} \right.$$

$$- \frac{1}{\mu_{00}^{(f)}} \sum_{\substack{n=0 \\ n+m<p+q-k-j}}^{p-k} \sum_{m=0}^{q-j} \binom{p-k}{n}\binom{q-j}{m}$$

$$\left. C(p-n-k, q-m-j)^{(f)} \mu_{nm}^{(f)} \right).$$

Using a shorter notation we can rewrite the last equation in the form

$$C(p, q)^{(g)} = C(p, q)^{(f)} + \sum_{\substack{k=0 \\ 0<k+j}}^{p} \sum_{j=0}^{q} \binom{p}{k}\binom{q}{j} \mu_{kj}^{(h)} \cdot D_{kj}, \qquad (5.19)$$

where

$$D_{kj} = \mu_{p-k,q-j}^{(f)} - \frac{1}{\mu_{00}^{(f)}} \sum_{\substack{n=0 \\ n+m<p+q-k-j}}^{p-k} \sum_{m=0}^{q-j} \binom{p-k}{n}\binom{q-j}{m}$$

$$C(p-n-k, q-m-j)^{(f)} \mu_{nm}^{(f)}.$$

If $k + j$ is odd, then, due to the centrosymmetry, $\mu_{kj}^{(h)} = 0$. If $k + j$ is even, then it follows from the definition (5.4) that

$$C(p-k, q-j) = \mu_{p-k,q-j} - \frac{1}{\mu_{00}} \sum_{n=0}^{p-k} \sum_{\substack{m=0 \\ 0<n+m<p+q-k-j}}^{q-j} \binom{p-k}{n} \binom{q-j}{m}$$

$$C(p-k-n, q-j-m) \cdot \mu_{nm}.$$

Consequently,

$$D_{kj} = C(p-k, q-j)^{(f)} - \frac{1}{\mu_{00}^{(f)}} C(p-k, q-j)^{(f)} \mu_{00}^{(f)} = 0.$$

Thus, equation (5.19) implies that $C(p, q)^{(g)} = C(p, q)^{(f)}$ for every p and q. □

Appendix 5.C

Proof of Theorem 5.3

Expanding the Fourier transform of $f(x, y)$ as a power series

$$F(u, v) = \sum_{k=0}^{\infty} \sum_{j=0}^{\infty} \frac{(-2\pi i)^{k+j}}{k! \cdot j!} m_{kj}^{(f)} \cdot u^k v^j$$

implies that

$$\mathcal{R}e(F(u, v)) = \sum_{\substack{k=0 \\ (k+j) \text{ even}}}^{\infty} \sum_{j=0}^{\infty} \frac{(-2\pi i)^{k+j}}{k! \cdot j!} m_{kj} \cdot u^k v^j$$

$$= \sum_{\substack{k=0 \\ (k+j) \text{ even}}}^{\infty} \sum_{j=0}^{\infty} \frac{(-1)^{(k+j)/2} (-2\pi)^{k+j}}{k! \cdot j!} m_{kj} \cdot u^k v^j$$

and

$$i \cdot \mathcal{I}m(F(u, v)) = \sum_{\substack{k=0 \\ (k+j) \text{ odd}}}^{\infty} \sum_{j=0}^{\infty} \frac{(-2\pi i)^{k+j}}{k! \cdot j!} m_{kj} \cdot u^k v^j$$

$$= i \cdot \sum_{\substack{k=0 \\ (k+j) \text{ odd}}}^{\infty} \sum_{j=0}^{\infty} \frac{(-1)^{(k+j-1)/2} (-2\pi)^{k+j}}{k! \cdot j!} m_{kj} \cdot u^k v^j.$$

Thus, $\tan(\text{ph} F(u, v))$ is a ratio of two absolutely convergent power series and therefore it can also be expressed as a power series

$$\tan(\text{ph} F(u, v)) = \frac{\mathcal{I}m(F(u, v))}{\mathcal{R}e(F(u, v))} = \sum_{k=0}^{\infty} \sum_{j=0}^{\infty} a_{kj} u^k v^j,$$

which must fulfill the condition

$$\sum_{\substack{k=0 \\ (k+j)\text{ odd}}}^{\infty} \sum_{j=0}^{\infty} \frac{(-1)^{(k+j-1)/2}(-2\pi)^{k+j}}{k! \cdot j!} m_{kj} \cdot u^k v^j$$

$$= \sum_{\substack{k=0 \\ (k+j)\text{ even}}}^{\infty} \sum_{j=0}^{\infty} \frac{(-1)^{(k+j)/2}(-2\pi)^{k+j}}{k! \cdot j!} m_{kj} \cdot u^k v^j \cdot \sum_{k=0}^{\infty} \sum_{j=0}^{\infty} a_{kj} u^k v^j. \qquad (5.20)$$

Let us prove by induction that a_{kj} has the form (5.10) if $k + j$ is odd.

- $k + j = 1$

  It follows from (5.20) that
  $$a_{10} = \frac{-2\pi m_{10}}{m_{00}}.$$

  On the other hand,
  $$\frac{(-2\pi)(-1)^0}{1! \cdot 0! \cdot \mu_{00}} M(1, 0) = \frac{-2\pi m_{10}}{m_{00}} = a_{10}.$$

- Let us suppose that the assertion has been proven for all k and $j, k + j \leq r$, where r is an odd integer and let $p + q = r + 2$. It follows from (5.20) that

  $$\frac{(-1)^{(p+q-1)/2}(-2\pi)^{p+q}}{p! \cdot q!} m_{pq}$$

  $$= m_{00} a_{pq} + \sum_{\substack{n=0 \\ 0<n+m}}^{p} \sum_{m=0}^{q} \frac{(-1)^{(n+m)/2}(-2\pi)^{n+m}}{n! \cdot m!} m_{nm} a_{p-n,q-m}.$$

  Introducing (5.10) into the right side, we obtain

  $$\frac{(-1)^{(p+q-1)/2}(-2\pi)^{p+q}}{p! \cdot q!} m_{pq}$$

  $$= m_{00} a_{pq} + \sum_{\substack{n=0 \\ 0<n+m<p+q}}^{p} \sum_{m=0}^{q} \frac{(-1)^{(p+q-1)/2}(-2\pi)^{p+q}}{n! \cdot m! \cdot (p-n)! \cdot (q-m)! \cdot \mu_{00}}$$

  $$\cdot M(p-n, q-m) \cdot m_{nm}$$

  and, consequently,

  $$a_{pq} = \frac{(-1)^{(p+q-1)/2}(-2\pi)^{p+q}}{p! \cdot q! \cdot m_{00}}$$

  $$\cdot \left(m_{pq} - \frac{1}{\mu_{00}} \sum_{\substack{n=0 \\ 0<n+m<p+q}}^{p} \sum_{m=0}^{q} \binom{p}{n}\binom{q}{m} M(p-n, q-m) \cdot m_{nm} \right).$$

Finally, it follows from (5.4) that

$$a_{pq} = \frac{(-1)^{(p+q-1)/2}(-2\pi)^{p+q}}{p! \cdot q! \cdot \mu_{00}} M(p, q).$$ □

Appendix 5.D

Proof of Theorem 5.6

Assuming that $p \geq q$, the proof is via induction with respect to q. If $q = 0$, then

$$K_G(p, q)^{(f*h)} = c_{pq}^{(f*h)} = c_{pq}^{(f)} c_{00}^{(h)} = c_{pq}^{(f)} = K_G(p, q)^{(f)}.$$

Now let $q > 0$.

$$K_G(p, q)^{(f*h)} - K_G(p, q)^{(f)}$$

$$= c_{pq}^{(f*h)} - \sum_{k=1}^{q} \binom{p}{k}\binom{q}{k} \left(\frac{c_{11}^{(f*h)}}{c_{00}^{(f*h)}}\right)^k k! K_G(p-k, q-k)^{(f*h)} - K_G(p, q)^{(f)}$$

$$= \sum_{j=0}^{q} \binom{p}{j}\binom{q}{j}(2\sigma^2)^j j! c_{p-j,q-j}^{(f)} - \sum_{k=0}^{q} \binom{p}{k}\binom{q}{k}\left(\frac{c_{11}^{(f)}}{c_{00}^{(f)}} + 2\sigma^2\right)^k k! K_G(p-k, q-k)^{(f)}.$$

Now we show that the two sums in the last formula are equal, hence that $K_G(p, q)^{(f*h)} - K_G(p, q)^{(f)} = 0$.

$$\sum_{k=0}^{q} \binom{p}{k}\binom{q}{k}\left(\frac{c_{11}^{(f)}}{c_{00}^{(f)}} + 2\sigma^2\right)^k k! K_G(p-k, q-k)^{(f)}$$

$$= \sum_{k=0}^{q} \binom{p}{k}\binom{q}{k} \sum_{j=0}^{k} \binom{k}{j}\left(\frac{c_{11}^{(f)}}{c_{00}^{(f)}}\right)^{k-j}(2\sigma^2)^j k! K_G(p-k, q-k)^{(f)}$$

$$= \sum_{j=0}^{q} \sum_{k=j}^{q} \frac{p!q!}{(p-k)!(q-k)!j!(k-j)!}\left(\frac{c_{11}^{(f)}}{c_{00}^{(f)}}\right)^{k-j}(2\sigma^2)^j K_G(p-k, q-k)^{(f)}$$

$$= \sum_{j=0}^{q} \sum_{k=0}^{q-j} \frac{p!q!}{(p-k-j)!(q-k-j)!j!k!}\left(\frac{c_{11}^{(f)}}{c_{00}^{(f)}}\right)^{k}(2\sigma^2)^j K_G(p-k-j, q-k-j)^{(f)}$$

$$= \sum_{j=0}^{q} \binom{p}{j}\binom{q}{j} j!(2\sigma^2)^j \sum_{k=0}^{q-j} \binom{p-j}{k}\binom{q-j}{k}\left(\frac{c_{11}^{(f)}}{c_{00}^{(f)}}\right)^{k} k! K_G(p-k-j, q-k-j)^{(f)}$$

$$= \sum_{j=0}^{q} \binom{p}{j}\binom{q}{j} j!(2\sigma^2)^j (K_G(p-j, q-j)^{(f)} - (K_G(p-j, q-j)^{(f)} - c_{p-j,q-j}^{(f)}))$$

$$= \sum_{j=0}^{q} \binom{p}{j}\binom{q}{j} j!(2\sigma^2)^j c_{p-j,q-j}^{(f)}.$$ □

Appendix 5.E

Proof of equation (5.16)

Assuming $p \geq q$, the proof is via induction with respect to q. If $q = 0$, then $K_G(p, q)^{(f)} = c_{pq}^{(f)}$. Now let $q > 0$.

$$K_G(p, q)^{(f)}$$

$$= c_{pq}^{(f)} - \sum_{k=1}^{q} \binom{p}{k}\binom{q}{k}\left(\frac{c_{11}^{(f)}}{c_{00}^{(f)}}\right)^k k! K_G(p-k, q-k)^{(f)}$$

$$= c_{pq}^{(f)} - \sum_{k=1}^{q} \binom{p}{k}\binom{q}{k}\left(\frac{c_{11}^{(f)}}{c_{00}^{(f)}}\right)^k k! \sum_{j=0}^{q-k}\binom{p-k}{j}\binom{q-k}{j}\left(-\frac{c_{11}^{(f)}}{c_{00}^{(f)}}\right)^j j! c_{p-k-j,q-k-j}^{(f)}$$

$$= c_{pq}^{(f)} - \sum_{k=1}^{q}\sum_{j=0}^{q-k}\frac{p!q!}{k!j!(p-k-j)!(q-k-j)!}\left(\frac{c_{11}^{(f)}}{c_{00}^{(f)}}\right)^{k+j}(-1)^j c_{p-k-j,q-k-j}^{(f)}$$

$$= c_{pq}^{(f)} - \sum_{k=1}^{q}\sum_{j=k}^{q}\frac{p!q!}{k!(j-k)!(p-j)!(q-j)!}\left(\frac{c_{11}^{(f)}}{c_{00}^{(f)}}\right)^{j}(-1)^{j-k} c_{p-j,q-j}^{(f)}$$

$$= c_{pq}^{(f)} - \sum_{j=1}^{q}\sum_{k=1}^{j}\frac{p!q!}{k!(j-k)!(p-j)!(q-j)!}\left(\frac{c_{11}^{(f)}}{c_{00}^{(f)}}\right)^{j}(-1)^{j-k} c_{p-j,q-j}^{(f)}$$

$$= c_{pq}^{(f)} - \sum_{j=1}^{q}\binom{p}{j}\binom{q}{j}\left(\frac{c_{11}^{(f)}}{c_{00}^{(f)}}\right)^{j}(-1)^j j! c_{p-j,q-j}^{(f)} \sum_{k=1}^{j}\binom{j}{k}(-1)^k$$

$$= c_{pq}^{(f)} - \sum_{j=1}^{q}\binom{p}{j}\binom{q}{j}\left(-\frac{c_{11}^{(f)}}{c_{00}^{(f)}}\right)^{j} j! c_{p-j,q-j}^{(f)}((-1+1)^j - 1)$$

$$= \sum_{j=0}^{q}\binom{p}{j}\binom{q}{j}\left(-\frac{c_{11}^{(f)}}{c_{00}^{(f)}}\right)^{j} j! c_{p-j,q-j}^{(f)}. \qquad \Box$$

Appendix 5.F

Proof of Theorem 5.7

Since $I(C(0, 0), \ldots, C(P, Q))$ is a function of blur invariants $C(p, q)$ only, it is also a blur invariant. To prove its invariance to affine transform, it is sufficient to prove that:

$$\frac{\partial I(C(0,0), \ldots, C(P, Q))}{\partial a} \equiv \sum_{p=0}^{P}\sum_{q=0}^{Q}\frac{\partial I(C(0,0), \ldots, C(P, Q))}{\partial C(p, q)} \cdot \frac{\partial C(p, q)}{\partial a} = 0$$

$$(5.21)$$

for each parameter $a \in \{a_0, a_1, a_2, b_0, b_1, b_2\}$ of the affine transform.

The affine invariance of $I(\mu_{00}, \ldots, \mu_{pq})$ implies that

$$\frac{\partial I(\mu_{00}, \ldots, \mu_{pq})}{\partial a} \equiv \sum_{p=0}^{P} \sum_{q=0}^{Q} \frac{\partial I(\mu_{00}, \ldots, \mu_{pq})}{\partial \mu_{pq}} \cdot \frac{\partial \mu_{pq}}{\partial a} = 0. \tag{5.22}$$

Thus, it is sufficient to prove that the partial derivatives of $C(p, q)$ are identical to the derivatives of moments μ_{pq}, when substituting $C(p, q)$ for μ_{pq}, i.e.

$$\frac{\partial C(p, q)}{\partial a} = \left. \frac{\partial \mu_{pq}}{\partial a} \right|_{\mu_{pq}=C(p,q)} \tag{5.23}$$

for each parameter a.

To prove this, we decompose the affine transform into six one-parametric transformations (similarly to in Chapter 3 when we solved the Cayley–Aronhold equation):

horizontal translation:

$$u = x + \alpha,$$
$$v = y \tag{5.24}$$

vertical translation:

$$u = x,$$
$$v = y + \beta \tag{5.25}$$

uniform scaling:

$$u = sx,$$
$$v = sy \tag{5.26}$$

stretching:

$$u = rx,$$
$$v = \frac{y}{r} \tag{5.27}$$

horizontal skewing:

$$u = x + ty,$$
$$v = y \tag{5.28}$$

vertical skewing:

$$u = x,$$
$$v = y + zx. \tag{5.29}$$

We prove that the constraint (5.23) holds for each of these simple transformations.

Equation (5.23) holds trivially for translations (5.24) and (5.25). Without loss of generality, we assume in the rest of the proof that $(x_c, y_c) = (0, 0)$. For uniform scaling (5.26), we have

$$\frac{\partial \mu_{pq}}{\partial s} = \frac{\partial}{\partial s} \int_{-\infty}^{\infty} \int_{-\infty}^{\infty} u^p v^q f'(u, v) \, du \, dv = \frac{\partial}{\partial s} \int_{-\infty}^{\infty} \int_{-\infty}^{\infty} s^{p+q+2} x^p y^q f(x, y) \, dx \, dy$$

$$= \int_{-\infty}^{\infty} \int_{-\infty}^{\infty} (p+q+2) s^{p+q+1} x^p y^q f(x, y) \, dx \, dy$$

$$= \frac{p+q+2}{s} \int_{-\infty}^{\infty} \int_{-\infty}^{\infty} u^p v^q f'(u, v) \, du \, dv$$

$$= \frac{p+q+2}{s} \mu_{pq}$$

and

$$\frac{\partial C(p, q)}{\partial s} = \frac{p+q+2}{s} \mu_{pq} - \sum_{n=0}^{p} \sum_{\substack{m=0 \\ 0<n+m<p+q}}^{q} \binom{p}{n}\binom{q}{m}$$

$$\times \left[\frac{\mu_{nm}}{\mu_{00}} \cdot \frac{\partial C(p-n, q-m)}{\partial s} + C(p-n, q-m) \cdot \frac{\partial}{\partial s} \frac{\mu_{nm}}{\mu_{00}} \right].$$

At this moment we use the induction principle. Clearly, (5.23) holds for $p+q=3$ because in that case $C(p, q) = \mu_{pq}$ (if $p+q < 3$, then (5.23) holds trivially). Let us assume the validity of (5.23) for all orders less than $p+q$. Using this assumption and performing the derivatives, we obtain

$$\frac{\partial C(p, q)}{\partial s} = \frac{p+q+2}{s} \mu_{pq} - \sum_{n=0}^{p} \sum_{\substack{m=0 \\ 0<n+m<p+q}}^{q} \binom{p}{n}\binom{q}{m}$$

$$\cdot \left[\frac{(p+q-n-m+2)\mu_{nm}}{s\mu_{00}} \cdot C(p-n, q-m) + \frac{(n+m+2)\mu_{nm}}{s\mu_{00}} \right.$$

$$\left. \cdot C(p-n, q-m) - \frac{2\mu_{nm}}{s\mu_{00}} \cdot C(p-n, q-m) \right]$$

$$= \frac{p+q+2}{s} \mu_{pq} - \sum_{n=0}^{p} \sum_{\substack{m=0 \\ 0<n+m<p+q}}^{q} \binom{p}{n}\binom{q}{m} \frac{(p+q+2)\mu_{nm}}{s\mu_{00}}$$

$$\cdot C(p-n, q-m)$$

$$= \frac{p+q+2}{s} \cdot C(p, q),$$

which proves the validity of (5.23) in case of scaling.

Similarly, for stretching (5.27) we have

$$\frac{\partial \mu_{pq}}{\partial r} = \frac{\partial}{\partial r} \int_{-\infty}^{\infty} \int_{-\infty}^{\infty} r^{p-q} x^p y^q f(x, y) \, dx \, dy$$

$$= \int_{-\infty}^{\infty} \int_{-\infty}^{\infty} (p-q) r^{p-q-1} x^p y^q f(x, y) \, dx \, dy = \frac{p-q}{r} \mu_{pq}$$

and

$$\frac{\partial C(p, q)}{\partial r} = \frac{p-q}{r} \mu_{pq} - \frac{1}{\mu_{00}} \sum_{n=0}^{p} \sum_{m=0}^{q} \binom{p}{n} \binom{q}{m}$$
$$\phantom{\frac{\partial C(p, q)}{\partial r} =} \scriptstyle 0 < n+m < p+q$$

$$\cdot \left[\mu_{nm} \frac{\partial C(p-n, q-m)}{\partial r} + \frac{n-m}{r} C(p-n, q-m) \mu_{nm} \right].$$

Similarly to the previous case, we obtain by induction

$$\frac{\partial C(p, q)}{\partial r} = \frac{p-q}{r} C(p, q).$$

Finally, for horizontal skewing (5.28) we have

$$\frac{\partial \mu_{pq}}{\partial t} = \frac{\partial}{\partial t} \int_{-\infty}^{\infty} \int_{-\infty}^{\infty} (x + ty)^p y^q f(x, y) \, dx \, dy$$

$$= \int_{-\infty}^{\infty} \int_{-\infty}^{\infty} p(x + ty)^{p-1} y^{q+1} f(x, y) \, dx \, dy = p \mu_{p-1, q+1}$$

and

$$\frac{\partial C(p, q)}{\partial t} = p \mu_{p-1, q+1} - \frac{1}{\mu_{00}} \sum_{n=0}^{p} \sum_{m=0}^{q} \binom{p}{n} \binom{q}{m}$$
$$\phantom{\frac{\partial C(p, q)}{\partial t} =} \scriptstyle 0 < n+m < p+q$$

$$\cdot \left[\mu_{nm} \frac{\partial C(p-n, q-m)}{\partial t} + C(p-n, q-m) \frac{\partial \mu_{nm}}{\partial t} \right].$$

Employing the induction principle, we have

$$\frac{\partial C(p, q)}{\partial t} = p \mu_{p-1, q+1} - \frac{1}{\mu_{00}} \sum_{n=0}^{p} \sum_{m=0}^{q} \binom{p}{n} \binom{q}{m}$$
$$\phantom{\frac{\partial C(p, q)}{\partial t} =} \scriptstyle 0 < n+m < p+q$$

$$\cdot [(p-n) \mu_{nm} C(p-n-1, q-m+1) + n C(p-n, q-m) \mu_{n-1, m+1}].$$

Shifting the indices and using the identities

$$(p-n) \binom{p}{n} = p \binom{p-1}{n} = (n+1) \binom{p}{n+1},$$

$$\binom{q}{m} + \binom{q}{m-1} = \binom{q+1}{m}$$

we obtain the final relation

$$\frac{\partial C(p, q)}{\partial t} = p\mu_{p-1,q+1}$$

$$- \frac{1}{\mu_{00}} \Bigg[\sum_{\substack{n=0 \\ 0<n+m<p+q}}^{p-1} \sum_{m=0}^{q} p \binom{p-1}{n} \binom{q}{m} C(p-n-1, q-m+1)\mu_{nm}$$

$$+ \sum_{\substack{n=0 \\ 0<n+m<p+q}}^{p-1} \sum_{m=1}^{q+1} (n+1) \binom{p}{n+1} \binom{q}{m-1} C(p-n-1, q-m+1)\mu_{nm} \Bigg]$$

$$= p\mu_{p-1,q+1}$$

$$- \frac{1}{\mu_{00}} \sum_{\substack{n=0 \\ 0<n+m<p+q}}^{p-1} \sum_{m=0}^{q+1} p \binom{p-1}{n} \binom{q+1}{m} C(p-n-1, q-m+1)\mu_{nm}$$

$$= pC(p-1, q+1).$$

Because of symmetry, the same is true for vertical skewing (5.29). Thus, the constraint (5.23) holds for each affine parameter and, consequently, $I(C(0, 0), \ldots, C(P, Q))$ is a combined invariant. $\qquad\square$

References

[1] Kundur D. and Hatzinakos, D. (1996) "Blind image deconvolution," *IEEE Signal Processing Magazine*, vol. 13, no. 3, pp. 43–64.

[2] Flusser, J., Suk, T. and Saic, S. (1996) "Recognition of blurred images by the method of moments," *IEEE Transactions on Image Processing*, vol. 5, no. 3, pp. 533–8.

[3] Flusser, J. and Suk, T. (1998) "Degraded image analysis: An invariant approach," *IEEE Transactions Pattern Analysis and Machine Intelligence*, vol. 20, no. 6, pp. 590–603.

[4] Ojansivu, V. and Heikkilä, J. (2007) "Image registration using blur-invariant phase correlation," *IEEE Signal Processing Letters*, vol. 14, no. 7, pp. 449–52.

[5] Zitová, B. and Flusser, J. (2003) "Image registration methods: A survey," *Image and Vision Computing*, vol. 21, no. 11, pp. 977–1000.

[6] Cash, G. L. and Hatamian, M. (1987) "Optical character recognition by the method of moments," *Computer Vision, Graphics, and Image Processing*, vol. 39, no. 3, pp. 291–310.

[7] Flusser, J., Suk, T. and Saic, S. (1996) "Recognition of images degraded by linear motion blur without restoration," *Computing Supplement*, vol. 11, pp. 37–51.

[8] Stern, A., Kruchakov, I., Yoavi, E. and Kopeika, S. (2002) "Recognition of motion-blurred images by use of the method of moments," *Applied Optics*, vol. 41, pp. 2164–72.

[9] Flusser, J., Boldyš, J. and Zitová, B. (2000) "Invariants to convolution in arbitrary dimensions," *Journal of Mathematical Imaging and Vision*, vol. 13, no. 2, pp. 101–13.

[10] Bentoutou, Y., Taleb, N., Kpalma, K. and Ronsin, J. (2005) "An automatic image registration for applications in remote sensing," *IEEE Transactions Geoscience and Remote Sensing*, vol. 43, no. 9, pp. 2127–37.

[11] Bentoutou, Y., Taleb, N., Chikr El Mezouar, M., Taleb, M. and Jetto, J. (2002) "An invariant approach for image registration in digital subtraction angiography," *Pattern Recognition*, vol. 35, no. 12, pp. 2853–65.

[12] Bentoutou, Y. and Taleb, N. (2005) "Automatic extraction of control points for digital subtraction angiography image enhancement," *IEEE Trans. Nuclear Science*, vol. 52, no. 1, pp. 238–46.

[13] Zhang, Y., Wen, C. and Zhang, Y. (2000) "Estimation of motion parameters from blurred images," *Pattern Recognition Letters*, vol. 21, no. 5, pp. 425–33.

[14] Zhang, Y., Wen, C., Zhang, Y. and Soh, Y. C. (2002) "Determination of blur and affine combined invariants by normalization," *Pattern Recognition*, vol. 35, no. 1, pp. 211–21.

[15] Lu, J. and Yoshida, Y. (1999) "Blurred image recognition based on phase invariants," *IEICE Transactions Fundamentals of Electronics, Communications and Computer Sciences*, vol. E82A, no. 8, pp. 1450–5.

[16] Wang, L. and Healey, G. (1998) "Using Zernike moments for the illumination and geometry invariant classification of multispectral texture," *IEEE Transactions Image Processing*, vol. 7, no. 2, pp. 196–203.

[17] Shen, X.-J. and Pan, J.-M. (2004) "Monocular visual servoing based on image moments," *IEICE Transactions Fundamentals of Electronics, Communications and Computer Sciences*, vol. E87-A, no. 7, pp. 1798–803.

[18] Zhang, Y., Zhang, Y. and Wen, C. (2000) "A new focus measure method using moments," *Image and Vision Computing*, vol. 18, no. 12, pp. 959–965.

[19] Flusser, J. and Zitová, B. (2004) "Invariants to convolution with circularly symmetric PSF," in *Proceedings of the 17th International Conference on Pattern Recognition ICPR'04* (Cambridge, UK), pp. 11–14, IEEE Computer Society.

[20] Liu, J. and Zhang, T. (2005) "Recognition of the blurred image by complex moment invariants," *Pattern Recognition Letters*, vol. 26, no. 8, pp. 1128–38.

[21] Flusser, J. and Zitová, B. (1999) "Combined invariants to linear filtering and rotation," *International Journal of Pattern Recognition and Artificial Intelligence*, vol. 13, no. 8, pp. 1123–36.

[22] Flusser, J., Zitová, B. and Suk, T. (1999) "Invariant-based registration of rotated and blurred images," in *Proceedings of the International Geoscience and Remote Sensing Symposium IGARSS'99* (Hamburg, Germany) (I. S. Tammy, ed.), vol. 2, pp. 1262–4, IEEE.

[23] Zitová, B. and Flusser, J. (2002) "Estimation of camera planar motion from blurred images," in *Proceedings of the International Conference on Image Processing ICIP'02* (Rochester, New York), vol. 2, pp. 329–32, IEEE Computer Society.

[24] Zitová, B. and Flusser, J. (2000) "Invariants to convolution and rotation," in *Invariants for Pattern Recognition and Classification* (M. A. Rodrigues, ed.), pp. 23–46, World Scientific.

[25] Flusser, J., Boldyš, J. and Zitová, B. (2003) "Moment forms invariant to rotation and blur in arbitrary number of dimensions," *IEEE Transactions Pattern Analysis and Machine Intelligence*, vol. 25, no. 2, pp. 234–46.

[26] Suk, T. and Flusser, J. (2003) "Combined blur and affine moment invariants and their use in pattern recognition," *Pattern Recognition*, vol. 36, no. 12, pp. 2895–907.

[27] Li, Y., Chen, H., Zhang, J. and Qu, P. (2003) "Combining blur and affine moment invariants in object recognition," in *Proceedings of the Fifth International Symposium on Instrumentation and Control Technology ISICT'03* (Beijing, China), vol. 5253, SPIE.

[28] Gao, Y., Song, H., Tian, X. and Chen, Y. (2007) "Identification algorithm of winged insects based on hybrid moment invariants," in *First International Conference on Bioinformatics and Biomedical Engineering ICBBE'07*, pp. 531–4, IEEE.

[29] Ojansivu, V. and Heikkilä, J. (2008) "A method for blur and affine invariant object recognition using phase-only bispectrum," in *Image Analysis and Recognition*, vol. 5112, pp. 527–36, Springer.

[30] Boldyš, J. and Flusser, J. (2008) "Extension of moment features' invariance to blur," *Journal of Mathematical Imaging and Vision*, vol. 32, no. 3, pp. 227–38.

6

Orthogonal moments

6.1   Introduction

It is well known from linear algebra that an orthogonal basis of a vector space has many favorable numerical properties compared to other bases. This applies also to bases of polynomials, and, consequently, to moments, that are "projections" of an image onto a basis.

All existing bases of a given vector space are, of course, theoretically equivalent, because each vector from one basis can be expressed as a linear combination of vectors from the other basis. When dealing with various polynomial bases up to a certain degree and with corresponding moments, any moment (with respect to any basis) can be expressed as a function of moments of the same or fewer orders with respect to an arbitrary basis. In particular, any moments can be expressed in terms of geometric moments. Taking this into consideration, there is seemingly no reason for introducing other than geometric moments.

The above reasoning is true on a theoretical level only. When considering numerical properties and implementation issues, the motivation for introducing *OG moments* becomes apparent. OG moments have already been briefly introduced in Chapter 1. Let us recall that they are moments to an orthogonal or weighted-orthogonal polynomial basis. The main reasons for introducing them are: stable and fast numerical implementation (OG polynomials can be evaluated by recurrent relations which can be efficiently implemented by means of multiplication with special matrices); avoidance of high dynamic range of moment values that may lead to the loss of precision due to overflow or underflow (unlike standard powers, the values of OG polynomials lie inside a narrow interval such as $\langle -1, 1 \rangle$ or similar); and a higher robustness to random noise reached. From this point of view it is a serious mistake to evaluate OG polynomials by expanding them into standard powers. Even if these expansions are commonly presented in literature (and we also do this in this chapter) it should serve for illustration and/or theoretical considerations only. They should never be used for evaluation of polynomials and for OG moment calculation.

Another reason for using OG moments frequently mentioned in literature is for better image reconstruction. As we already saw in Chapter 1, OG moments are, unlike geometric and all other moments, coordinates of f in the polynomial basis in the common sense used in linear algebra. Therefore, the image reconstruction from OG moments can be performed

Moments and Moment Invariants in Pattern Recognition   Jan Flusser, Tomáš Suk and Barbara Zitová
© 2009 John Wiley & Sons, Ltd

easily as

$$f(x, y) = \sum_{k,j} M_{kj} \cdot p_{kj}(x, y).$$

This reconstruction is "optimal" because it minimizes the mean-square error when using only a finite set of moments. On the other hand, image reconstruction from geometric moments is carried out in the Fourier domain using the fact that geometric moments form Taylor coefficients of the Fourier transform $F(u, v)$

$$F(u, v) = \sum_p \sum_q \frac{(-2\pi i)^{p+q}}{p!q!} m_{pq} u^p v^q.$$

This argument for OG moments is correct but at the same time we emphasize that the task of image reconstruction from its moments is rather artificial, very rarely appearing in practice. The reconstruction problem appears most frequently in image compression when trying to recover the original from a lossy representation. However, moments are not a good tool for image compression, providing a much worse compression rate than JPEG or wavelets, and are hardly ever used for compression purposes.

Numerous types of OG polynomial, both in 1D and 2D, have been described in traditional mathematical literature. In this chapter we present a survey of OG moments that are of importance in image analysis. The literature on OG moments is very broad, namely in the area of practical applications, and our survey has no claim on completeness.

We divide OG polynomials and OG moments into two basic groups. The polynomials *orthogonal on a rectangle* originate from 1D OG polynomials whose 2D versions were created as products of 1D polynomials in x and y. The main advantage of the moments orthogonal on a rectangle is that they preserve the orthogonality even on the sampled image. They can be made scale-invariant but creating rotation invariants from them is very complicated. The polynomials *orthogonal on a disk* are intrinsically 2D functions. They are constructed as products of a radial factor (usually a 1D OG polynomial) and angular factor which is usually a kind of harmonic function. When implementing these moments, an image must be mapped into a disk of orthogonality which creates certain resampling problems. On the other hand, moments orthogonal on a disk can easily be used for construction of rotation invariants because they change under rotation in a simple way.

6.2   Moments orthogonal on a rectangle

OG moments were introduced in Chapter 1 as moments, whose kernel functions satisfy a relation of orthogonality (1.6) or weighted orthogonality (1.7). 2D polynomials orthogonal on a rectangle are mostly constructed as products of 1D OG polynomials $p_k(x)$.[1] Thus, the corresponding moments have a form

$$v_{pq} = n_p n_q \iint_{\Omega} p_p(x) p_q(y) f(x, y) \, dx \, dy \quad p, q = 0, 1, 2, \ldots, \tag{6.1}$$

where n_p, n_q are some normalizing factors and Ω is the area of orthogonality. The image $f(x, y)$ should be scaled such that its support is contained in Ω. The polynomials $p_k(x)$

---

[1] Theoretically, we can use 2D kernel functions that cannot be decomposed to the product $p_p(x) p_q(y)$, but it is impractical and, to the authors' knowledge, nobody has used it when working on a rectangular domain.

satisfy a relation of orthogonality

$$\int_{\Omega_1} w(x) p_p(x) p_q(x) \, dx = h_p \delta_{pq}, \tag{6.2}$$

where $w(x)$ is a weight function, $h_p = \|p_p\|^2$, and δ_{pq} is a *Kronecker delta*, $\delta_{pq} = 1$ if $p = q$ and $\delta_{pq} = 0$ otherwise. Ω_1 is the interval of orthogonality, $\Omega = \Omega_1 \times \Omega_1$.

6.2.1 Hypergeometric functions

Orthogonal polynomials are related to a concept of hypergeometric series. If it is convergent, it defines a *hypergeometric function*. The hypergeometric functions include such special functions as Bessel functions, the incomplete gamma function, the error function and the elliptic integrals. They are also often used for definition of the orthogonal polynomials. The *hypergeometric series* $_pF_q$ is defined as

$$_pF_q(a_1, \ldots, a_p; b_1, \ldots, b_q; z) = \sum_{n=0}^{\infty} \frac{(a_1)_n (a_2)_n \cdots (a_p)_n}{(b_1)_n (b_2)_n \cdots (b_q)_n} \frac{z^n}{n!}, \tag{6.3}$$

where $(a)_n$ is a *Pochhammer symbol* (also *rising factorial*)

$$(a)_n = a(a+1)\cdots(a+n-1) = (a+n-1)!/(a-1)! = \Gamma(a+n)/\Gamma(a). \tag{6.4}$$

Particularly,

$$_2F_1(a, b; c; z) = \sum_{n=0}^{\infty} \frac{(a)_n (b)_n}{(c)_n} \frac{z^n}{n!} \quad \text{or} \quad _3F_2(a, b, c; d, e; z) = \sum_{n=0}^{\infty} \frac{(a)_n (b)_n (c)_n}{(d)_n (e)_n} \frac{z^n}{n!}. \tag{6.5}$$

Some elementary functions can be expressed by means of the hypergeometric function $_2F_1$ [1]

$$(1+z)^k = {}_2F_1(-k, 1; 1; -z),$$

$$\log(1+z) = z \, {}_2F_1(1, 1; 2; -z),$$

$$\log\left(\frac{1+z}{1-z}\right) = 2z \, {}_2F_1\left(\frac{1}{2}, 1; \frac{3}{2}; z^2\right),$$

$$\arcsin(z) = z \, {}_2F_1\left(\frac{1}{2}, \frac{1}{2}; \frac{3}{2}; z^2\right),$$

$$\arctan(z) = z \, {}_2F_1\left(\frac{1}{2}, 1; \frac{3}{2}; -z^2\right). \tag{6.6}$$

Expressing orthogonal polynomials by hypergeometric functions is of theoretical rather than practical importance. It may provide us with an idea concerning complexity of the orthogonal polynomials but recurrent formulas are more efficient for enumeration.

6.2.2   Legendre moments[2]

There have been many papers describing the use of the Legendre moments in image processing, e.g. references [2] and [3], among many others. The *Legendre moments* are defined as

$$\lambda_{mn} = \frac{(2m+1)(2n+1)}{4} \int_{-1}^{1}\int_{-1}^{1} P_m(x)P_n(y)f(x,y)\,\mathrm{d}x\,\mathrm{d}y \quad m, n = 0, 1, 2, \ldots, \quad (6.7)$$

where $P_n(x)$ is the nth degree *Legendre polynomial* (expression by the so-called Rodrigues' formula)

$$P_n(x) = \frac{1}{2^n n!}\frac{\mathrm{d}^n}{\mathrm{d}x^n}(x^2-1)^n \qquad (6.8)$$

and the image $f(x, y)$ is mapped into the square $\langle -1, 1\rangle \times \langle -1, 1\rangle$. The Legendre polynomials of low degrees expressed in terms of x^n are

$$P_0(x) = 1,$$

$$P_1(x) = x,$$

$$P_2(x) = \tfrac{1}{2}(3x^2-1),$$

$$P_3(x) = \tfrac{1}{2}(5x^3-3x), \qquad (6.9)$$

$$P_4(x) = \tfrac{1}{8}(35x^4-30x^2+3),$$

$$P_5(x) = \tfrac{1}{8}(63x^5-70x^3+15x),$$

$$P_6(x) = \tfrac{1}{16}(231x^6-315x^4+105x^2-5).$$

The graphs of the Legendre polynomials are in Figure 6.1. We can compare them with the kernel functions of the geometric moments x^n for $n = 0, 1, \ldots, 6$ in Figure 6.2, where the phenomenon of high values near the interval border can be observed even on $\langle -1.5, 1.5\rangle$. The kernel functions of the 2D Legendre moments on $\langle -1, 1\rangle \times \langle -1, 1\rangle$ are shown in Figure 6.3.

The relation of orthogonality is

$$\int_{-1}^{1} P_m(x)P_n(x)\,\mathrm{d}x = \frac{2}{2n+1}\delta_{mn}, \qquad (6.10)$$

on the interval of orthogonality $\Omega_1 = \langle -1, 1\rangle$.

The recurrence relation, which can be used for efficient computation of the Legendre polynomials, are

$$P_0(x) = 1,$$

$$P_1(x) = x,$$

$$P_{n+1}(x) = \frac{2n+1}{n+1}x P_n(x) - \frac{n}{n+1}P_{n-1}(x). \qquad (6.11)$$

---

[2] A. M. Legendre (1752–1833), a French mathematician.

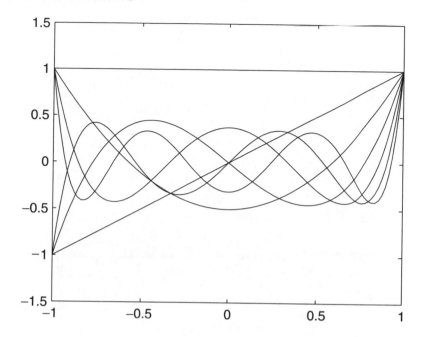

Figure 6.1 The graphs of the Legendre polynomials up to the sixth degree.

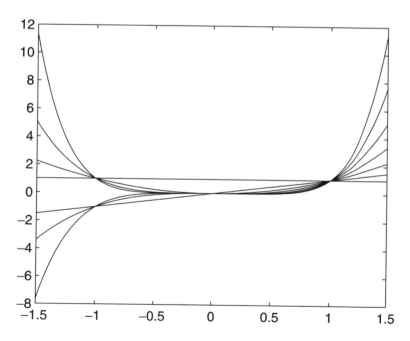

Figure 6.2 The graphs of the kernel functions x^n of the geometric moments up to $n = 6$.

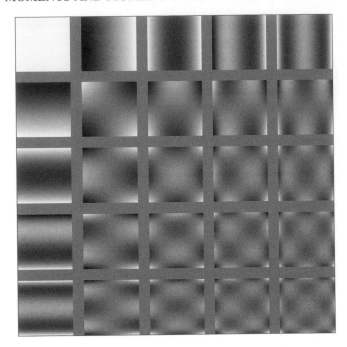

Figure 6.3 The graphs of 2D kernel functions of the Legendre moments up to the fourth degree. Black $= -1$, white $= 1$.

The Legendre polynomials can be expressed in terms of the hypergeometric functions

$$P_n(x) = {}_2F_1\left(-n, n+1; 1; \frac{1-x}{2}\right),$$

$$P_n(x) = \binom{2n}{n}\left(\frac{x-1}{2}\right)^n {}_2F_1\left(-n, -n; -2n; \frac{2}{1-x}\right),$$

$$P_n(x) = \binom{2n}{n}\left(\frac{x}{2}\right)^n {}_2F_1\left(-\frac{n}{2}, \frac{1-n}{2}; \frac{1}{2}-n; \frac{1}{x^2}\right).$$

(6.12)

In the case of a digital image $M \times N$, we have

$$\lambda_{mn} = \frac{(2m+1)(2n+1)}{4MN} \sum_{x=0}^{M-1} \sum_{y=0}^{N-1} P_m\left(-1+2\frac{x}{M-1}\right) P_n\left(-1+2\frac{y}{N-1}\right) f(x, y).$$

(6.13)

If we denote the coefficients of the Legendre polynomials a_{np}

$$P_n(x) = \sum_{p=0}^{n} a_{np} x^p,$$

(6.14)

then we can express the Legendre moments in terms of geometric moments

$$\lambda_{mn} = \frac{(2m+1)(2n+1)}{4} \sum_{p=0}^{m} \sum_{q=0}^{n} a_{mp} a_{nq} m_{pq}.$$

(6.15)

The coefficients a_{mp} are

$$a_{mp} = \begin{cases} (-1)^{(m-p)/2} \dfrac{1}{2^m} \dfrac{(m+p)!}{((m-p)/2)!((m+p)/2)!p!} & m-p \text{ even} \\ 0 & m-p \text{ odd.} \end{cases} \quad (6.16)$$

The first four Legendre moments expressed in terms of geometric moments are

$$\lambda_{00} = \tfrac{1}{4}m_{00},$$
$$\lambda_{10} = \tfrac{3}{4}m_{10},$$
$$\lambda_{20} = \tfrac{5}{8}(3m_{20} - m_{00}),$$
$$\lambda_{11} = \tfrac{9}{4}m_{11}. \quad (6.17)$$

6.2.3 Chebyshev moments[3]

The Chebyshev moments were introduced into image processing by Mukundan, who published a series of papers [4–7].

The nth degree *Chebyshev polynomial* of the first kind can be expressed by the recurrence relations

$$T_0(x) = 1,$$
$$T_1(x) = x, \quad (6.18)$$
$$T_n(x) = 2xT_{n-1}(x) - T_{n-2}(x).$$

The Chebyshev polynomials of the second kind differ from them by the initial values

$$U_0(x) = 1,$$
$$U_1(x) = 2x, \quad (6.19)$$
$$U_n(x) = 2xU_{n-1}(x) - U_{n-2}(x).$$

The Chebyshev polynomials of the first kind are more important in image processing. They can be expressed by means of trigonometric functions

$$T_n(x) = \cos(n \arccos x). \quad (6.20)$$

The Chebyshev polynomials are solutions of the Chebyshev differential equation

$$(1 - x^2)\frac{d^2 y}{dx^2} - x\frac{dy}{dx} + n^2 y = 0. \quad (6.21)$$

---

[3]P. L. Chebyshev (1821–1894), a Russian mathematician. There exist several Latin transcriptions of his name (П. Л. Чебышёв in Russian), e.g. Tchebichef and Tschebischeff.

They can also be expressed by explicit formulas

$$T_n(x) = \frac{(x + \sqrt{x^2 - 1})^n + (x - \sqrt{x^2 - 1})^n}{2}$$

$$= \sum_{k=0}^{[n/2]} \binom{n}{2k}(x^2 - 1)^k x^{n-2k}$$

$$= \frac{n}{2} \sum_{m=0}^{[n/2]} (-1)^m \frac{(n - m - 1)!}{m!(n - 2m)!}(2x)^{n-2m}. \tag{6.22}$$

The last version cannot be used for $n = 0$.
   The orthogonality relation is

$$\int_{-1}^{1} T_n(x)T_m(x) \frac{1}{\sqrt{1 - x^2}} \, dx = \begin{cases} 0 & : \ n \neq m \\ \pi & : \ n = m = 0 \\ \pi/2 & : \ n = m \neq 0, \end{cases} \tag{6.23}$$

on the interval of orthogonality $\Omega_1 = \langle -1, 1 \rangle$. The Chebyshev polynomials can be expressed in terms of hypergeometric functions

$$T_n(x) = {}_2F_1\left(-n, n; \frac{1}{2}; \frac{1 - x}{2}\right). \tag{6.24}$$

Below, we present a few examples of the Chebyshev polynomials of low degrees:

$$\begin{aligned}
T_0(x) &= 1, \\
T_1(x) &= x, \\
T_2(x) &= 2x^2 - 1, \\
T_3(x) &= 4x^3 - 3x, \\
T_4(x) &= 8x^4 - 8x^2 + 1, \\
T_5(x) &= 16x^5 - 20x^3 + 5x, \\
T_6(x) &= 32x^6 - 48x^4 + 18x^2 - 1.
\end{aligned} \tag{6.25}$$

The graphs of the Chebyshev polynomials of the first kind are in Figure 6.4, that of the second kind are in Figure 6.5. The kernel functions of the 2D Chebyshev moments on $\langle -1, 1 \rangle \times \langle -1, 1 \rangle$ are shown in Figure 6.6.
   The *Chebyshev moments* are defined in the basic version as

$$\tau_{mn} = \iint_{\Omega} T_m(x)T_n(y)f(x, y) \, dx \, dy \quad m, n = 0, 1, 2, \ldots. \tag{6.26}$$

The conversion of the geometric moments to the Chebyshev moments is similar to that in the case of the Legendre moments (6.14) and (6.15). We can take the coefficients of the

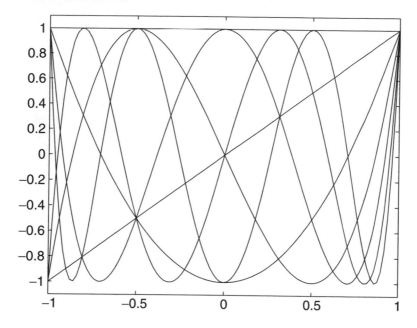

Figure 6.4 The graphs of the Chebyshev polynomials of the first kind up to the sixth degree.

Chebyshev polynomials from (6.22)

$$
t_{n,n-2m} = \begin{cases} 0 & m < 0 \text{ or } m > n/2 \\ 1 & n = 0 \\ \dfrac{n}{2}(-1)^m \dfrac{(n - m - 1)!}{m!(n - 2m)!} 2^{n-2m} & n \neq 0, \end{cases} \tag{6.27}
$$

then

$$
\tau_{mn} = \sum_{p=0}^{m} \sum_{q=0}^{n} t_{mp} t_{nq} m_{pq} \quad m, n = 0, 1, 2, \dots. \tag{6.28}
$$

An important property of the Chebyshev moments, that is useful particularly in image processing, is that they can be modified to preserve the orthogonality even on discrete images; see the paragraph "Discrete Chebyshev polynomials" in Section 6.2.5.

6.2.4 Other moments orthogonal on a rectangle

There are various orthogonal polynomials that can be used for definition of OG moments. Recently, many of them have been introduced into the image processing area. Here, we survey them briefly, including only moments with certain relations to image analysis. We refer to reference [1] for more details.

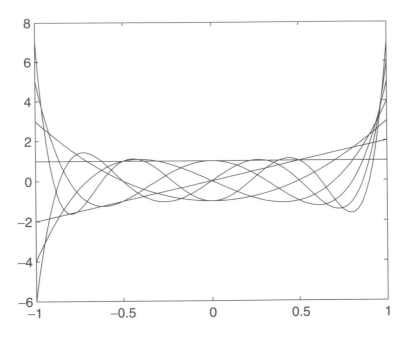

Figure 6.5 The graphs of the Chebyshev polynomials of the second kind up to the sixth degree.

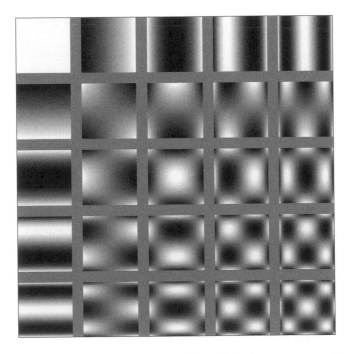

Figure 6.6 The graphs of 2D kernel functions of the Chebyshev moments of the first kind up to the fourth order. Black $= -1$, white $= 1$.

Gegenbauer polynomials of the nth degree with parameter λ (also *ultraspherical poly-nomials*, see also references [3] and [8]) can be expressed by the recurrence formula

$$C_0^{(\lambda)}(x) = 1,$$
$$C_1^{(\lambda)}(x) = 2\lambda x,$$
$$(n+1)C_{n+1}^{(\lambda)}(x) = 2(n+\lambda)xC_n^{(\lambda)}(x) - (n+2\lambda-1)C_{n-1}^{(\lambda)}(x). \tag{6.29}$$

$C_n^{(\lambda)}(x)$ would be zero for $\lambda = 0$ and $n > 0$, so the different initial values are used in this case

$$C_1^{(0)}(x) = 2x,$$
$$C_2^{(0)}(x) = x^2 - \frac{1}{2}, \tag{6.30}$$

where $C_0^{(0)}(x) = 1$ is defined separately.

The corresponding differential equation is

$$(1-x^2)\frac{d^2y}{dx^2} - (2\lambda+1)x\frac{dy}{dx} + n(2\lambda+n)y = 0, \tag{6.31}$$

where $\lambda > -1/2$. The explicit expression

$$C_n^{(\lambda)}(x) = \frac{1}{\Gamma(\lambda)}\sum_{m=0}^{[n/2]}(-1)^m\frac{\Gamma(\lambda+n-m)}{m!\,(n-2m)!}(2x)^{n-2m} \tag{6.32}$$

cannot be used for $\lambda = 0$. The relation of orthogonality is

$$\int_{-1}^{1}(1-x^2)^{\lambda-1/2}C_n^{(\lambda)}(x)C_m^{(\lambda)}(x)\,dx = \begin{cases} \dfrac{\pi 2^{1-2\lambda}\Gamma(n+2\lambda)}{(n+\lambda)\,n!\,\Gamma^2(\lambda)}\delta_{nm}: & \lambda \neq 0 \\ \dfrac{2\pi}{n^2}\delta_{nm}: & \lambda = 0. \end{cases} \tag{6.33}$$

The Gegenbauer polynomials can be expressed in terms of the hypergeometric functions as

$$C_n^{(\lambda)}(x) = \frac{\Gamma(n+2\lambda)}{n!\Gamma(2\lambda)}\,{}_2F_1\left(-n, n+2\lambda; \lambda+\frac{1}{2}; \frac{1-x}{2}\right). \tag{6.34}$$

The Gegenbauer polynomials can be understood as a generalization of the previous polynomials. We obtain the Legendre polynomials for $\lambda = 1/2$ (compare (6.29) with (6.11)) and the Chebyshev polynomials of the second kind for $\lambda = 1$. If we use special initial values (6.30) in the case of $\lambda = 0$, then we obtain $2/n$ multiple of the Chebyshev polynomials of the first kind. Therefore, the graphs in Figure 6.7 for $\lambda = 0.75$ are "something between" the Legendre polynomials and the Chebyshev polynomials of the second kind.

Jacobi polynomials of the nth degree with parameters α and β can be obtained as a solution of the Jacobi differential equation

$$(1-x^2)\frac{d^2y}{dx^2} + (\beta-\alpha-(\alpha+\beta+2)x)\frac{dy}{dx} + n(\alpha+\beta+n+1)y = 0, \tag{6.35}$$

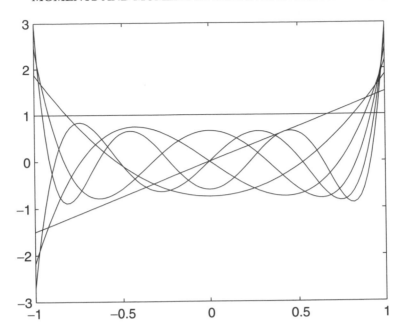

Figure 6.7 The graphs of the Gegenbauer polynomials for $\lambda = 0.75$ up to the sixth degree.

where $\alpha > -1$, $\beta > -1$. Equivalently, they can be expressed by the recurrence formula

$$2(n+1)(n+\alpha+\beta+1)(2n+\alpha+\beta)P_{n+1}^{(\alpha,\beta)}(x)$$

$$= ((2n+\alpha+\beta+1)(\alpha^2-\beta^2)+(2n+\alpha+\beta)_3 x)P_n^{(\alpha,\beta)}(x) \qquad (6.36)$$

$$- 2(n+\alpha)(n+\beta)(2n+\alpha+\beta+2)P_{n-1}^{(\alpha,\beta)}(x),$$

where $(a)_n$ is the Pochhammer symbol (6.4), $P_0^{(\alpha,\beta)} = 1$ and $P_1^{(\alpha,\beta)} = \frac{1}{2}(\alpha - \beta + (\alpha + \beta + 2)x)$ or in the explicit form

$$P_n^{(\alpha,\beta)}(x) = \frac{1}{2^n} \sum_{m=0}^{n} \binom{n+\alpha}{m}\binom{n+\beta}{n-m}(x-1)^{n-m}(x+1)^m. \qquad (6.37)$$

The relation of orthogonality is

$$\int_{-1}^{1} (1-x)^\alpha (1+x)^\beta P_n^{(\alpha,\beta)}(x) P_m^{(\alpha,\beta)}(x)\,dx = \frac{2^{\alpha+\beta+1}\Gamma(n+\alpha+1)\Gamma(n+\beta+1)}{(2n+\alpha+\beta+1)\,n!\,\Gamma(n+\alpha+\beta+1)}\delta_{nm}.$$

$$(6.38)$$

The expression of the Jacobi polynomials in terms of the hypergeometric functions is

$$P_n^{(\alpha,\beta)}(x) = \binom{n+\alpha}{n} {}_2F_1\left(-n, n+\alpha+\beta+1; \alpha+1; \frac{1-x}{2}\right). \qquad (6.39)$$

This version of the Jacobi polynomials is orthogonal on the interval $\langle -1, 1\rangle$. We can find another expression of the Jacobi polynomials in the literature that is more suitable for conversion from geometric moments. On the other hand, comparison with the previous polynomials is more difficult. *Jacobi polynomials* of the nth degree with parameters p and q are orthogonal on the interval $\langle 0, 1 \rangle$. They can be expressed by the recurrence formula

$$(2n + p - 2)_4 (2n + p - 1) G_{n+1}^{(p,q)}(x)$$

$$= ((2n + p - 2)_4 (2n + p - 1)x - (2n(n + p) + q(p - 1))(2n + p - 2)_3) G_n^{(p,q)}(x)$$

$$- n(n + p - 1)(n + q - 1)(n + p - q)(2n + p + 1) G_{n-1}^{(p,q)}(x), \tag{6.40}$$

where $(a)_n$ is again the Pochhammer symbol (6.4), $G_0^{(p,q)} = 1$ and $G_1^{(p,q)} = x - (q/(p + 1))$ or in the explicit form

$$G_n^{(p,q)}(x) = \frac{\Gamma(q + n)}{\Gamma(p + 2n)} \sum_{m=0}^{n} (-1)^m \binom{n}{m} \frac{\Gamma(p + 2n - m)}{\Gamma(q + n - m)} x^{n-m}. \tag{6.41}$$

The parameters must satisfy the constraint $p - q > -1$ and $q > 0$. The relation of orthogonality is

$$\int_0^1 (1 - x)^{p-q} x^{q-1} G_n^{(p,q)}(x) G_m^{(p,q)}(x)\, \mathrm{d}x = \frac{n!\Gamma(n + p)\Gamma(n + q)\Gamma(n + p - q + 1)}{(2n + p)\Gamma^2(2n + p)} \delta_{nm}. \tag{6.42}$$

The relations between both expressions are

$$G_n^{(p,q)}(x) = \frac{n!\Gamma(n + p)}{\Gamma(2n + p)} P_n^{(p-q,q-1)}(2x - 1) \tag{6.43}$$

and in the opposite direction

$$P_n^{(\alpha,\beta)}(x) = \frac{\Gamma(2n + \alpha + \beta + 1)}{n!\Gamma(n + \alpha + \beta + 1)} G_n^{(\alpha+\beta+1,\beta+1)}\left(\frac{x + 1}{2}\right). \tag{6.44}$$

The Jacobi polynomials are a further generalization of Gegenbauer polynomials; they can be obtained from them for $\alpha = \beta = \lambda - 1/2$

$$C_n^{(\lambda)}(x) = \frac{\Gamma(2\lambda + n)\Gamma(\lambda + 1/2)}{\Gamma(2\lambda)\Gamma(\lambda + n + 1/2)} P_n^{(\lambda-1/2,\lambda-1/2)}(x). \tag{6.45}$$

Both the Jacobi and the Gegenbauer polynomials can be used for definition of the corresponding moments [9, 10].

Generalized Laguerre polynomials of the nth degree with parameter α are defined as

$$L_n^{(\alpha)}(x) = \frac{x^{-\alpha} \mathrm{e}^x}{n!} \frac{\mathrm{d}^n}{\mathrm{d}x^n} x^{n+\alpha} \mathrm{e}^{-x} \tag{6.46}$$

or explicitly

$$L_n^{(\alpha)}(x) = \sum_{m=0}^{n} (-1)^m \binom{n + \alpha}{n - m} \frac{1}{m!} x^m, \tag{6.47}$$

where the parameter $\alpha > -1$. We can obtain *Laguerre polynomials* (nongeneralized) from them for $\alpha = 0$. The Laguerre polynomials are orthogonal on the interval $\langle 0, \infty \rangle$; the relation of orthogonality is

$$\int_0^\infty L_n^{(\alpha)}(x) L_m^{(\alpha)}(x) e^{-x} x^\alpha \, dx = \delta_{nm} \frac{\Gamma(n+\alpha+1)}{n!}. \tag{6.48}$$

The area of orthogonality is not bounded but it is not necessary to rescale the image onto the whole interval $\langle 0, \infty \rangle$. It is sufficient to map the image, whose support is always bounded, without scaling anywhere inside this interval [11].

Hermite polynomials of the nth degree are defined as

$$H_n(x) = (-1)^n e^{x^2} \frac{d^n}{dx^n} e^{-x^2}. \tag{6.49}$$

The relation of orthogonality is

$$\int_{-\infty}^\infty H_n(x) H_m(x) e^{-x^2} dx = \delta_{nm} n! 2^n \sqrt{\pi}. \tag{6.50}$$

The Hermite polynomials are orthogonal on the interval $(-\infty, \infty)$, so they are generally less suitable for image-processing purposes. Nevertheless, Wu and Shen [12] proposed so-called *orthogonal Gaussian–Hermite moments*

$$M_{pq}(x, y, f(x, y)) = \int_{-\infty}^\infty \int_{-\infty}^\infty g(u, v, \sigma) H_p(u/\sigma) H_q(v/\sigma) f(x+u, y+v) \, du \, dv \tag{6.51}$$

that depend on the image coordinates (x, y). The $g(u, v, \sigma)$ is a Gaussian function

$$g(u, v, \sigma) = \frac{1}{2\pi\sigma^2} e^{-(u^2+v^2)/(2\sigma^2)}, \tag{6.52}$$

where σ is its standard deviation. The authors used these moments for change detection in video sequences.

6.2.5 OG moments of a discrete variable

There is a group of orthogonal polynomials defined directly on a series of points and therefore they are especially suitable for digital images. The relation of discrete orthogonality of the kernel function has a form

$$\sum_{\ell=1}^N p_m(x_\ell) \, p_n(x_\ell) = 0 \tag{6.53}$$

or in case of the weighted orthogonality

$$\sum_{\ell=1}^N w_\ell^2 \, p_m(x_\ell) \, p_n(x_\ell) = 0 \tag{6.54}$$

for $m \neq n$. The point coordinates are usually ordered and always distinct

$$x_1 < x_2 < \cdots < x_N. \tag{6.55}$$

An overview of polynomials of a discrete variable can be found in reference [13]. In the following text, we briefly review those which have found applications in image processing.

Discrete Chebyshev polynomials [5] are defined as

$$T_n(x) = (1 - N)_n \, {}_3F_2(-n, -x, 1 + n; 1, 1 - N; 1), \quad n, x = 0, 1, 2, \ldots, N - 1. \quad (6.56)$$

They can also be written as

$$T_n(x) = n! \sum_{k=0}^{n} (-1)^{n-k} \binom{N-1-k}{n-k} \binom{n+k}{n} \binom{x}{k}. \quad (6.57)$$

They satisfy the relation of orthogonality

$$\sum_{x=0}^{N-1} T_n(x) T_m(x) = \varrho(n, N) \delta_{mn} \quad (6.58)$$

with the normalizing factor

$$\varrho(n, N) = \frac{N(N^2 - 1^2)(N^2 - 2^2) \cdots (N^2 - n^2)}{2n + 1}$$

$$= (2n)! \binom{N+n}{2n+1}, \quad n = 0, 1, 2, \ldots, N - 1. \quad (6.59)$$

The recurrence formula is

$$(n + 1)T_{n+1}(x) = (2n + 1)(2x - N + 1)T_n(x) - n(N^2 - n^2)T_{n-1}(x), \quad n = 1, 2, \ldots \quad (6.60)$$

with the initial values $T_0(x) = 1$ and $T_1(x) = 2x + 1 - N$.

It is advantageous to work with *orthonormal* polynomials $\hat{T}_n(x)$ whose norm equals one:

$$\sum_{x=0}^{N-1} (\hat{T}_n(x))^2 = 1. \quad (6.61)$$

Mukundan *et al.* [5] and Mukundan [7] proposed first to normalize the polynomials to the magnitude by N^n and then to normalize them by $\varrho(n, N)$ to obtain the orthonormal polynomials. It resulted in the modified definition of the discrete Chebyshev polynomials

$$\hat{T}_0(x) = \frac{1}{\sqrt{N}}$$

$$\hat{T}_1(x) = (2x + 1 - N)\sqrt{\frac{3}{N(N^2 - 1)}} \quad (6.62)$$

$$\hat{T}_n(x) = (\alpha_1 x + \alpha_2)\hat{T}_{n-1}(x) - \alpha_3 \hat{T}_{n-2}(x),$$

where

$$\alpha_1 = \frac{2}{n}\sqrt{\frac{4n^2 - 1}{N^2 - n^2}}$$

$$\alpha_2 = \frac{1 - N}{n}\sqrt{\frac{4n^2 - 1}{N^2 - n^2}} \tag{6.63}$$

$$\alpha_3 = \frac{n - 1}{n}\sqrt{\frac{2n + 1}{2n - 3}}\sqrt{\frac{N^2 - (n - 1)^2}{N^2 - n^2}}.$$

Then the *discrete Chebyshev moments* for digital images are

$$\tau_{pq} = \sum_{x=0}^{M-1}\sum_{y=0}^{N-1} \hat{T}_m(x)\hat{T}_n(y)f(x, y)\,dx\,dy. \tag{6.64}$$

Discrete generalized Laguerre polynomials of the nth degree with parameter α are defined as

$$L_n^{(\alpha)}(x) = \frac{(\alpha + 1)_n}{n!}\,{}_1F_1(-n; \alpha + 1; x) \quad n, x = 0, 1, 2, \ldots, N - 1, \tag{6.65}$$

where the parameter $\alpha > -1$. If α is an integer, then it can also be expressed

$$L_n^{(\alpha)}(x) = \sum_{m=0}^{n}(-1)^m\binom{\alpha + 1}{n - m}\frac{1}{m!}x^m. \tag{6.66}$$

They satisfy the relation of orthogonality

$$\sum_{x=0}^{N-1} L_n^{(\alpha)}(x)L_m^{(\alpha)}(x)e^{-x}x^\alpha = \delta_{nm}\left(\frac{\Gamma(n + \alpha + 1)}{\Gamma(\alpha + 1)n!}\right)^2 N, \quad n = 0, 1, 2, \ldots, N - 1. \tag{6.67}$$

The recurrence formula suitable for evaluation is

$$(n + 1)L_{n+1}^{(\alpha)}(x) = (2n + 1 + \alpha - x)L_n^{(\alpha)}(x) - (n + \alpha)L_{n-1}^{(\alpha)}(x), \quad n = 1, 2, \ldots \tag{6.68}$$

with initial values $L_0^{(\alpha)}(x) = 1$ and $L_1^{(\alpha)}(x) = \alpha + 1 - x$.

The *Laguerre moments* of the image $f(x, y)$ are then [11]

$$\lambda_{mn} = \frac{\Gamma^2(\alpha + 1)m!n!}{\Gamma(m + \alpha + 1)\Gamma(n + \alpha + 1)\sqrt{MN}}$$

$$\times \sum_{x=0}^{M-1}\sum_{y=0}^{N-1} e^{-(x+y)/2}x^{\alpha/2}y^{\alpha/2}L_m^{(\alpha)}(x)L_n^{(\alpha)}(y)f(x, y). \tag{6.69}$$

The nth degree *Krawtchouk polynomials* with parameter p are defined as

$$K_n^{(p)}(x, N) = {}_2F_1\left(-n, -x; -N; \frac{1}{p}\right), \tag{6.70}$$

where $n = 0, 1, \ldots, N$ is the degree, x is a discrete variable with values $0, 1, \ldots, N$, i.e. the number of samples is $N + 1$, $N > 0$, and the parameter $p \in (0, 1)$. Their use in image processing was reported in reference [14].

The relation of orthogonality is

$$\sum_{x=0}^{N} w(x; p, N) K_m^{(p)}(x, N) K_n^{(p)}(x, N) = \varrho(n; p, N) \delta_{mn}, \tag{6.71}$$

where $m, n = 0, 1, \ldots, N$, the weight function

$$w(x; p, N) = \binom{N}{x} p^x (1 - p)^{N-x} \tag{6.72}$$

and the norm

$$\varrho(n; p, N) = \left(\frac{1-p}{p}\right)^n \frac{1}{\binom{N}{n}}. \tag{6.73}$$

We can also derive the recurrence formula

$$K_0^{(p)}(x, N) = 1,$$

$$K_1^{(p)}(x, N) = 1 - \frac{x}{Np}, \tag{6.74}$$

$$K_{n+1}^{(p)}(x, N) = \frac{Np - 2np + n - x}{(N - n)p} K_n^{(p)}(x, N) - \frac{n(1 - p)}{(N - n)p} K_{n-1}^{(p)}(x, N).$$

Graphs of the Krawtchouk polynomials for $p = 0.5$ are shown in Figure 6.8.

The Krawtchouk polynomials have better numerical stability when normalized both to the weight and the norm as

$$\hat{K}_n^{(p)}(x, N) = K_n^{(p)}(x, N) \sqrt{\frac{w(x; p, N)}{\varrho(n; p, N)}}. \tag{6.75}$$

Then the modified recurrence formula is

$$\hat{K}_0^{(p)}(x, N) = \sqrt{\frac{w(x; p, N)}{\varrho(0; p, N)}},$$

$$\hat{K}_1^{(p)}(x, N) = \left(1 - \frac{x}{Np}\right) \sqrt{\frac{w(x; p, N)}{\varrho(1; p, N)}}, \tag{6.76}$$

$$\hat{K}_{n+1}^{(p)}(x, N) = A \frac{Np - 2np + n - x}{(N - n)p} \hat{K}_n^{(p)}(x, N) - B \frac{n(1 - p)}{(N - n)p} \hat{K}_{n-1}^{(p)}(x, N),$$

where

$$A = \sqrt{\frac{p}{1 - p} \frac{N - n}{n + 1}},$$

$$B = \frac{p}{1 - p} \sqrt{\frac{(N - n)(N - n + 1)}{n(n + 1)}}.$$

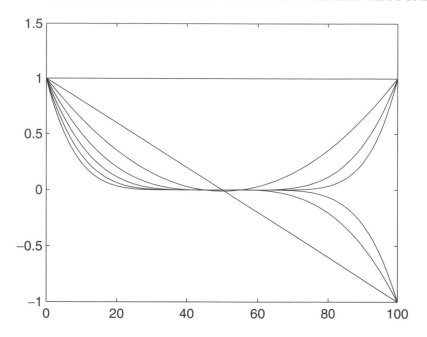

Figure 6.8 The graphs of the Krawtchouk polynomials up to the sixth degree.

The relation of orthogonality changes to

$$\sum_{x=0}^{N} \hat{K}_m^{(p)}(x, N) \hat{K}_n^{(p)}(x, N) = \delta_{mn}. \tag{6.77}$$

The parameter p determines the "localization" of the polynomial in the interval $\langle 0, N \rangle$. If we know a priori where in the image is our area of interest located, we can choose such p that the central part of the polynomial is shifted to this area. Having no prior knowledge, a common choice is $p = 0.5$. Graphs of the weighted Krawtchouk polynomials for $p = 0.5$ can be seen in Figure 6.9 and for $p = 0.2$ in Figure 6.10.

Dual Hahn polynomials are defined on a nonuniform lattice $x(s) = s(s + 1)$, where s is the order of a sample and x is its distance from the origin. Zhu *et al.* [15] adapted them for use in image processing such that s stands for a traditional coordinate in a digital image. Here, we use their approach although their definition differs from other papers, e.g. from reference [13].

The nth degree dual Hahn polynomials with parameter c on the interval $\langle a, b - 1 \rangle$ are then defined as

$$W_n^{(c)}(s, a, b) = \frac{(a - b + 1)_n (a + c + 1)_n}{n!}$$

$$\times \, {}_3F_2(-n, a - s, a + s + 1; a - b + 1, a + c + 1; 1), \tag{6.78}$$

where $(a)_n$ is the Pochhammer symbol (6.4), $n = 0, 1, \ldots, N - 1$ and $s = a, a + 1, \ldots, b - 1$. It is supposed that

$$-\tfrac{1}{2} < a < b, \quad |c| < 1 + a, \quad b = a + N. \tag{6.79}$$

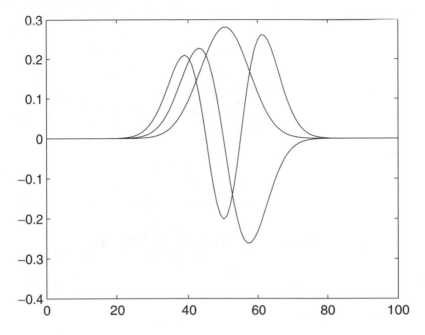

Figure 6.9 The graphs of the weighted Krawtchouk polynomials for $p = 0.5$ up to the second degree.

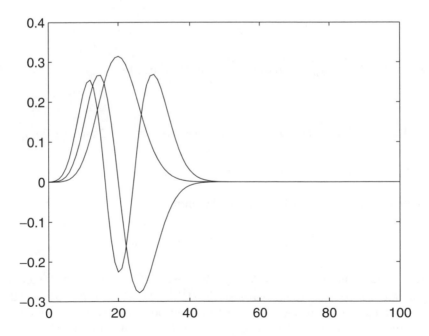

Figure 6.10 The graphs of the weighted Krawtchouk polynomials for $p = 0.2$ up to the second degree.

The dual Hahn polynomials satisfy the relation of orthogonality

$$\sum_{s=a}^{b-1} W_m^{(c)}(s, a, b) W_n^{(c)}(s, a, b) \varrho(s)(2s + 1) = d_n^2 \delta_{mn}, \qquad (6.80)$$

where $m, n = 0, 1, \ldots, N - 1$. The weight function is

$$\varrho(s) = \frac{\Gamma(a + s + 1)\Gamma(c + s + 1)}{\Gamma(s - a + 1)\Gamma(b - s)\Gamma(b + s + 1)\Gamma(s - c + 1)} \qquad (6.81)$$

and the norm

$$d_n^2 = \frac{\Gamma(a + c + n + 1)}{n!(b - a - n - 1)!\Gamma(b - c - n)}. \qquad (6.82)$$

Dual Hahn polynomials also have a normalized version with better numerical stability

$$\hat{W}_n^{(c)}(s, a, b) = W_n^{(c)}(s, a, b)\sqrt{\frac{\varrho(s)}{d_n^2}(2s + 1)}. \qquad (6.83)$$

The recurrence formula is then

$$\hat{W}_0^{(c)}(s, a, b) = \sqrt{\frac{\varrho(s)}{d_0^2}(2s + 1)},$$

$$\hat{W}_1^{(c)}(s, a, b) = -\frac{\varrho(s + 1)(s + 1 - a)(s + 1 + b)(s + 1 - c) - \varrho(s)(s - a)(s + b)(s - c)}{\varrho(s)(2s + 1)}$$

$$\cdot \sqrt{\frac{\varrho(s)}{d_1^2}(2s + 1)},$$

$$\hat{W}_{n+1}^{(c)}(s, a, b) = A\frac{d_n}{d_{n+1}}\hat{W}_n^{(c)}(s, a, b) - B\frac{d_{n-1}}{d_{n+1}}\hat{W}_{n-1}^{(c)}(s, a, b), \qquad (6.84)$$

where

$$A = \frac{1}{n + 1}(s(s + 1) - ab + ac - bc - (b - a - c - 1)(2n + 1) + 2n^2),$$

$$B = \frac{1}{n + 1}(a + c + n)(b - a - n)(b - c - n).$$

The relation of orthogonality then changes to the form

$$\sum_{s=a}^{b-1} \hat{W}_m^{(c)}(s, a, b)\hat{W}_n^{(c)}(s, a, b) = \delta_{mn}. \qquad (6.85)$$

If the parameters have values $a = (\alpha + \beta)/2$, $b = a + N$, $c = (\beta - \alpha)/2$, then the dual Hahn polynomials become *Hahn polynomials* $H_n^{(\alpha,\beta)(x,N)}$. If $\alpha = \beta = 0$, then the dual Hahn polynomials reduce to the discrete Chebyshev polynomials. If we set $\beta = pt$ and $\alpha = (1 - p)t$ in the Hahn polynomials, then for $t \to \infty$ it tends to the Krawtchouk polynomials [15].

Racah polynomials have many common properties with the dual Hahn polynomials, including a nonuniform lattice $x(s) = s(s + 1)$. We use the adaptation from reference [16] here. The nth degree Racah polynomials with parameters α and β on the interval $\langle a, b - 1\rangle$ are defined as

$$U_n^{(\alpha,\beta)}(s, a, b)$$

$$= \frac{(a - b + 1)_n(\beta + 1)_n(a + b + \alpha + 1)_n}{n!}$$

$$\cdot {}_4F_3(-n, \alpha + \beta + n + 1, a - s, a + s + 1; \beta + 1, a - b + 1, a + b + \alpha + 1; 1),$$
$$(6.86)$$

where $(a)_n$ is the Pochhammer symbol (6.4), $n = 0, 1, \ldots, L - 1$ and $s = a, a + 1, \ldots, b - 1$. It is supposed that

$$-\tfrac{1}{2} < a < b, \quad \alpha > -1, \quad -1 < \beta < 2a + 1, \quad b = a + N. \tag{6.87}$$

The Racah polynomials satisfy the relation of orthogonality

$$\sum_{s=a}^{b-1} U_m^{(\alpha,\beta)}(s, a, b)U_n^{(\alpha,\beta)}(s, a, b)\varrho(s)(2s + 1) = d_n^2 \delta_{mn}, \tag{6.88}$$

where $m, n = 0, 1, \ldots, L - 1$. The weight function is

$$\varrho(s) = \frac{\Gamma(a + s + 1)\Gamma(s - a + \beta + 1)\Gamma(b + \alpha - s)\Gamma(b + \alpha + s + 1)}{\Gamma(a - \beta + s + 1)\Gamma(s - a + 1)\Gamma(b - s)\Gamma(b + s + 1)} \tag{6.89}$$

and the norm

$$d_n^2 = \frac{\Gamma(\alpha + n + 1)\Gamma(\beta + n + 1)\Gamma(b - a + \alpha + \beta + n + 1)\Gamma(a + b + \alpha + n + 1)}{(\alpha + \beta + 2n + 1)n!(b - a - n - 1)!\Gamma(\alpha + \beta + n + 1)\Gamma(a + b - \beta - n)}. \tag{6.90}$$

As in the case of the dual Hahn polynomials, the Racah polynomials have a normalized version having better numerical properties

$$\hat{U}_n^{(\alpha,\beta)}(s, a, b) = U_n^{(\alpha,\beta)}(s, a, b)\sqrt{\frac{\varrho(s)}{d_n^2}(2s + 1)}. \tag{6.91}$$

The recurrence formula is then

$$\hat{U}_0^{(\alpha,\beta)}(s, a, b) = \sqrt{\frac{\varrho(s)}{d_0^2}(2s + 1)},$$

$$\hat{U}_1^{(\alpha,\beta)}(s, a, b) = -\left(\frac{\varrho(s + 1)(s + 1 - a)(s + 1 + b)(s + 1 + a - \beta)(b + \alpha - s - 1)}{\varrho(s)(2s + 1)}\right.$$

$$\left. - \frac{\varrho(s)(s - a)(s + b)(s + a - \beta)(b + \alpha - s)}{\varrho(s)(2s + 1)}\right)\sqrt{\frac{\varrho(s)}{d_1^2}(2s + 1)},$$

$$A\,\hat{U}_{n+1}^{(\alpha,\beta)}(s, a, b) = B\frac{d_n}{d_{n+1}}\hat{U}_n^{(\alpha,\beta)}(s, a, b) - C\frac{d_{n-1}}{d_{n+1}}\hat{U}_{n-1}^{(\alpha,\beta)}(s, a, b), \tag{6.92}$$

where

$$A = \frac{(n+1)(\alpha+\beta+n+1)}{(\alpha+\beta+2n+1)(\alpha+\beta+2n+2)},$$

$$B = s(s+1) - \frac{a^2+b^2+(a-\beta)^2+(b+\alpha)^2-2}{4} + \frac{(\alpha+\beta+2n)(\alpha+\beta+2n+2)}{8}$$

$$- \frac{(\beta^2-\alpha^2)((b+\alpha/2)^2-(a-\beta/2)^2)}{2(\alpha+\beta+2n)(\alpha+\beta+2n+2)},$$

$$C = \frac{(\alpha+n)(\beta+n)}{(\alpha+\beta+2n)(\alpha+\beta+2n+1)}$$

$$\cdot \left(\left(a+b+\frac{\alpha-\beta}{2}\right)^2-\left(n+\frac{\alpha+\beta}{2}\right)^2\right)\left(\left(b-a+\frac{\alpha+\beta}{2}\right)^2-\left(n+\frac{\alpha+\beta}{2}\right)^2\right).$$

The weighted Racah polynomials satisfy the relation of orthogonality

$$\sum_{s=a}^{b-1} \hat{U}_m^{(\alpha,\beta)}(s,a,b)\hat{U}_n^{(\alpha,\beta)}(s,a,b) = \delta_{mn}. \tag{6.93}$$

6.3   Moments orthogonal on a disk

One of the main motivations when working with moments is an easy construction of rotation invariants. As already observed in Chapter 2, an efficient way is to work in polar coordinates, where rotation is transformed into a shift. In the case of complex moments, this shift causes a change of phase that is further eliminated by the multiplication of proper moments. The same idea applies when introducing moments orthogonal on a disk.

Such moments are mostly constructed in the form:

$$v_{pq} = n_{pq} \int_0^{2\pi}\int_0^1 R_{pq}(r)e^{-iq\varphi}f(r,\varphi)r\,dr d\varphi \quad p=0,1,2,\ldots,q=-p,\ldots,p, \tag{6.94}$$

where n_{pq} is some normalizing factor, r,φ are polar coordinates (2.8), R_{pq} is a radial part and $e^{iq\varphi}$ is an angular part of the respective polynomial. The area of orthogonality is a unit disk $\Omega = \{(x,y) \mid x^2+y^2 \le 1\}$. To calculate the moments, the image f must be scaled such that it is fully contained in Ω.

6.3.1   Zernike and Pseudo-Zernike moments[4]

Zernike moments (ZMs) were introduced into image analysis about 30 years ago by Teague [17] who used ZMs to construct rotation invariants. He used the fact that the ZMs keep their magnitude under arbitrary rotation. He also showed that the Zernike invariants of the second and third orders are equivalent to the Hu invariants when expressed in terms of geometric moments. He presented the invariants up to the eighth order in explicit form but no general rule about how to derive them was given. Later, Wallin [18] described an

---

[4]F. Zernike (1888–1966), a Dutch physicist.

algorithm for the formation of rotation invariants of any order. Numerical properties and possible applications of ZMs in image processing were described in references [19], [2], [3] and [20], among others.

ZMs of the nth order with repetition ℓ are defined as

$$A_{n\ell} = \frac{n+1}{\pi} \int_0^{2\pi} \int_0^1 V_{n\ell}^*(r, \varphi) f(r, \varphi) r \, dr \, d\varphi \quad n = 0, 1, 2, \ldots, \ell = -n, -n+2, \ldots, n,$$

(6.95)

i.e. the difference $n - |\ell|$ is always even. The asterisk means the complex conjugate. The *Zernike polynomials* are defined as products

$$V_{n\ell}(r, \varphi) = R_{n\ell}(r) e^{i\ell\varphi},$$

(6.96)

where the radial part is

$$R_{n\ell}(r) = \sum_{s=0}^{(n-|\ell|)/2} (-1)^s \frac{(n-s)!}{s!((n+|\ell|)/2 - s)!((n-|\ell|)/2 - s)!} r^{n-2s}$$

$$= \sum_{k=|\ell|,|\ell|+2,\ldots}^{n} B_{n\ell k} r^k.$$

(6.97)

The coefficients

$$B_{n\ell k} = \frac{(-1)^{(n-k)/2} ((n+k)/2)!}{((n-k)/2)!((k+\ell)/2)!((k-\ell)/2)!}$$

(6.98)

can be used for conversion from the geometric moments:

$$A_{n\ell} = \frac{n+1}{\pi} \sum_{k=|\ell|,|\ell|+2,\ldots}^{n} \sum_{j=0}^{(k-|\ell|)/2} \sum_{m=0}^{|\ell|} \binom{(k-|\ell|)/2}{j} \binom{|\ell|}{m} w^m B_{n\ell k} m_{k-2j-m, 2j+m},$$

(6.99)

where

$$w = \begin{cases} -i & \ell > 0 \\ i & \ell \leq 0. \end{cases}$$

(6.100)

The Zernike polynomials satisfy the relation of orthogonality

$$\int_0^{2\pi} \int_0^1 V_{n\ell}^*(r, \varphi) V_{mk}(r, \varphi) r \, dr d\varphi = \frac{\pi}{n+1} \delta_{mn} \delta_{k\ell},$$

(6.101)

particularly

$$\int_0^1 R_{n\ell}(r) R_{m\ell}(r) r \, dr = \frac{1}{2(n+1)} \delta_{mn}.$$

(6.102)

The recurrence formula for enumeration of the radial part (sometimes called the Prata method) is

$$R_{n\ell}(r) = \frac{2rn}{n+\ell} R_{n-1,\ell-1}(r) - \frac{n-\ell}{n+\ell} R_{n-2,\ell}(r).$$

(6.103)

Computation of the Zernike polynomials by this formula must commence with

$$R_{nn}(r) = r^n, \quad R_{n,-n}(r) = r^n, \quad n = 0, 1, \ldots \tag{6.104}$$

and then the Zernike polynomials can be computed for $n = 2, 3, \ldots, \ell = -n + 2, -n + 4, \ldots, n - 2$. Other, formally more complicated but more computationally efficient recurrence formulas can be found in the next chapter.

There is a relation between the Jacobi and a radial part of Zernike polynomials [21]

$$R_{m+2s,|m|}(r) = (-1)^{m+2s} \binom{m+s}{s} r^m G_s^{(m+1,m+1)}(r^2) \tag{6.105}$$

and between the Legendre and a radial part of Zernike polynomials

$$R_{2n,0}(r) = P_n(2r^2 - 1). \tag{6.106}$$

Examples of selected radial polynomials:

$R_{00}(r) = 1,$

$R_{11}(r) = r,$

$R_{20}(r) = 2r^2 - 1,$

$R_{22}(r) = r^2,$

$R_{31}(r) = 3r^3 - 2r,$

$R_{33}(r) = r^3,$

$R_{40}(r) = 6r^4 - 6r^2 + 1,$

$R_{42}(r) = 4r^4 - 3r^2,$

$R_{44}(r) = r^4,$

$R_{51}(r) = 10r^5 - 12r^3 + 3r,$

$R_{53}(r) = 5r^5 - 4r^3,$

$R_{55}(r) = r^5,$

$R_{60}(r) = 20r^6 - 30r^4 + 12r^2 - 1,$

$R_{62}(r) = 15r^6 - 20r^4 + 6r^2,$

$R_{64}(r) = 6r^6 - 5r^4,$

$R_{66}(r) = r^6,$

$R_{71}(r) = 35r^7 - 60r^5 + 30r^3 - 4r,$

$R_{73}(r) = 21r^7 - 30r^5 + 10r^3,$

$R_{75}(r) = 7r^7 - 6r^5,$

$R_{77}(r) = r^7,$

$R_{80}(r) = 70r^8 - 140r^6 + 90r^4 - 20r^2 + 1,$

$R_{82}(r) = 56r^8 - 105r^6 + 60r^4 - 10r^2,$

$R_{84}(r) = 28r^8 - 42r^6 + 15r^4,$

$R_{86}(r) = 8r^8 - 7r^6,$

$R_{88}(r) = r^8,$

$R_{91}(r) = 126r^9 - 280r^7 + 210r^5 - 60r^3 + 5r,$

$R_{93}(r) = 84r^9 - 168r^7 + 105r^5 - 20r^3,$

$R_{95}(r) = 36r^9 - 56r^7 + 21r^5,$

$R_{97}(r) = 9r^9 - 8r^7,$

$R_{99}(r) = r^9. \tag{6.107}$

The radial function is symmetric with respect to repetition, i.e. $R_{n,-\ell}(r) = R_{n\ell}(r)$. The graphs of the radial polynomials are in Figure 6.11. The Zernike polynomials $V_{n\ell}(r, \varphi)$ on the unit disk $\{r \mid r \leq 1\}$ are visualized in Figure 6.12.

Digital images must be mapped onto the unit disk before the Zernike moments can be calculated. ZMs are not natural scaling invariants. The scaling invariance should be provided by this mapping; therefore, the correct mapping of objects into the unit disk is a crucial step. If two objects differ by the TRS transform, the corresponding pixels should be mapped onto

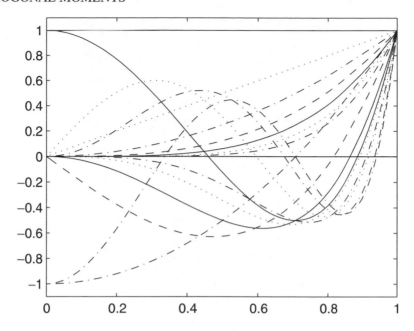

Figure 6.11 The graphs of the Zernike radial functions up to the sixth degree. The graphs of the polynomials of the same degree are drawn by the same type of line.

the same distance from the center of the unit disk. If the image has the size $M \times N$, then

$$A_{n\ell} = \frac{n+1}{\pi} \sum_{x=0}^{M-1} \sum_{y=0}^{N-1} V_{n\ell}^*(r, \varphi) f(x, y), \quad n = 0, 1, 2, \ldots, \ell = -n, -n+2, \ldots, n.$$

(6.108)

There are several methods for converting Cartesian coordinates x, y to polar coordinates r, φ and their suitability depends on the specific application. One of the most common supposes the center of rotation in the centroid of the image

$$r = \frac{\sqrt{(x - x_c)^2 + (y - y_c)^2}}{r_{max}}, \quad \text{where } r_{max} = \frac{\sqrt{m_{00}}}{2} \sqrt{\frac{M}{N} + \frac{N}{M}}$$

(6.109)

$$\varphi = \arctan\left(\frac{y - y_c}{x - x_c}\right).$$

The image brightness is mapped into the interval $\langle 0, 1 \rangle$. Another method uses the center of the image $(M - 1)/2$, $(N - 1)/2$ instead of the centroid x_c, y_c and MN instead of m_{00}.

The values of ZMs should also be normalized to the density of sampling. We can divide the values by the number of pixels in the unit disk, but their division by the A_{00} yields slightly better results.

The mapping to the unit disk and the normalization to the sampling density can also be provided by the formula (6.99), where we substitute the geometric moments m_{pq} by the central moments normalized to scaling

$$\nu_{pq} = \frac{\mu_{pq}}{\mu_{00}^{(p+q)/2+1}}.$$

(6.110)

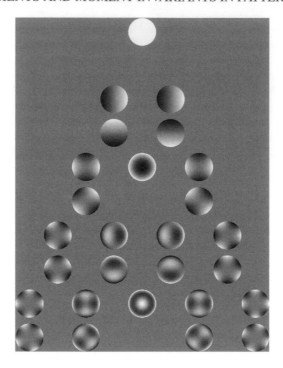

Figure 6.12 The graphs of the Zernike polynomials up to the fourth degree. Black $= -1$, white $= 1$. Real parts: 1st row: $n = 0$, $\ell = 0$, 3rd row: $n = 1$, $\ell = -1$, 1, 5th row: $n = 2$, $\ell = -2, 0, 2$, 7th row: $n = 3$, $\ell = -3, -1, 1, 3$, 9th row: $n = 4$, $\ell = -4, -2, 0, 2, 4$. Imaginary parts: 2nd, 4th, 6th, 8th and 10th rows, respectively. The indices are the same as above.

The behavior of ZMs under rotation is similar to that of the complex moments. If we rotate the image by an angle θ, the values of the ZMs change as

$$A'_{n\ell} = A_{n\ell} e^{-i\ell\theta}. \tag{6.111}$$

Thus, rotation invariants can be constructed either by taking the magnitude $|A_{n\ell}|$, that leads to an incomplete system with limited recognition power, or by taking proper moment products as in the case of the complex moments.

Alternatively, we can apply the normalization approach. The algorithm for determining a proper nonzero normalizing moment is analogous to that from Chapter 2. First, we search the ZMs with repetition 1, i.e. $A_{31}, A_{51}, \ldots A_{n_{max}1}$. If they are all zero (i.e. under-threshold), we suppose the object to be rotationally symmetric and then search the moments with successively increased repetitions 2, 3, etc. For nonsymmetric objects, the suitable normalizing moment is A_{31}. If the normalizing moment $A_{m_r \ell_r}$ has a phase

$$\phi = \frac{1}{\ell_r} \arctan \left[\frac{\mathcal{I}m(A_{m_r \ell_r})}{\mathcal{R}e(A_{m_r \ell_r})} \right], \tag{6.112}$$

then

$$Z_{m\ell} = A_{m\ell} e^{-i\ell\phi} \tag{6.113}$$

is a rotation invariant. It is an analogy to equation (2.22). Another method is to use the ratios

$$\bar{Z}_{m\ell} = A_{m\ell}/(A_{m_r\ell_r})^{\ell/\ell_r}, \tag{6.114}$$

which is equivalent to the phase cancellation by multiplication according to Theorem 2.2.

If we want to avoid using one specific ZM for normalization to the rotation, we can use the method from reference [22]. If the ZMs of two images I and J are to be compared, we look for an angle ϕ minimizing the sum of square differences

$$d_{I,J}^2(\phi) = \sum_{n=0}^{n_{\max}} \sum_{\ell=-n,-n+2,\dots}^{n} \frac{\pi}{n+1} |A_{n\ell}^{(I)} - A_{n\ell}^{(J)} e^{i\ell\phi}|^2. \tag{6.115}$$

The value of $d_{I,J}^2(\phi)$ in this minimum is used as an image dissimilarity measure.

The ZMs can also be normalized to the affine transform. The method is similar to the affine normalization from Chapter 3. This approach was used, for example, in reference [23] for pseudo-Zernike moments.

Pseudo-Zernike moments (PZMs) were derived from the ZMs by releasing from the condition $n - |\ell|$ is even. It means that the definition (6.95) is valid also for PZMs, but with $\ell = -n, -n+1, \dots, n$. The formula (6.97) for the radial part is then rewritten as[5]

$$R_{n\ell}(r) = \sum_{s=0}^{n-|\ell|} (-1)^s \frac{(2n+1-s)!}{s!(n+|\ell|+1-s)!(n-|\ell|-s)!} r^{n-s} = \sum_{k=|\ell|}^{n} S_{n\ell k} r^k. \tag{6.116}$$

The consequence is that the number of Zernike moments up to the nth order $(n+1)(n+2)/2$ increases to $(n+1)^2$ for the PZMs. The radius r appears also in odd powers in (6.116), so we must use a special method for conversion of the geometric moments to the PZMs [23]

$$
\begin{aligned}
A_{n\ell} = {} & \frac{n+1}{\pi} \sum_{\substack{s=0 \\ n-\ell-s \text{ even}}}^{n-|\ell|} S_{n\ell s} \sum_{j=0}^{(n-|\ell|-s)/2} \sum_{m=0}^{|\ell|} \binom{(n-|\ell|-s)/2}{j} \\
& \cdot \binom{|\ell|}{m} (-i)^m m_{n-s-2j-m,2j+m} \\
{} + {} & \frac{n+1}{\pi} \sum_{\substack{s=0 \\ n-\ell-s \text{ odd}}}^{n-|\ell|} S_{n\ell s} \sum_{j=0}^{(n-|\ell|-s+1)/2} \sum_{m=0}^{|\ell|} \binom{(n-|\ell|-s+1)/2}{j} \\
& \cdot \binom{|\ell|}{m} (-i)^m g_{n-s+1-2j-m,2j+m},
\end{aligned}
\tag{6.117}
$$

where g_{pq} are *radial geometric moments*

$$g_{pq} = \int_{-\infty}^{\infty} \int_{-\infty}^{\infty} x^p y^q \sqrt{x^2+y^2} f(x,y) \, \mathrm{d}x \, \mathrm{d}y, \quad p, q = 0, 1, 2, \dots. \tag{6.118}$$

The radius $\sqrt{x^2+y^2}$ can be expanded as a power series. Therefore, if we have an infinite number of moments of all orders, we can express the radial geometric moments in terms of

---

[5] Some papers use a wrong formula with a factor $(n-|\ell|+1-s)!$ instead of $(n+|\ell|+1-s)!$

geometric moments. From this point of view, the independence of the PZMs claimed in [2] is questionable. On the other hand, if we had only a limited number of geometric moments, we would be able to compute the radial geometric moments only approximately.

Several authors tested the robustness of PZMs to additive noise and claimed that it is better than the robustness of the ZMs. This is considered to be the main advantage of PZMs.

6.3.2   Orthogonal Fourier–Mellin moments

Orthogonal Fourier–Mellin moments were proposed by Sheng and Shen [21]. They are defined as

$$\Phi_{n\ell} = \frac{n+1}{\pi} \int_0^{2\pi} \int_0^1 Q_n(r) e^{-i\ell\varphi} f(r, \varphi)\, r\, dr\, d\varphi \quad n = 0, 1, 2, \ldots, \ell = 0, \pm 1, \pm 2, \ldots,$$

(6.119)

where the radial function $Q_n(r)$ is

$$Q_n(r) = \sum_{s=0}^{n} (-1)^{n+s} \frac{(n+s+1)!}{(n-s)!\,s!\,(s+1)!}\, r^s.$$

(6.120)

There is a relation to the Jacobi polynomials

$$Q_n(r) = (-1)^n \binom{n+1}{n} G_n^{(2,2)}(r).$$

(6.121)

The relation of orthogonality of the radial functions is

$$\int_0^1 Q_n(r) Q_m(r)\, r\, dr = \frac{1}{2(n+1)}\, \delta_{mn}.$$

(6.122)

The first six radial polynomials have the form

$$
\begin{aligned}
Q_0(r) &= 1, \\
Q_1(r) &= -2 + 3r, \\
Q_2(r) &= 3 - 12r + 10r^2, \\
Q_3(r) &= -4 + 30r - 60r^2 + 35r^3, \\
Q_4(r) &= 5 - 60r + 210r^2 - 280r^3 + 126r^4, \\
Q_5(r) &= -6 + 105r - 560r^2 + 1260r^3 - 1260r^4 + 462r^5.
\end{aligned}
$$

(6.123)

Their graphs are in Figure 6.13. The kernel functions $Q_n(r)e^{-i\ell\varphi}$ of the orthogonal Fourier–Mellin moments on the unit disk $\{r \mid r \le 1\}$ are in Figure 6.14.

The name "Fourier–Mellin" moments is associated with their relation to the Fourier–Mellin transform for the integers s and ℓ

$$M_f(s, \ell) = \frac{1}{2\pi} \int_0^\infty \int_0^{2\pi} r^s e^{-i\ell\varphi} f(r, \varphi)\, d\varphi\, \frac{dr}{r}.$$

(6.124)

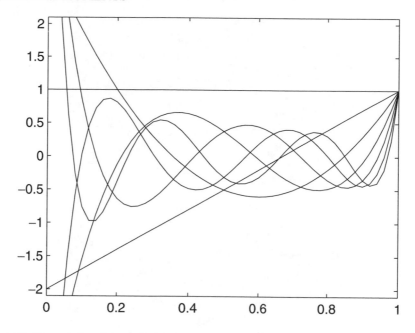

Figure 6.13 The graphs of the radial functions of the orthogonal Fourier–Mellin moments up to the sixth degree.

The relation between these two quantities is

$$\Phi_{n\ell} = 2(n + 1) \sum_{s=0}^{n} (-1)^{n+s} \frac{(n + s + 1)!}{(n - s)!s!(s + 1)!} M_f(s + 2, \ell).$$ (6.125)

In reference [21], the values $\chi_{s\ell} = 2\pi M_f(s + 2, \ell)$ are called *Fourier–Mellin moments* (without the adjective "orthogonal"). Nevertheless, a relation similar to (6.125) can also be derived for other moments orthogonal on the unit disk and from this point of view the name "orthogonal Fourier–Mellin moments" is not quite accurate. On the other hand, the angular part

$$\int_{0}^{2\pi} e^{-i\ell \varphi} f(r, \varphi) \, d\varphi$$ (6.126)

can really be understood as a Fourier transform of the angular coordinate of the image.

The main difference between the orthogonal Fourier–Mellin moments and the ZMs is the absence of the constraint $\ell \leq n$ and the independence of the radial function $Q_n(r)$ on the repetition factor ℓ. The different distribution of zeros over the unit disk is also interesting; compare Figure 6.13 with Figure 6.11. There is a direct relation to the Zernike radial function

$$r Q_n(r^2) = R_{2n+1,1}(r).$$ (6.127)

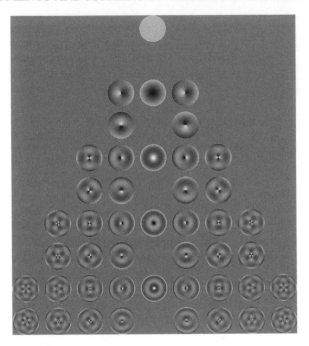

Figure 6.14 The graphs of 2D kernel functions of the orthogonal Fourier–Mellin moments up to the fourth order. Black $= -2$, white $= 2$. Real parts: 1st row: $n = 0$, $\ell = 0$, 3rd row: $n = 1$, $\ell = -1, 0, 1$, 5th row: $n = 2$, $\ell = -2, -1, 0, 1, 2$, 7th row: $n = 3$, $\ell = -3, -2, -1, 0, 1, 2, 3$, 9th row: $n = 4$, $\ell = -4, -3, -2, -1, 0, 1, 2, 3, 4$. Imaginary parts: 2nd, 4th, 6th, 8th and 10th rows. Indices are the same as above.

The orthogonal Fourier–Mellin moments can be formally expressed by means of the complex moments

$$\Phi_{n\ell} = \frac{n+1}{\pi} \sum_{s=0}^{n} (-1)^{n+s} \frac{(n+s+1)!}{(n-s)!s!(s+1)!} c_{(s-\ell)/2,(s+\ell)/2}.$$ (6.128)

We cannot avoid a formal use of the complex moments with half-integer or even negative indices in this conversion, i.e. if we compute the orthogonal Fourier–Mellin moments from the geometric moments, we would not avoid the radial geometric moments (6.118). It illustrates that there is a similar problem with independency as in the PZMs. Nevertheless, we can still choose such a range of the orders n and the repetitions ℓ in a specific application that the orthogonal Fourier–Mellin moments can be considered independent. Comparing (6.125) and (6.128) we can easily find that it holds for the Fourier–Mellin moments

$$\chi_{s\ell} = c_{(s-\ell)/2,(s+\ell)/2}.$$

6.3.3   Other moments orthogonal on a disk

There is a number of moments in the form (6.94), where $R_{pq}(r)$ are certain traditional 1D orthogonal polynomials. Here we present a short survey of such moments. It is worth mentioning that according to Ping *et al.* [24] moments of a different form providing rotation invariance do not exist.

Jacobi–Fourier moments with parameters p and q are defined as

$$\Phi_{n\ell}^{(p,q)} = \int_0^{2\pi} \int_0^1 J_n^{(p,q)}(r) e^{-i\ell\varphi} f(r, \varphi)\, r\, dr\, d\varphi, \tag{6.129}$$

the n and ℓ are arbitrary integers. $J_n^{(p,q)}(x)$ are normalized Jacobi polynomials

$$J_n^{(p,q)}(r) = \sqrt{\frac{w(p, q, r)}{b(n, p, q)}}\, G_n^{(p,q)}(r), \tag{6.130}$$

where the normalization constant is

$$b(n, p, q) = \frac{n!\Gamma(n + p)\Gamma(n + q)\Gamma(n + p - q + 1)}{(2n + p)\Gamma^2(2n + p)} \tag{6.131}$$

and the weight function is

$$w(p, q, r) = (1 - r)^{p-q} r^{q-1}. \tag{6.132}$$

Note that in reference [9] a slightly different normalization was used.

Ping *et al.* [24] introduced also *Chebyshev–Fourier moments*

$$\Psi_{n\ell} = \int_0^{2\pi} \int_0^1 R_n(r) e^{-i\ell\varphi} f(r, \varphi)\, r\, dr\, d\varphi, \tag{6.133}$$

where $R_n(r)$ are normalized shifted Chebyshev polynomials of the second kind

$$\begin{aligned}
R_n(r) &= \sqrt{\frac{8}{\pi}}\, U_n^*(r)(r - r^2)^{1/4} r^{-1/2} \\
&= \sqrt{\frac{8}{\pi}} \left(\frac{1-r}{r}\right)^{1/4} \sum_{k=0}^{(n+2)/2} (-1)^k \frac{(n-k)!}{k!(n-2k)!} (2(2r-1))^{n-2k}.
\end{aligned} \tag{6.134}$$

$U_n^*(r)$ are shifted Chebyshev moments of the second kind

$$U_n^*(r) = U_n(2r - 1) \tag{6.135}$$

that are orthogonal on the interval $\langle 0, 1 \rangle$ with relation

$$\int_0^1 U_m^*(r) U_n^*(r)(r - r^2)^{1/2}\, dr = \frac{8}{\pi}\, \delta_{mn}. \tag{6.136}$$

Another type of OG moments called *radial harmonic Fourier moments* was proposed in reference [25]:

$$\Phi_{n\ell} = \int_0^{2\pi} \int_0^1 T_n(r) e^{-i\ell\varphi} f(r, \varphi)\, r\, dr\, d\varphi, \tag{6.137}$$

where $T_n(r)$ is

$$
T_n(r) = \begin{cases}
\dfrac{1}{\sqrt{r}} & \text{if } n = 0 \\[2ex]
\sqrt{\dfrac{2}{r}}\,\sin((n+1)\pi r) & \text{if } n \text{ is odd} \\[2ex]
\sqrt{\dfrac{2}{r}}\,\cos(n\pi r) & \text{if } n > 0 \text{ is even.}
\end{cases}
\tag{6.138}
$$

The radial harmonic Fourier moments satisfy the relation of orthogonality

$$
\int_0^{2\pi} \int_0^1 T_n(r)e^{i\ell\varphi} T_m(r)e^{ik\varphi} r \, dr \, d\varphi = \delta_{mn}\delta_{k\ell}.
\tag{6.139}
$$

We can see that $T_n(r)$ is not a polynomial in r, so the radial harmonic Fourier moments are not moments in the strict sense of the second definition in section 1.3. Nevertheless, they originate from similar ideas as other OG moments and the authors have used them in a similar way.

6.4   Object recognition by ZMs

The goal of this experiment is to demonstrate the recognition power of ZMs. The experiment was performed on real data, i.e. on real photographs and real rotations; no computer simulation was used. However, we intentionally choose circular objects to avoid problems with the definition domain of Zernike polynomials.

We photographed a set of children playing cards from the popular "Ferdy the Ant" collection;[6] see Figure 6.15, to create our database.

We put the cards on the floor and rotated the hand-held camera. Each card was captured eight times with different rotation angles. The cards were photographed on a dark background allowing an easy segmentation. The first snapshot of each card was used as a representative of its class and the other seven images were recognized by the minimum-distance classifier.

We tried to recognize graylevel versions of the images first. There are several parameters which can influence the recognition rate – the number of invariants, the precision of their calculation and the way the invariants of different orders are normalized to the same dynamic range. Normalization to the range is in fact the same as the setting of weights when using weighted Euclidean distance in the space of invariants. In practice, this step is of great importance and the features are usually normalized by their standard deviations. If we did not normalize the values, the features with a high range would override the others which is not desirable.

We tested both normalized ZMs and Zernike rotation invariants. As expected, the differences between them were not significant. When using normalized ZMs (6.113) by the phase of A_{31} and normalizing their values by standard deviations of the features over the whole test set, we obtained correct recognition when we took the moments up to the third order. A 2D subspace of the feature space is shown in Figure 6.16. Note that two features are too few to provide enough separability. When using Zernike invariants (6.114) with range normalization by standard deviations, we had two misclassifications. Avoiding range

---

[6]Fairy tales by Ondřej Sekora, a Czech writer.

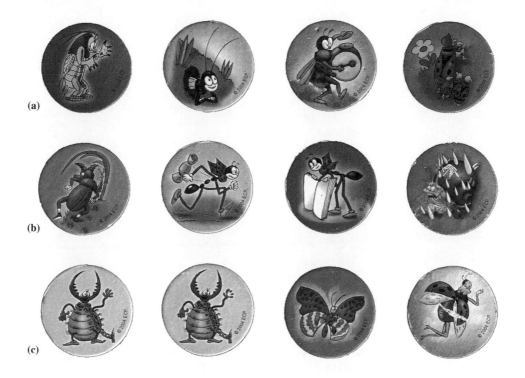

Figure 6.15 The playing cards: (from left) (a) Mole cricket, Cricket, Bumblebee and Heteropter; (b) Poke the Bug, Ferdy the Ant 1, Ferdy the Ant 2 and Snail; (c) Ant-lion 1, Ant-lion 2, Butterfly and Ladybird.

normalization led to 2–5 misclassifications in both cases, even if we increased the number of the invariants. We classified the images also by the complex moment method for comparison. The best result achieved was one misclassification when using invariants up to the fifth order. This illustrates the slightly higher discrimination power of ZMs due to their better numerical properties.

We repeated this experiment on color images. The features were computed from each color channel separately, which yielded three times the features of each order. The recognition rates were much better than before. Misclassifications occurred only if no normalization to the range was performed. For all other settings mentioned above, including the complex moment method, we obtained the correct results. Moreover, fewer features were sufficient in most cases (see Figure 6.17 showing separable clusters in the space of two Zernike normalized moments).

6.5 Image reconstruction from moments

Reconstruction of the original image from a set of its moments has been discussed in the literature quite often. Although it has attracted considerable attention, its practical importance is not very high because moments in general are not a good tool for image

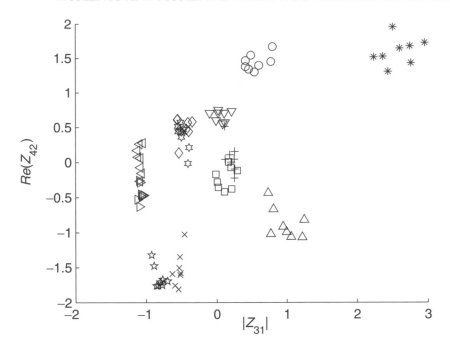

Figure 6.16 The feature space of two Zernike normalized moments $|Z_{31}|$ and $\mathcal{R}e(Z_{42})$. ▽ – Ferdy the Ant 1, △ – Ferdy the Ant 2, □ – Ladybird, ◇ – Poke the Bug, ▷ – Ant-lion 1, ◁ – Ant-lion 2, × – Mole cricket, + – Snail, ☆ – Butterfly, ✳ – Cricket, ✿ – Bumblebee, ○ – Heteropter.

compression, so the reconstruction problem does not appear very frequently. On the other hand, reconstruction abilities of moments are connected with their recognition power. Thus, studying the reconstruction problem provides an insight into those moment properties that are important for classification.

In a continuous domain we have a uniqueness theorem saying that each image function (which is supposed to be compactly supported and piecewise continuous by definition[7]) can be precisely reconstructed from its geometric moments. Using equivalence among different polynomial bases, this theorem holds for other moments also.

The reconstruction problem can be formulated in a more general way: can any series of numbers be considered a set of moments of an image? This question is well known from statistics and probability as the Hausdorff problem and the general answer is "no."[8]

In a discrete domain, the uniqueness theorem says that the image can be precisely reconstructed from the same number of its moments as the number of pixels. In applied digital image processing, the reconstruction problem converts into two questions: what is the optimal number of moments for the reconstruction and what is the error of this reconstruction?

In image processing, reconstruction is sometimes considered not directly from moments but from certain moment invariants. This task can be easily transformed to the previous one,

---

[7] The reconstruction problem on noncompact supports is not considered in this book.

[8] A sufficient condition is a so-called complete monotonicity of the moment values; see reference [3] for details.

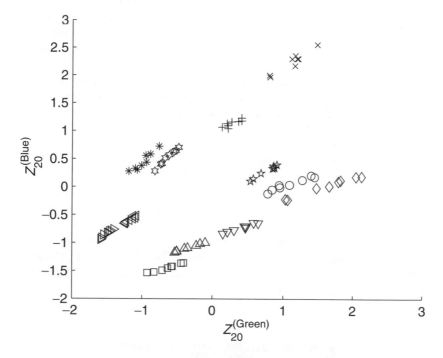

Figure 6.17 The feature space of two Zernike normalized moments $Z_{20}^{(Green)}$ and $Z_{20}^{(Blue)}$. $\triangledown$ – Ferdy the Ant 1, $\triangle$ – Ferdy the Ant 2, $\square$ – Ladybird, $\diamondsuit$ – Poke the Bug, $\triangleright$ – Ant-lion 1, $\triangleleft$ – Ant-lion 2, $\times$ – Mole cricket, $+$ – Snail, $\star$ – Butterfly, $*$ – Cricket, $\maltese$ – Bumblebee, $\bigcirc$ – Heteropter.

bearing in mind that we never can "reconstruct" those image properties to which the invariants are insensitive (e.g. it is impossible to recover the orientation of the object from its rotation invariants).

In the sequel, we review three approaches to the image reconstruction from moments – direct calculation, reconstruction in a Fourier domain and reconstruction from OG moments. By experiment we show their major advantages and limitations.

6.5.1 Reconstruction by the direct calculation

Let us have a discrete image of a size $M \times N$. Its discrete geometric moments are defined as

$$m_{pq} = \sum_{x=0}^{M-1} \sum_{y=0}^{N-1} x^p y^q f(x, y), \quad p = 0, 1, 2, \ldots, M-1, q = 0, 1, 2, \ldots, N-1$$

(6.140)

(this is exactly equivalent to the second Definition in section 1.3 if we imagine the discrete image as an array of Dirac δ-functions). We can consider (6.140) to be a system of linear equations for unknowns $f(x, y)$, where the moments arranged to a vector of size MN create the right-hand side and kernel functions $x^p y^q$ arranged to the matrix $MN \times MN$ create the matrix of the system. For very small images, this system can be resolved by Gaussian elimination and in this way an original image is obtained. For larger images, the system becomes ill-conditioned. Double-precision floating point arithmetic is no longer sufficient, higher-order moments as well as the kernel functions $x^p y^q$ lose their precision and the whole algorithm collapses.

We can slightly improve the algorithm by translation of the origin of coordinates to the center of the image, but the same problem appears a little later. Of course we can use variables with higher precision, but neither is that a principal improvement. Moreover, the set of moments would demand much more memory.

Summarizing, the direct calculation is convenient for very small images only. To demonstrate this, we carried out the following experiment. We created a series of images of the capital letter E of the size $N \times N$ for $N = 7, 8, \ldots, 14$, computed the geometric moments m_{pq} for $p, q = 0, 1, \ldots, N - 1$ of them and then solved the system (6.140). The reconstruction of images up to 11×11 was precise; Figure 6.18 catches the collapse of the algorithm. The original image 12×12 is top left, its reconstruction (bottom left) is almost precise (the maximum error is 5 graylevels from 255). The reconstruction of the image 13×13 (top right) totally collapsed (bottom right), the range of values is actually $\langle -857, 1460 \rangle$ instead of the original $\langle 0, 255 \rangle$. The difference between these two intervals is an attribute of the depth of the error.

Figure 6.18 The collapse of the image reconstruction from moments by the direct calculation. Top row: original images 12×12 and 13×13; bottom row: the reconstructed images.

6.5.2 Reconstruction in the Fourier domain

The kernel function of the discrete Fourier transform

$$F(u, v) = \sum_{x=0}^{M} \sum_{y=0}^{N} e^{-2\pi i ((xu)/M + (yv)/N)} f(x, y) \qquad (6.141)$$

can be expanded as a power series. Integrating it termwise, we obtain

$$F(u, v) = \sum_{p=0}^{\infty} \sum_{q=0}^{\infty} \frac{(-2\pi i)^{p+q}}{p!q!} \left(\frac{u}{M}\right)^{p} \left(\frac{v}{N}\right)^{q} m_{pq}. \tag{6.142}$$

We can compute the Fourier transform from the moments by (6.142) and then obtain the original image by inverse Fourier transform from $F(u, v)$. The main problem of this method is that the power series is only an approximation of the exponential function. If we take the same number of moments as the number of pixels, then the error of this approximation is still too high. We can fill in the higher order moments either by extrapolation or, if possible, calculate them from the original image. In both cases we face the problem well known from the previous method; the overflow of the factors $x^{p}y^{q}$. We can see that there must be a certain optimal moment order and higher order moments cannot improve the image reconstruction.

If the number of moments is not sufficient, the most affected frequencies are the highest ones. We must erase the higher frequencies from the spectrum $F(u, v)$ to obtain an acceptable result. Therefore, we must search for two parameters in this method: the optimal moment order and the optimal threshold in the frequency domain.

An analogous method can be used when reconstructing the image from the complex moments instead of the geometric ones [26]. Then, the formula (6.142) reads

$$F(u, v) = \sum_{p=0}^{\infty} \frac{(-\pi i)^{p}}{p!} \sum_{k=0}^{p} \binom{p}{k} \left(\frac{u+iv}{N}\right)^{p-k} \left(\frac{u-iv}{M}\right)^{k} c_{k, p-k}. \tag{6.143}$$

We tested the Fourier domain reconstruction in the following experiment, similar to the previous one. We used an image of the capital letter E of size 32×32 pixels, computed its geometric moments, reconstructed its spectrum by (6.142) and computed the inverse Fourier transform. We used the moments up to the order 21, 32, 43, 54, 65, 76 and 87, respectively. Higher orders were meaningless because the loss of precision was so high that the reconstruction failed completely. We found optimal frequency thresholds θ_f as 0.12, 0.18, 0.21, 0.25, 0.34, 0.4 and 0.43, for the respective orders, and set $F(u, v) = 0$ for such u, v that $\sqrt{(u/M)^2 + (v/N)^2} > \theta_f$. The reconstructions are visualized in Figure 6.19. One can see that this method does not yield very good results even if a redundant number of moments was used. The reasons are that neither the implementation nor the higher-order moments themselves can be calculated with sufficient precision.

6.5.3   Reconstruction from OG moments

Another approach to the image reconstruction uses OG moments. Having discrete moments v_{pq} orthogonal on a rectangle

$$v_{pq} = \sum_{x=0}^{M-1} \sum_{y=0}^{N-1} P_p(x) P_q(y) f(x, y), \quad p = 0, 1, 2, \ldots, M-1, q = 0, 1, 2, \ldots, N-1,$$

$$\tag{6.144}$$

we can reconstruct the original image as

$$f(x, y) = \sum_{p=0}^{M-1} \sum_{q=0}^{N-1} v_{pq} P_p(x) P_q(y), \quad x = 0, 1, 2, \ldots, M-1, y = 0, 1, 2, \ldots, N-1.$$

$$\tag{6.145}$$

Figure 6.19 Image reconstruction from moments in the Fourier domain. The original image 32×32 and the reconstructed images with maximum moment orders 21, 32, 43, 54, 65, 76 and 87, respectively.

For precise reconstruction, the orthogonality of the polynomial basis should be a discrete one, i.e.

$$\sum_{x=0}^{N-1} p_p(x)p_q(x) = \delta_{pq}. \tag{6.146}$$

A similar formula can be written for the moments orthogonal on a disk. For instance, the reconstruction from the Zernike moments is accomplished by

$$f(x, y) = \sum_{n=0}^{\infty} \sum_{\ell=-n,-n+2,\dots}^{n} A_{n\ell} V_{n\ell}(x, y), \tag{6.147}$$

where $V_{n\ell}(x, y)$ was obtained from $V_{n\ell}(r, \varphi)$ by the same mapping (6.109) as in the computation of the ZMs. In reality, the first sum is truncated to a finite range.

The above idea can also be used when reconstructing the image from geometric moments. If we express the orthogonal polynomial in powers

$$p_p(x) = \sum_{s=0}^{p} a_{ps}x^s, \tag{6.148}$$

then for the corresponding OG moment holds

$$v_{pq} = \sum_{s=0}^{p} \sum_{t=0}^{q} a_{ps}a_{qt}m_{st}. \tag{6.149}$$

Now we can use (6.145) directly for the reconstruction. In this way we avoid any numerically unstable operations that appeared in the direct solution of (6.140). Nevertheless, as the geometric moments already lost their precision, then the OG moments are also imprecise and the reconstruction is not perfect.

In spite of this progress, the image reconstruction is still a numerically delicate process, we must pay attention to the numerical stability in each step of the algorithm. Therefore, it is better to use the moments orthogonal exactly on a discrete set of points rather than the discrete versions of moments with "continuous" orthogonality. The moments orthogonal on a rectangle are more suitable in a rectangular raster and the moments orthogonal on a disk are more suitable in a polar raster (Figure 6.20). The problems with the polar raster and the errors of the reconstruction are analyzed in detail in reference [3].

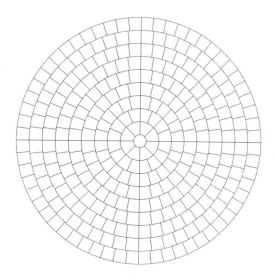

Figure 6.20 An example of the polar raster.

Let us explain on an example of Chebyshev moments how to organize the calculations. Suppose $N = M$ for simplicity and suppose the discrete Chebyshev polynomials (6.56) to be already orthonormalized.

We could employ the formulas (6.62) and (6.63) separately for each x for the computation of the values of the polynomials, but repeated multiplication by large x $(x = N - 1$ in maximum) would cause large magnitude variations of the $\hat{T}_n(x)$. This can violate the numerical stability for N somewhere about 100 and the reconstruction fails.

We can avoid this problem by using formulas recurrent to x. The initial values are

$$\hat{T}_0(x) = \frac{1}{\sqrt{N}}, \quad x = 0, 1, \ldots, N - 1, \tag{6.150}$$

then

$$\hat{T}_n(0) = -\sqrt{\frac{N - n}{N + n}}\sqrt{\frac{2n + 1}{2n - 1}}\hat{T}_{n-1}(0), \quad n = 1, 2, \ldots, N - 1, \tag{6.151}$$

and from that

$$\hat{T}_n(1) = \left(1 + \frac{n(n + 1)}{1 - N}\right)\hat{T}_n(0), \quad n = 1, 2, \ldots, N - 1. \tag{6.152}$$

Other values are computed by the recurrence

$$\hat{T}_n(x) = \gamma_1 \hat{T}_n(x-1) + \gamma_2 \hat{T}_n(x-2), \quad n = 1, 2, \ldots, N-1; \quad x = 2, 3, \ldots, N/2, \tag{6.153}$$

where

$$\gamma_1 = \frac{-n(n+1) - (2x-1)(x-N-1) - x}{x(N-x)}, \quad \gamma_2 = \frac{(x-1)(x-N-1)}{x(N-x)}. \tag{6.154}$$

The values for higher x can be computed using the symmetry condition

$$\hat{T}_n(N-1-x) = (-1)^n \hat{T}_n(x). \tag{6.155}$$

Small numerical errors can be corrected by renormalization

$$\hat{T}_n(x) \leftarrow \frac{\hat{T}_n(x)}{\sqrt{\sum_{i=0}^{N-1} (\hat{T}_n(i))^2}}, \quad x = 0, 1, \ldots, N-1. \tag{6.156}$$

This algorithm can be used for precise reconstruction of images up to about 1024×1024 pixels. If we continue to increase the size of the image, then we face an underflow.[9] When N exceeds 1078, the values of the polynomials are less than the "machine epsilon" which violates the orthogonality and, consequently, leads to an imperfect reconstruction.

6.5.4 Reconstruction from noisy data

In practice, the moment values are often calculated from a noisy version of the original image. In such cases, even if we had a "perfect" reconstruction algorithm, we never obtain the noise-free original. If we minimize the mean-square error between the reconstructed and noise-free original images, we find that an optimal order of moments exists. Moments of higher orders contribute more to reconstruction of the noise rather than of the image. The optimal order decreases as the noise increases and may be surprisingly low for heavy noise. Since the problem is ill-posed, the reconstruction error from noisy data may be very high, even if we use this optimal moment order. If we replace the sequence of original moments $\{m_{pq}\}$ by $\{m_{pq} + \epsilon_{pq}\}$, where $\{\epsilon_{pq}\}$ are arbitrarily small numbers, the reconstruction result $f_\epsilon(x, y)$ may be arbitrarily far from $f(x, y)$. For detailed analysis of this phenomenon, we refer to reference [3].

6.5.5 Numerical experiments with image reconstruction from OG moments

In the first experiment, the image 32×32 of the capital E (Figure 6.19 top left) was reconstructed from three kinds of OG. The goal of this experiment was to show an impact of using moments orthogonal in a continuous domain. First, we computed the OG moments of the original image up to 31st order[10] (in the case of the Zernike moments up to 44th order), so the precise reconstruction is theoretically possible. Then we reconstructed the image.

---

[9] Since computers have certain minimum positive floating-point numbers that can be stored, all lower values are treated as zeros.

[10] Solely in this section, we take the moment order as a maximum of $\{p, q\}$.

We subsequently used the Legendre moments normalized to be orthonormal (Figure 6.21 left), the continuous Chebyshev moments of the first kind without any normalization (Figure 6.21 middle), and the Zernike moments (Figure 6.21 right). Finally, we used the discrete Chebyshev moments and the reconstruction was correct, without any error. The actual range of values of the results was $\langle 44, 700 \rangle$ for the Legendre moments, $\langle 13, 374 \rangle$ for the Chebyshev moments and $\langle 37, 138 \rangle$ for the Zernike moments. The errors are exclusively caused by using continuous OG moments on the discrete image, in the case of the Zernike moments, this error is larger because the sampling error on the unit disk is larger than that on the square. We can also see "roundness" of the result in the case of the ZMs in comparison with the other two results. It is interesting to note that the normalization of the Legendre moments brings almost no improvement over the nonnormalized Chebyshev moments.

Figure 6.21 Image reconstruction from the OG moments. From left to right: Legendre moments, Chebyshev moments of the first kind, and ZMs.

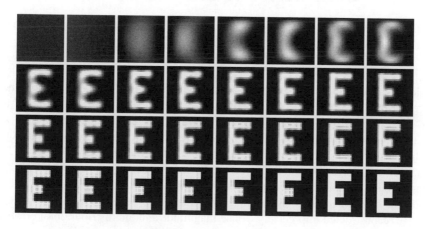

Figure 6.22 The image reconstruction from the incomplete set of OG moments. The discrete Chebyshev moments with the order limit $0, 1, \ldots, 31$ were used. The last image (bottom right) is precise reconstruction of the original image.

The aim of the next experiment is to examine the content of the image information in the moments of low orders. We took the image of the letter E from the previous experiment, computed the discrete Chebyshev moments up to the order 31, and then successively reconstructed the image. The result is shown in Figure 6.22. We can reconstruct only a constant image from the zero-order moment, linear function from the first-order moments, etc. When the number of moments reaches the number of pixels (the last image bottom right), the precise reconstruction is possible.

This experiment illustrates the fact that if we use moments up to e.g. the third order for recognition, we cannot distinguish among the fourth, fifth and further images in Figure 6.22 (even if they are easy to discriminate visually) since their third-order moments are the same. On the other hand, using 31st-order moments for recognition would not be a good choice because of its low robustness. To find an optimal order of moments (or invariants) for object recognition, we have to perform a discriminative analysis of the given training set.

Figure 6.23   The test image for the reconstruction experiment from discrete Chebyshev moments (Astronomical Clock, Prague, Czech Republic).

The last experiment was performed on a graylevel image. Its aim was to explore the limits of the reconstruction abilities of OG moments. We used a standard photograph of the size 2048×1536 pixels (Figure 6.23) converted to grayscale. We cropped it to 1024×1024, then to 1025×1025, 1026×1026, etc. For each crop we calculated the normalized discrete Chebyshev moments with all available algorithmic improvements, (6.150) to (6.156), and reconstructed the image. Up to 1076×1076, the reconstruction was precise. From 1077×1077, the errors started appearing due to floating point underflow. For 1077×1077 the errors are negligible but they increase as the image size increases. For 1079×1079, about 80% of pixels were reconstructed imprecisely.[11] The magnitude and the distribution of errors can be seen in Figures 6.24 and 6.25.

6.6   Three-dimensional OG moments

The generalization of the moments orthogonal on a rectangle into 3D is relatively straightforward. Having a polynomial basis $\{G_k(x)\}$ in 1D that is orthogonal on Ω_1, we can define 3D

---

[11]These values may be slightly different for other images but in general, this behavior is determined by the fact that both in C and MATLAB, the floating-point variables use a 48-bit mantissa that underflows as the size of the image reaches a certain limit.

(a)   (b)

Figure 6.24 The crop 1079×1079 of the test image; (a) the original image, and (b) the error map. The range of errors from -6 to 6 is visualized in the range black–white.

Figure 6.25 The detail of the central part of the error map with typical oscillations.

OG moments as

$$v_{pqr} = n_p n_q n_r \iiint\limits_{\Omega} G_p(x) G_q(y) G_r(z) f(x, y, z) \, \mathrm{d}x \, \mathrm{d}y \, \mathrm{d}z \quad p, q, r = 0, 1, 2, \ldots ,$$

(6.157)

where the area of the orthogonality $\Omega = \Omega_1 \times \Omega_1 \times \Omega_1$. The image $f(x, y, z)$ should again be scaled in such a way that it is fully contained in Ω. This can be performed for any OG polynomials mentioned in the previous sections. In particular, the equations for 3D Legendre moments can be found in reference [27].

The generalization of the moments orthogonal on a disk is more difficult. In 3D the area of orthogonality can be either a cylinder or a sphere. On a cylinder, we may define

$$v_{pqr} = n_{pqr} \int_0^{2\pi} \int_0^1 \int_{-1}^1 G_r(z) R_{pq}(\varrho) e^{-iq\varphi} f(\varrho, \varphi, z) \varrho \, dz \, d\varrho \, d\varphi$$

$$p, r = 0, 1, 2, \ldots, q = -p, \ldots, p, \tag{6.158}$$

where $R_{pq}(\varrho) e^{-iq\varphi}$ is the kernel function of some moments orthogonal on a disk and $G_r(z)$ are 1D orthogonal polynomials.

Probably more common is a generalization to a sphere [28, 29]. When using spherical harmonics (3.95), the *3D Zernike polynomials* have the form

$$V_{n\ell}^m(\varrho, \theta, \varphi) = R_{n\ell}(\varrho) Y_\ell^m(\theta, \varphi), \tag{6.159}$$

where Y_ℓ^m is the spherical harmonics of degree ℓ and order m, and ϱ, θ, φ are the spherical coordinates

$$x = \varrho \sin \theta \cos \varphi$$

$$y = \varrho \sin \theta \sin \varphi \tag{6.160}$$

$$z = \varrho \cos \theta.$$

The *3D Zernike moments* are then defined as

$$A_{n\ell}^m = \int_0^{2\pi} \int_0^\pi \int_0^1 (V_{n\ell}^m(\varrho, \theta, \varphi))^* f(\varrho, \theta, \varphi) \varrho \, d\varrho \, d\theta \, d\varphi \tag{6.161}$$

$$n = 0, 1, 2, \ldots, \ell = -n, -n+2, \ldots, n, m = -\ell, \ldots, \ell.$$

The relation of orthogonality in the interior of the unit sphere is

$$\frac{3}{4\pi} \int_0^{2\pi} \int_0^\pi \int_0^1 V_{n\ell}^m(\varrho, \theta, \varphi)(V_{n'\ell'}^{m'}(\varrho, \theta, \varphi))^* \varrho^2 \sin \theta \, d\varrho \, d\theta \, d\varphi = \delta_{nn'} \delta_{\ell\ell'} \delta_{mm'}. \tag{6.162}$$

Similar to in 2D, the 3D Zernike moments can be employed to construct 3D rotation invariants.

In the same way we can generalize other moments orthogonal on a disk to a sphere:

$$v_{n\ell}^m = n_{n\ell m} \int_0^{2\pi} \int_0^\pi \int_0^1 R_{n\ell}(\varrho)(Y_\ell^m(\theta, \varphi))^* f(\varrho, \theta, \varphi) \varrho \, d\varrho \, d\theta \, d\varphi \tag{6.163}$$

$$n = 0, 1, 2, \ldots, \ell = -n, \ldots, n, m = -\ell, \ldots, \ell$$

where $R_{n\ell}$ is a radial function of moments orthogonal on a disk. The area of orthogonality is the unit sphere $\Omega = \{(x, y, z) \mid x^2 + y^2 + z^2 \leq 1\}$, inside which the image $f(\varrho, \theta, \varphi)$ is supposed to be mapped.

6.7 Conclusion

In this chapter we presented a quick overview of OG moments with respect to their applications in image analysis. OG moments are a useful tool, overcoming certain drawbacks of the geometric or complex moments. The primary motivation for using them is their stable numerical implementation and improved reconstruction abilities.

From the theoretical point of view, all moment systems up to the given order are equivalent. This equivalence implies that one cannot expect any principal differences between geometric and OG moments when used for recognition. However, certain distinctions may appear in the experiments owing to better numerical properties of the OG moments or if we use incomplete sets of moments of the given order. In that case the sets of OG moments and geometric moments are not equivalent and even if the number of elements is the same, the OG moments capture more information about the image than the geometric moments, as they are highly correlated due to the standard powers being nearly dependent.

On the other hand, derivation of invariants from OG moments is typically more complicated than from the geometric or complex moments (for instance, constructing rotation invariants from the moments orthogonal on a rectangle is possible but difficult) and the image must always be mapped into the area of orthogonality. To work with OG polynomials outside their area of orthogonality where their numerical behavior is worse than that of the geometric moments would be a serious mistake.

Practical applications of OG moments in digital watermarking and optical flow estimation can be found in Chapter 8.

References

[1] Abramowitz, M. and Stegun, I. A. (1964) *Handbook of Mathematical Functions With Formulas, Graphs and Mathematical Tables*. Washington, DC: National Bureau of Standards.

[2] Teh, C.-H. and Chin, R. T. (1988) "On image analysis by the method of moments," *IEEE Transactions on Pattern Analysis and Machine Intelligence*, vol. 10, no. 4, pp. 496–513.

[3] Pawlak, M. (2006) *Image Analysis by Moments: Reconstruction and Computational Aspects*. Wrocław: Oficyna Wydawnicza Politechniki Wrocławskiej.

[4] Mukundan, R. and Ramakrishnan, K. R. (1998) *Moment Functions in Image Analysis*. Singapore: World Scientific.

[5] Mukundan, R., Ong, S. H. and Lee, P. A. (2001) "Image analysis by Tchebichef moments," *IEEE Transactions on Image Processing*, vol. 10, no. 9, pp. 1357–64.

[6] Mukundan, R. (2003) "Improving image reconstruction accuracy using discrete orthonormal moments," in *Proceedings of the International Conference on Imaging Systems, Science and Technology CISST'03* (Las Vegas, Nevada), pp. 287–93, CSREA Press.

[7] Mukundan, R. (2004) "Some computational aspects of discrete orthonormal moments," *IEEE Transactions on Image Processing*, vol. 13, no. 8, pp. 1055–9.

[8] Liao, S. X., Chiang, A., Lu, Q. and Pawlak, M. (2002) "Chinese character recognition via Gegenbauer moments," in *Proceedings of the 16th International Conference on Pattern Recognition ICPR'02* (Québec City, Canada), vol. 3, pp. 485–8, IEEE Computer Society.

[9] Ping, Z., Ren, H., Zou, J., Sheng, Y. and Bo, W. (2007) "Generic orthogonal moments: Jacobi-Fourier moments for invariant image description," *Pattern Recognition*, vol. 40, no. 4, pp. 1245–54.

[10] Yap, P.-T. and Paramesran, R. (2004) "Jacobi moments as image features," in *Proceedings of the Region 10 Conference on TENCON'04*, vol. 1, pp. 594–7, IEEE.

[11] Qjidaa, H. (2006) "Image reconstruction by Laguerre moments," in *Proceedings of the Second International Symposium on Communications, Control and Signal Processing ISCCSP'06* (Marrakech, Morocco), IEEE.

[12] Wu, Y. and Shen, J. (2005) "Properties of orthogonal Gaussian–Hermite moments and their applications," *EURASIP Journal on Applied Signal Processing*, no. 4, pp. 588–99.

[13] Koekoek, R. and Swarttouw, R. F. (1996) "The askey-scheme of hypergeometric orthogonal polynomials and its q-analogue," Report 98–17, Technische Universiteit Delft, Faculty of Technical Mathematics and Informatics.

[14] Yap, P.-T., Paramesran, R. and Ong, S.-H. (2003) "Image analysis by Krawtchouk moments," *IEEE Transactions Image Processing*, vol. 12, no. 11, pp. 1367–77.

[15] Zhu, H., Shu, H., Zhou, J., Luo, L. and Coatrieux, J.-L. (2007) "Image analysis by discrete orthogonal dual Hahn moments," *Pattern Recognition Letters*, vol. 28, no. 13, pp. 1688–1704.

[16] Zhu, H., Shu, H., Liang, J., Luo, L. and Coatrieux, J.-L. (2007) "Image analysis by discrete orthogonal Racah moments," *Signal Processing*, vol. 87, no. 4, pp. 687–708.

[17] Teague, M. R. (1980) "Image analysis via the general theory of moments," *Journal of the Optical Society of America*, vol. 70, no. 8, pp. 920–30.

[18] Wallin, A. and Kübler, O. (1995) "Complete sets of complex Zernike moment invariants and the role of the pseudoinvariants," *IEEE Transactions Pattern Analysis and Machine Intelligence*, vol. 17, no. 11, pp. 1106–10.

[19] Khotanzad, A. and Hong, Y. H. (1990) "Invariant image recognition by Zernike moments," *IEEE Transactions Pattern Analysis and Machine Intelligence*, vol. 12, no. 5, pp. 489–97.

[20] Hwang, S.-K. and Kim, W.-Y. (2006) "A novel approach to the fast computation of Zernike moments," *Pattern Recognition*, vol. 39, no. 11, pp. 2065–76.

[21] Sheng, Y. and Shen, L. (1994) "Orthogonal Fourier-Mellin moments for invariant pattern recognition," *Journal of the Optical Society of America A*, vol. 11, no. 6, pp. 1748–57.

[22] Revaud, J., Lavoué, G. and Baskurt, A. (2009) "Improving Zernike moments comparison for optimal similarity and rotation angle retrieval," *IEEE Transactions on Pattern Analysis and Machine Intelligence*, vol. 31, no. 4, pp. 627–36.

[23] Haddadnia, J., Ahmadi, M. and Faez, K. (2003) "An efficient feature extraction method with pseudo-Zernike moment in RBF neural network-based human face recognition system," *EURASIP Journal on Applied Signal Processing*, vol. 9, pp. 890–901.

[24] Ping, Z., Wu, R. and Sheng, Y. (2002) "Image description with Chebyshev-Fourier moments," *Journal of the Optical Society of America A*, vol. 19, no. 9, pp. 1748–54.

[25] Ren, H., Ping, Z., Bo, W., Wu, W. and Sheng, Y. (2003) "Multidistortion-invariant image recognition with radial harmonic Fourier moments," *Journal of the Optical Society of America A*, vol. 20, no. 4, pp. 631–7.

[26] Ghorbel, F., Derrode, S., Mezhoud, R., Bannour, T. and Dhahbi, S. (2006) "Image reconstruction from a complete set of similarity invariants extracted from complex moments," *Pattern Recognition Letters*, vol. 27, no. 12, pp. 1361–9.

[27] Zhang, H., Shu, H., Luo, L. and Dillenseger, J. L. (2005) "A Legendre orthogonal moment based 3D edge operator," *Science in China Series G: Physics, Mechanics and Astronomy*, vol. 48, no. 1, pp. 1–13.

[28] Canterakis, N. (1999) "3D Zernike moments and Zernike affine invariants for 3D image analysis and recognition," in *Proceedings of the 11th Scandinavian Conference on Image Analysis SCIA'99* (Greenland) (B. K. Ersbøll and P. Johansen, eds), DSAGM.

[29] Novotni, M. and Klein, R. (2003) "3D Zernike descriptors for content based shape retrieval," in *Proceedings of the Eighth Symposium on Solid Modeling and Applications SM'03* (Seattle, Washington), pp. 216–25, ACM.

7

Algorithms for moment computation

7.1 Introduction

This chapter is devoted to the computational aspects of moments of digital images. Until now, we have been dealing mostly with moments and moment invariants in a continuous domain. In digital image processing, all quantities have to be converted from the continuous to the discrete domain and efficient algorithms for dealing with discrete quantities must be developed. Sometimes, the discrete algorithms follow their continuous ancestors in a straightforward manner; sometimes, to develop a computationally efficient algorithm requires a new invention. The same holds true for the moments.

We present primarily the algorithms for calculating geometric moments. Central moments, normalized moments and complex moments can be evaluated by their straightforward modifications. Theoretically, orthogonal moments can also be calculated via geometric moments but as we have already pointed out in Chapter 6, a numerically stable calculation requires special algorithms based on recurrent formulas. We do not speak explicitly about the calculation of invariants because the computing complexity of all invariants is determined by the complexity of moment computation. Having the moments, we can calculate any invariant in $\mathcal{O}(1)$ time.

First, we show how to define moments of digital images. Then we review two basic groups of algorithms for the computation of geometric moments of binary images – the decomposition methods and the boundary-based methods. We also present a few methods for speeding up the computation of the geometric moments of graylevel images. Finally, we discuss efficient algorithms for the computation of discrete orthogonal moments.

7.2 Moments in a discrete domain

Although the notion of a digital or discrete image is intuitive, we start with its formal definition.

Moments and Moment Invariants in Pattern Recognition   Jan Flusser, Tomáš Suk and Barbara Zitová
© 2009 John Wiley & Sons, Ltd

Definition 7.1 By a *digital image* we understand any $N \times M$ matrix $\mathbf{f}$, the elements of which are from the finite set of integers $\{0, 1, 2, \ldots, L-1\}$, where $L > 1$.

In practice, usually $L = 256$ and each pixel value is stored in one byte. If $L = 2$, the image is called *binary*.

The conversion from a "continuous" image $f(x, y)$ into a digital image $\mathbf{f}$ is accomplished by *sampling* (spatial discretization) followed by *quantization* (graylevel discretization). The model of these processes in real imaging systems is very complicated and comprises many device-dependent parameters. However, an approximative model of digital image formation is well known from standard textbooks [1,2].

Now we can start thinking of the moments of digital images. A philosophic question arises immediately: the moments of which image shall we calculate?

If we adopt a common sampling model as a multiplication of $f(x, y)$ with an infinite "comb" of Dirac δ-functions [1], then we can apply the "continuous" definition of geometric moments

$$m_{pq} = \int\limits_{-\infty}^{\infty} \int\limits_{-\infty}^{\infty} x^p y^q f(x, y) \, \mathrm{d}x \, \mathrm{d}y \qquad (7.1)$$

to the sampled image. We obtain, without any approximations, the formula for a moment of a discrete image

$$m_{pq} = \sum_{i=1}^{N} \sum_{j=1}^{M} i^p j^q f_{ij}, \qquad (7.2)$$

where i, j are coordinates of the pixel centers and f_{ij} is the graylevel of the pixel (i, j) (see Figure 7.1 (a)).

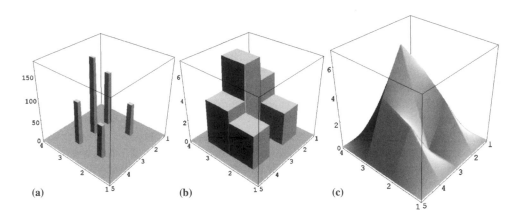

(a)   (b)   (c)

Figure 7.1 Digital image (a) as a sum of Dirac δ-functions, (b) nearest neighbor interpolation, and (c) bilinear interpolation.

We can, however, also think differently. We may want to calculate (or at least estimate) the moments of the *original* image from its digital version. To do that, we interpolate the discrete values by a proper interpolation function to obtain an image defined in a continuous domain.

Nearest-neighbor interpolation leads to a piecewise constant image (see Figure 7.1 (b)); bilinear interpolation produces a more complicated surface (Figure 7.1 (c)). (Let us recall that an optimal but impractical interpolant is a *sinc* function which leads, under Nyquist sampling, to a precise reconstruction of $f(x, y)$.) Let us denote the geometric moments of the interpolated image as $\bar{m}_{pq}$. For the nearest-neighbor interpolation and for an approximation of the monomials $x^p y^q$ by constants we obtain essentially the same formula as (7.2)

$$\bar{m}_{pq} = \sum_{i=1}^{N} \sum_{j=1}^{M} i^p j^q f_{ij} \tag{7.3}$$

but its interpretation is now different – it is a double zero-order approximation of moments of the original image $f(x, y)$.

We can, however, obtain a more precise estimation originally proposed by Lin and Wang [3] and later generalized by Flusser [4]. Consider again the nearest-neighbor interpolation and integrate the monomials $x^p y^q$ exactly by the Newton–Leibnitz formula on each pixel. In this way we obtain

$$\hat{m}_{pq} = \sum_{i=1}^{N} \sum_{j=1}^{M} f_{ij} \iint_{S_{ij}} x^p y^q \, dx \, dy$$

$$= \frac{1}{(p+1)(q+1)} \sum_{i=1}^{N} \sum_{j=1}^{M} f_{ij} \cdot ((i+0.5)^{p+1} - (i-0.5)^{p+1})$$

$$\times ((j+0.5)^{q+1} - (j-0.5)^{q+1}), \tag{7.4}$$

where S_{ij} denotes the area of the pixel (i, j). This formula is still a zero-order approximation of moments of the original image $f(x, y)$ but only the image is approximated, the monomials are integrated exactly.

We can continue this process by taking higher-order image approximations which should yield smaller approximation errors but this is meaningless for practical purposes. It should be noted that most of the papers on moment computation skip the above analysis and use only (7.3) without any reasoning.

Since the direct evaluation of both (7.3) and (7.4) is time-consuming (it requires $\mathcal{O}(MN)$ operations), much effort has been spent in recent years to develop more efficient algorithms [5–15] both for binary as well as for graylevel images. In the sequel we review the main approaches.

7.3   Geometric moments of binary images

Particular attention has been paid to binary images because of their importance in practical pattern recognition applications. Since any binary object is fully determined by its boundary, which is supposed to consist of far fewer than $\mathcal{O}(MN)$ pixels (this assumption may not necessarily be true; see a chessboard), there is a significant potential for improvement.

The methods for fast computation of the moments of the binary images can be divided into two groups referred to as *decomposition methods* and *boundary-based* methods.

7.3.1 Decomposition methods for binary images

The idea behind all decomposition methods is quite simple. Having a binary object Ω (by a binary object, we understand a set of all pixels of a binary image whose values equal one), we decompose it into $K \geq 1$ blocks $B_1, B_2, \ldots, B_K$ such that $B_i \cap B_j = \emptyset$ for any $i \neq j$ and $\Omega = \bigcup_{k=1}^{K} B_k$. Then

$$m_{pq}^{\Omega} = \sum_{k=1}^{K} m_{pq}^{B_k}.$$

If we can calculate the moment of each block in $\mathcal{O}(1)$ time (as we can for rectangular blocks, for instance) then the overall complexity of m_{pq}^{Ω} is $\mathcal{O}(K)$. If $K \ll MN$ the speed-up may be significant.

The power of any decomposition method depends on our ability to decompose the object into a small number of blocks in a reasonable time. Individual decomposition methods differ from one another namely by the decomposition algorithms. Simple algorithms produce a relatively high number of components but perform fast, while more sophisticated decomposition methods end up with a small number of blocks but require more time. Some authors ignore the complexity of the decomposition itself and do not include it in the overall complexity estimation, claiming that "the fewer blocks the better method." This is a serious methodical mistake. Even if the decomposition is performed only once and can be used for the calculation of all moments, the time needed for decomposing the image is often so long that it substantially influences the efficiency of the whole method.

It should also be noted that some images cannot be efficiently decomposed by any algorithm and always $K \sim MN$. A chessboard is an extreme example.

The "delta" method and its modifications

The first attempt to speed up the moment calculation by image decomposition came from Zakaria *et al.* [5]. The basic idea of his "delta" method is to decompose the object into individual rows of pixels (as in the run-length encoding compression). The original Zakaria method worked for convex shapes only and dealt with moments up to the third order only that approximated (7.3). The lengths of the row segments were labeled by the Greek letter δ, from that the name of the method came. Dai *et al.* [9] extended the Zakaria method also to approximate (7.4) and Li [10] generalized it for nonconvex shapes. His improved decomposition scheme into rows and row segments is visualized in Figure 7.2. The decomposition algorithm is very fast but the number of blocks is higher than necessary.

To calculate the moments of a row segment, we can use (7.3) or (7.4). When using (7.3), we obtain for a row segment B_k of the length δ starting at the pixel (x_0, y_0)

$$\tilde{m}_{pq}^{B_k} = y_0^q \sum_{i=x_0}^{x_0+\delta-1} i^p.$$

To evaluate the sum, we have to employ formulas for the sums of the series of powers

$$\sum_{i=1}^{n} i = \frac{n(n+1)}{2}, \qquad \sum_{i=1}^{n} i^2 = \frac{n(n+1)(2n+1)}{6},$$

$$\sum_{i=1}^{n} i^3 = \frac{n^2(n+1)^2}{4}, \qquad \sum_{i=1}^{n} i^4 = \frac{n(n+1)(2n+1)(3n^2+3n+1)}{30}.$$

$$(7.5)$$

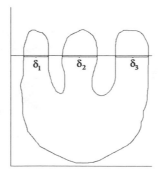

Figure 7.2 Decomposition of the object into row segments.

The formulas for higher powers are given recursively as

$$\sum_{t=1}^{m} \binom{m+1}{t} S_n^t = (n+1)((n+1)^m - 1),$$

(7.6)

where

$$S_n^t = \sum_{i=1}^{n} i^t.$$

(7.7)

One can see that the computational complexity of a row moment is not $\mathcal{O}(1)$ but rather, $\mathcal{O}(p)$. A substantial improvement can be reached when using (7.4) and when integrating the monomials not over separate pixels but on the whole row segment. In that way we obtain the formula

$$\hat{m}_{pq}^{Bk} = \int_{x_0-0.5}^{x_0+\delta-0.5} \int_{y_0-0.5}^{y_1+0.5} x^p y^q \, dx \, dy$$

$$= \frac{1}{(p+1)(q+1)}((x_0 + \delta - 0.5)^{p+1} - (x_0 - 0.5)^{p+1})$$

$$\times ((y_0 + 0.5)^{q+1} - (y_0 - 0.5)^{q+1}),$$

(7.8)

the complexity of which is $\mathcal{O}(1)$ independently of p and q.

Later on, Spiliotis and Mertzios [14] published an advanced modification of the delta method. Their algorithm employs a rectangularwise object representation instead of the row-wise one. The adjacent rows are compared and if there are some segments with the same beginning and end, they are unified into one rectangle (see Figure 7.3). The block moments can be calculated similarly as in the row-wise delta method, either by summation as in [14] in $\mathcal{O}(pq)$ steps or in $\mathcal{O}(1)$ time by integration as was proposed in reference [4].

Quadtree decomposition

Quadtree decomposition is a popular hierarchical decomposition scheme used in several image processing areas such as compression and representation. Wu et al. [16] proposed using

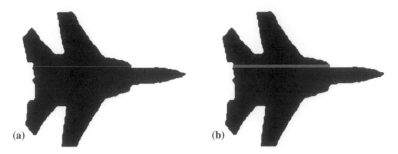

(a)   (b)

Figure 7.3 Decomposition into rectangular blocks of adjacent row segments of the same length. A segment (a) is unified with its neighbors into (b) a 5-pixel rectangle.

the quadtree decomposition as the first stage of their moment computation method. It works with square images, ideally with a size of a power of two. The image is iteratively divided into four quadrants. The homogeneity of each quadrant is checked and if the whole quadrant lies either in the object or in the background, it is not further divided. If it contains both object and background pixels, it is divided into quadrants and each process is repeated until all blocks are homogeneous. For an example of the quadtree decomposition, see Figure 7.4. A drawback of this decomposition algorithm is that the division of the blocks is fixed on each level. The division scheme is not adapted with respect to the content of the image. This may lead to absurd decompositions when, for instance, a large single square is uselessly decomposed into individual pixels. We may use other trees such as a bintree or a hexatree but it does not overcome this weakness.

After decomposition, the moments of each square block can be calculated as in the delta method in $\mathcal{O}(1)$ time by means of integration.

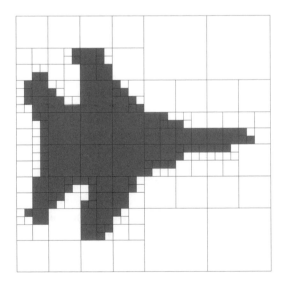

Figure 7.4 Quadtree decomposition of the image.

Morphological decomposition

In order to adapt the decomposition to the image content, Sossa-Azuela *et al.* [17] published an algorithm based on a *morphological erosion*. The algorithm works in an iterative manner: it finds the largest square inscribed in the object, removes it and looks for the largest square inscribed in the rest of the object. This "outer" loop is repeated until the object completely disappears.

In each outer loop, the center of the largest inscribed square is found by morphological erosion with a 3×3 square structure element. Morphological erosion is performed in "inner" loops. We move the structure element over the image and if the whole square lies inside the object, the central pixel of the structure element is left in the object, otherwise it is assigned to the background. After each inner loop, the object is shrunk by pixels along its boundary. This is repeated until the object disappears. The pixels disappearing in the last inner loop are the possible centers of the largest inscribed square (the task may not have a unique solution; in that case, we choose one of them randomly). The size of the square is determined by the number of inner loop runs.

The algorithm is very efficient because it yields a small number of blocks. It might be further improved by allowing rectangular blocks. On the other hand, compared to the previous decomposition algorithms, it performs much slower. After decomposition, the moments of each square block are calculated as in the previous methods.

An intermediate object decomposition after two outer loops can be seen in Figure 7.5.

Figure 7.5 Partial object decomposition after two outer loops of the morphological method.

7.3.2   Boundary-based methods for binary images

Boundary-based methods employ the property that the boundary of a binary object contains complete information on the object; all other pixels are redundant. Provided that the boundary consists of far fewer pixels than the whole object (which may be true for "normal" shapes but does not hold, in general), an algorithm that calculates the moments from the boundary pixels only should be more efficient than direct calculation by definition.

The methods based on Green's theorem

A majority of the boundary-based methods use *Green's theorem*, which evaluates the double integral over the object by means of a line integration around the object boundary. Green's theorem exists in several versions, an appropriate one (in the continuous domain) is

$$\iint_{\Omega} \frac{\partial}{\partial x} g(x, y) \, dx \, dy = \oint_{\partial \Omega} g(x, y) \, dy,$$

where $g(x, y)$ is an arbitrary function having continuous partial derivatives and $\partial \Omega$ is a piecewise smooth, simple closed-curve boundary of object Ω. By setting

$$g(x, y) = \frac{x^{p+1} y^q}{p+1},$$

we obtain a relation for the geometric moments

$$m_{pq}^{\Omega} = \frac{1}{p+1} \oint_{\partial \Omega} x^{p+1} y^q \, dy.$$

In the discrete domain, the individual methods differ from each other by the procedure used to discretize the boundary and calculate the line integral. Probably the first method of this kind was published by Li and Shen [7]. However, their results depend on the choice of discrete approximation of the boundary and differ from the theoretical values. Jiang and Bunke [6] approximated the object first by a polygon and then they applied Green's theorem. Therefore, they calculated only the single integral along line segments. Unfortunately, due to the two-stage approximation, their method produces inaccurate results.

Philips [12] proposed using discrete Green's theorem [18] instead of the discretized continuous one. The Philips method yields exact results as it does not use any approximation. Philips started from the continuous-domain Green's theorem in a different version from above

$$\oint_{\partial \Omega} h(x, y) \, dx + g(x, y) \, dy = \iint_{\Omega} \left(\frac{\partial g}{\partial x} - \frac{\partial h}{\partial y} \right) dx \, dy, \tag{7.9}$$

where $g(x, y)$ and $h(x, y)$ are arbitrary functions having continuous partial derivatives. By means of a discrete analogy of this theorem, Philips proved that the discrete moment (7.3) can be calculated from the boundary pixels as

$$\bar{m}_{pq}^{\Omega} = \sum_{(x,y) \in \partial \Omega+} y^q \sum_{i=1}^{x} i^p - \sum_{(x,y) \in \partial \Omega-} y^q \sum_{i=1}^{x} i^p, \tag{7.10}$$

where $\partial \Omega-$ and $\partial \Omega+$ are so-called *left-hand-side* and *right-hand-side boundaries*, respectively. They are defined as

$$\partial \Omega- = \{(x, y) | (x, y) \notin \Omega, (x + 1, y) \in \Omega\},$$

$$\partial \Omega+ = \{(x, y) | (x, y) \in \Omega, (x + 1, y) \notin \Omega\}.$$

The Philips method can also be used to calculate the moment approximation (7.4)

$$\hat{m}_{pq}^{\Omega} = \frac{1}{(p+1)(q+1)} \left\{ \sum_{(x,y) \in \partial \Omega+} (x + 0.5)^{p+1} ((y + 0.5)^{q+1} - (y - 0.5)^{q+1}) \right.$$

$$\left. - \sum_{(x,y) \in \partial \Omega-} (x + 0.5)^{p+1} ((y + 0.5)^{q+1} - (y - 0.5)^{q+1}) \right\}. \tag{7.11}$$

It is worth noting that for convex shapes the Philips approach leads to the same formulas as the delta method.

Yang and Albregtsen [13], Sossa-Azuela et al. [15] and Flusser [19] further improved the speed of the Philips method. A significant speed-up was achieved by precalculating certain variables which do not depend on the object and are used repeatedly, such as

$$\sum_{n=1}^{j} n^{i-1}$$

for all possible values of i and j and by storing them in properly organized matrices (see reference [19] for details).

The methods based on boundary approximations

Another group of boundary-based methods exploits approximation of the boundary by a polygon.[1] Any polygon is fully determined by its vertices only. Then its moments can be calculated by means of the corner points only [8, 11], that is particularly efficient for simple shapes with few corners. However, the results obtained by methods of this kind are only approximate. Their accuracy depends on the accuracy of the boundary approximation. Recall that the method by Jiang and Bunke [6], already mentioned in connection with Green's theorem, also approximates the boundary by a polygon and thus it belongs at least partially to this category.

Let us assume that the approximating polygon has n vertices $P_0, P_1, P_2, \ldots, P_{n-1}, P_n$, where $P_0 = P_n$, with coordinates $P_i = (x_i, y_i)$. We can create a set of triangles, each triangle with two vertices equal to two adjacent vertices of the polygon P_i, P_{i+1} and with the third vertex in the origin of the coordinates $O = (0, 0)$. The moment of the polygon is then

$$\hat{m}_{pq} = \sum_{i=0}^{n-1} \hat{m}_{pq}^{(i)}, \tag{7.12}$$

where $\hat{m}_{pq}^{(i)}$ is the moment of the triangle $P_i P_{i+1} O$. It can be calculated as

$$\hat{m}_{pq}^{(i)} = \alpha_i \sum_{k=0}^{p} \sum_{\ell=0}^{q} C_{k\ell}(p, q) x_i^k x_{i+1}^{p-k} y_i^\ell y_{i+1}^{q-\ell}, \tag{7.13}$$

where $\alpha_i = (x_i y_{i+1} - x_{i+1} y_i)/2$ is the area of the triangle $P_i P_{i+1} O$ and the coefficient $C_{k\ell}(p, q)$ is

$$C_{k\ell}(p, q) = \frac{2 \, p!q!(k+\ell)!(p+q-k-\ell)!}{(p-k)!(q-\ell)!k!\ell!(p+q+2)!}. \tag{7.14}$$

The proof of (7.13) can be found in reference [11]. The formula (7.13) can be evaluated efficiently by means of a proper recurrence.

7.3.3 Other methods for binary images

Of the methods that do not fall into the two above-mentioned categories, parallel algorithms play an important role. Since moment computation is easy to parallelize, several authors have

---

[1] Although other piecewise smooth curves can also be theoretically used, only polygons are of practical importance.

proposed parallel implementations on special processors that may be very efficient. Being dependent on the availability of a particular hardware, they are not reviewed here in detail. Two frequently cited representatives of this class are described in reference [20] and [21]. Chen [20] proposed a recursive algorithm for a SIMD processor array; Chung [21] presented a constant-time algorithm on reconfigurable meshes.

7.4   Geometric moments of graylevel images

When calculating moments of a graylevel image, we cannot hope for a significant speed-up over the direct calculation without loss of accuracy. Clearly, any correct algorithm must read each pixel value at least once, which is itself of the same complexity $\mathcal{O}(MN)$ as the direct calculation. Yet certain fast algorithms do still exist. However, they are meaningful for specific images only and/or do not yield exact results. In the following, we mention two basic groups – *intensity slicing* and *graylevel approximation*.

7.4.1   Intensity slicing

The intensity slicing method came from Papakostas *et al.* [22]. The idea behind this method is a decomposition of the image into so-called *slices*. A slice is a set of all pixels having the same intensity value. The original image can be considered to be a sum of slices multiplied by the respective intensity:

$$f(x, y) = \sum_{k=1}^{L-1} k\, f_k(x, y), \tag{7.15}$$

where L is the number of graylevels (usually $L = 256$) and the slice $f_k(x, y)$ contains pixels of the graylevel k only, i.e.

$$f_k(x, y) = \begin{cases} 1 & \text{if } f(x, y) = k \\ 0 & \text{if } f(x, y) \neq k. \end{cases} \tag{7.16}$$

In this way we transformed the problem of calculating graylevel moments to the previous task of calculating binary moments. Each slice $f_k(x, y)$ is nothing but a binary image and we have for the moments of $f(x, y)$

$$m_{pq}^{(f)} = \sum_{k=1}^{L-1} k\, m_{pq}^{(f_k)}.$$

For an example of the decomposition of an image into slices, see Figures 7.6 and 7.7. The moments of individual slices can be evaluated by any method used for binary images. In reference [22], the Spiliotis and Mertzios decomposition was used, but we can use an arbitrary method.

The computing complexity of intensity slicing depends greatly on the specific image. The method is appropriate for images with only few slices, where the slices are not "fragmented" and consist of large compact areas. In such cases we can save a significant amount of time. If these conditions are not met, the moment computation by intensity slicing may be even slower than a direct calculation from the definition.

The slicing method can be modified such that *bit planes* are used instead of the intensity slices. Then

$$m_{pq}^{(f)} = \sum_{k=0}^{\log L - 1} 2^k \, m_{pq}^{(f_k)},$$

where $f_k(x, y)$ is a binary image of the kth bit plane.

Figure 7.6 A picture having ten graylevels.

7.4.2 Approximation methods

Approximation methods are algorithms for computation of moments of graylevel images that reach low complexity at the expense of accuracy. They decompose the image into rectangular blocks such that the graylevels in each block can be approximated by a "simply integrable" function with a user-defined tolerance. There is a clear analogy between the approximation methods and the decomposition methods for binary images. Also, here the computing complexity of a moment is given by the number of blocks since the image moment is a sum of all block moments. The approximating function should be symbolically integrable – we can use a constant, linear/bilinear function, and also higher-degree polynomials. In such cases, the block moments can be obtained in $\mathcal{O}(1)$ time by means of the Newton–Leibnitz formula.

There is a tradeoff between the number of blocks and the degree of the approximating polynomial on one hand and the accuracy of the moment, on the other. For common images, higher-degree approximating polynomials generate fewer blocks than lower-degree ones at the same accuracy level. On the other hand, for a given degree, a request for higher accuracy leads to a higher number of blocks and, consequently, to a slower algorithm. Since both accuracy and the degree of approximating polynomials are user-defined parameters that significantly influence complexity, one has to choose them carefully with respect to the image and to the particular application.

Individual approximation methods differ from each other mainly in the decomposition scheme (in principle, any decomposition method for binary images can be adapted to this purpose) and by the approximating functions.

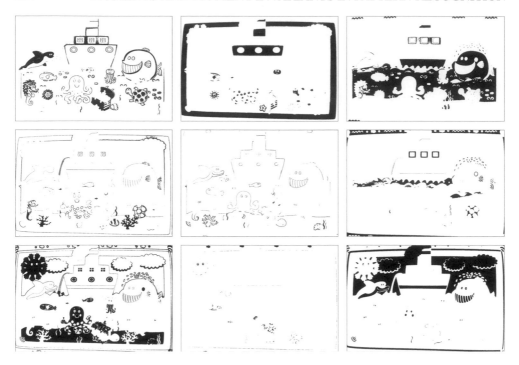

Figure 7.7   The picture decomposed into nine intensity slices (the slice of the zero intensity does not influence the moment computation).

Chung and Chen [23] proposed a method whereby the image is decomposed by a *bintree* to rectangular blocks. The bintree is similar to the quadtree, but the block is divided into two rectangles only in one step and the algorithm alternates between horizontal and vertical directions. The image function $f(x, y)$ in the block is approximated by the bilinear function $b(x, y) = a_0 + a_1 x + a_2 y + a_3 xy$ such that $f(x, y) = b(x, y)$ at the block corners (this is always possible for any block). If $|f(x, y) - b(x, y)| \leq e$ (where e is a user-defined tolerance parameter) for all pixels of the block, this block is not further divided. If $|f(x, y) - b(x, y)| > e$, the block is divided into two subblocks and the procedure repeats until the required approximation accuracy is reached. In reference [23], the block moments are then computed by the formulas for sums of the series of powers (7.5) which is readily inefficient. A better approach is to use integration over the block by means of the Newton–Leibnitz formula.

Chung's method can be improved in several ways. The bintree decomposition is not optimal for this purpose; it should be replaced by a method that would respect "natural" homogeneous blocks in the image. There is no theoretical reason to apply only bilinear approximation. In principle, any polynomial (and even any "reasonable" function) can be applied to obtain larger blocks.

Let $P_n(x, y)$ be a bivariate polynomial of degree n

$$P_n(x, y) = \sum_{\substack{k=0 \\ \ell+k \leq n}}^{n} \sum_{\ell=0}^{n} a_{k\ell} x^k y^\ell$$

and let $C = (n + 1)(n + 2)/2$ be the number of its coefficients. To find the coefficients of the approximating $P_n(x, y)$ in the block, we need C points (x_i, y_i, f_i) in which the interpolation constraints

$$P_n(x_i, y_i) = f(x_i, y_i)$$

are fulfilled. These points can be selected regularly or randomly within the block. Since polynomials are well known to be numerically unstable interpolants, it is better to release the interpolating conditions and to calculate the coefficients of $P_n(x, y)$ by a least-square fit over more than C points. The degree of the polynomials can be set up by the user in advance or it can be automatically selected during the decomposition (note that different blocks may be approximated by polynomials of different degrees). The moment of a block $M \times N$ is then

$$\hat{m}_{pq}^{B} = \int_{0}^{M} \int_{0}^{N} x^p y^q P_n(x, y) \, dx \, dy = \sum_{\substack{k=0 \\ \ell+k \leq n}}^{n} \sum_{\ell=0}^{n} a_{k\ell} \frac{M^{p+k+1} N^{q+\ell+1}}{(p+k+1)(q+\ell+1)}. \tag{7.17}$$

7.5 Efficient methods for calculating OG moments

7.5.1 Methods using recurrent relations

As already explained in Chapter 6, OG moments should not be calculated from geometric moments (although this is theoretically possible) for numerical reasons. If they are calculated directly from the respective definition, the kernel polynomials should be evaluated by recurrent relations rather than by expanding into powers. It was discovered that in some cases we can, by reordering the recurrent calculations, reach a certain speed-up of the algorithm or increase its numerical precision while preserving an acceptable computing complexity.

Let us illustrate this approach by an algorithm for calculating the discrete Chebyshev moments introduced in reference [24], where the original relations recurrent to the degree of the orthogonal polynomials are substituted by relations recurrent to the variable of the polynomials. If we have a system of orthogonal polynomials $T_n(x)$, then the usual form of the recurrence relation is

$$T_n(x) = (\alpha_1 x + \alpha_2) T_{n-1}(x) - \alpha_3 T_{n-2}(x) \tag{7.18}$$

for $n = 2, 3, \ldots, N - 1$, $x = 0, 1, \ldots, N - 1$, where $N \times N$ is the size of the image. The multiplications by big values of x near $N - 1$ lead to numerical instability that can be overcome by the recurrence relation in x

$$T_n(x) = \gamma_1 T_n(x - 1) + \gamma_2 T_n(x - 2). \tag{7.19}$$

For more details about these formulas see Chapter 6, equations (6.62) to (6.63) and equations (6.150) to (6.155). It is worth noting that similar recurrent formulas can also be derived for other orthogonal moments (e.g. dual Hahn [25] or Racah [26]). Nevertheless, the computing complexity stays $\mathcal{O}(N^2 p^2)$, where p is the maximum moment order.

Another group of methods is based on better utilization of the recurrence relations for ZMs [27,28]. The ZMs (together with the PZMs) differ from other moments orthogonal on a disk by dependency of their radial function $R_{n\ell}(r)$ not only on the degree n, but also on the repetition factor ℓ.

If we need to compute all ZMs up to the order p of an image, then the direct method using the definition (6.97) needs $\mathcal{O}(p^3)$ multiplications. The recurrent relation (6.103) and (6.104), referred to in reference [28] as the Prata method, needs $\mathcal{O}(p^2)$ multiplications only for the computation of all ZMs up to the order p. The computational flow of the Prata method is shown in Figure 7.8.

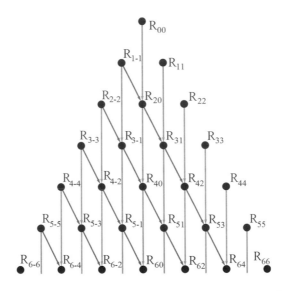

Figure 7.8 The computational flow of the Prata method. $R_{nn}(r) = r^n$, $R_{n,-n}(r) = r^n$ are initial values.

The Kintner method [28] consists of a recurrence formula

$$R_{n\ell}(r) = \frac{(K_2 r^2 + K_3)R_{n-2,\ell}(r) + K_4 R_{n-4,\ell}(r)}{K_1}, \tag{7.20}$$

where

$$
\begin{aligned}
K_1 &= (n+\ell)(n-\ell)(n-2)/2, \\
K_2 &= 2n(n-1)(n-2), \\
K_3 &= -\ell^2(n-1) - n(n-1)(n-2), \\
K_4 &= -n(n+\ell-2)(n-\ell-2)/2.
\end{aligned}
\tag{7.21}
$$

This formula cannot be used for $|\ell| = n$ and $|\ell| = n - 2$. Originally, the direct method (6.97) was proposed for these cases, but the computation can also start with the initial conditions

$$R_{nn}(r) = r^n,$$

$$R_{n+2,n}(r) = (n+2)R_{n+2,n+2}(r) - (n+1)R_{nn}(r),$$

(7.22)

that overcome this problem. The computational flow of the Kintner method is shown in Figure 7.9.

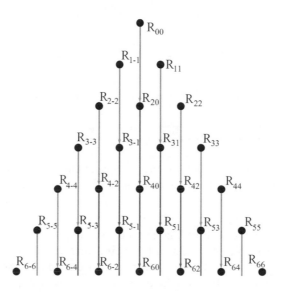

Figure 7.9   The computational flow of the Kintner method. The first two values in each sequence, i.e. $R_{n,-n}$, $R_{n+2,-n}$ and R_{nn}, $R_{n+2,n}$ must be computed directly from equation (7.22).

The Chong repetition–recursive method [28] uses a recurrence formula

$$R_{n,\ell-4}(r) = H_1 R_{n\ell}(r) + \left(H_2 + \frac{H_3}{r^2}\right) R_{n,\ell-2}(r),$$

(7.23)

where

$$H_3 = \frac{-4(\ell-2)(\ell-3)}{(n+\ell-2)(n-\ell+4)},$$

$$H_2 = \frac{H_3(n+\ell)(n-\ell+2)}{4(\ell-1)} + \ell - 2,$$

(7.24)

$$H_1 = \frac{\ell(\ell-1)}{2} - \ell H_2 + \frac{H_3(n+\ell+2)(n-\ell)}{8}.$$

The computation starts with

$$R_{nn}(r) = r^n,$$

$$R_{n,n-2}(r) = n R_{nn}(r) - (n-1)R_{n-2,n-2}(r).$$

(7.25)

The computational flow of the Chong method is shown in Figure 7.10.

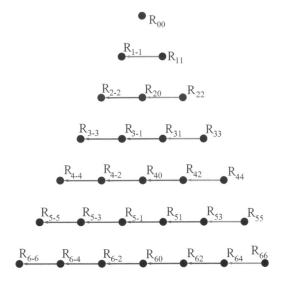

Figure 7.10 The computational flow of the Chong method. The first two values in each sequence, i.e. $R_{n,n-2}$, R_{nn} must be computed directly from equation (7.25).

Both the Kintner method and the Chong method require $\mathcal{O}(p^2)$ multiplications for the computation of all ZMs up to the order p, but their actual complexity is lower than that of the Prata method. The Chong method is appropriate if we need only the ZMs of a particular order, because the recurrence goes over the repetition. If we need all moments up to a certain order, then the Kintner method is more efficient.

7.5.2   Decomposition methods

There exist many methods for calculating OG moments that were inspired by decomposition methods and boundary-based methods for binary geometric moments and also analogous methods dealing with OG moments of graylevel images. For OG moments orthogonal on a rectangle we can, in principal, take any decomposition scheme used previously and simply replace the monomials $x^p y^q$ by respective OG polynomials. For ZMs (and for other moments orthogonal on a disk), this is not straightforward but there exist other possibilities of improvement. Here, we present a few examples of such methods.

Zhou *et al.* [29] proposed the generalization of the delta method for the computation of Legendre moments of binary images. The computation is organized row by row and in one row, it is split to separate line segments; see Figure 7.2. The formula for the Legendre moments (6.7) can be rewritten as

$$\lambda_{pq} = \frac{(2p+1)(2q+1)}{4} \int_{-1}^{1} P_q(y)\, dy \sum_{k=1}^{K} \int_{x_k(y_j)}^{x_k(y_j)+\delta_k(y_j)} P_p(x)\, dx \quad p, q = 0, 1, 2, \ldots,$$

(7.26)

where y_j is the y-coordinate of the jth-row, $x_k(y_j)$ is the x-coordinate of the beginning of the kth line segment in the jth row and $\delta_k(y_j)$ is the length of this segment. The integral of a

Legendre polynomial is again a function of Legendre polynomials

$$\int P_p(x)\,dx = \frac{1}{2p+1}(P_{p+1}(x) - P_{p-1}(x)),\qquad (7.27)$$

so the contribution of a line segment to a Legendre moment can be computed by integration, where we need simply to evaluate $P_{p+1}(x)$ and $P_{p-1}(x)$ at the first and the last pixels.

An example of a method that does not have a counterpart for geometric moments is the following algorithm for computation of ZMs [30]. The formula (6.95) can be rewritten as

$$A_{n\ell} = \frac{n+1}{\pi} \int_0^{2\pi}\!\!\int_0^1 \sum_{k=|\ell|,|\ell|+2,\ldots}^n B_{n\ell k} r^k e^{-i\ell\varphi} f(r,\varphi) r\,dr\,d\varphi$$

$$= \frac{n+1}{\pi} \sum_{k=|\ell|,|\ell|+2,\ldots}^n B_{n\ell k} \int_0^{2\pi}\!\!\int_0^1 r^k e^{-i\ell\varphi} f(r,\varphi) r\,dr\,d\varphi \qquad (7.28)$$

$$= \frac{n+1}{\pi} \sum_{k=|\ell|,|\ell|+2,\ldots}^n B_{n\ell k}\, \chi_{k\ell},$$

where $\chi_{k\ell}$ are Fourier–Mellin moments that do not depend on n. Thus, they can be stored and used repeatedly for various n.

Hwang and Kim [31] proposed another improvement of the computation of the ZMs that is based on their symmetry. Zernike polynomials are symmetrical not only across the x and y axes, but also with respect to the axes $x = y$ and $x = -y$. Thanks to this, it is sufficient to evaluate them in one octant only. If we denote $h_1 = f(x, y)$ and the values in the symmetric points as $h_2 = f(y, x)$, $h_3 = f(-y, x)$, $h_4 = f(-x, y)$, $h_5 = f(-x, -y)$, $h_6 = f(-y, -x)$, $h_7 = f(y, -x)$ and $h_8 = f(x, -y)$, we have for the ZM

$$A_{n\ell} = \frac{n+1}{\pi} \iint_{\substack{x^2+y^2\leq 1\\ 0\leq x\leq 1\\ 0\leq y\leq x}} R_{n\ell}(r)(g_\ell^r(x, y) - ig_\ell^i(x, y))\,dx\,dy, \qquad (7.29)$$

where $g_\ell^r(x, y)$ and $g_\ell^i(x, y)$ depend on the modulus m of ℓ after dividing by four:

$$g_{4k}^r(x, y) = (h_1 + h_2 + h_3 + h_4 + h_5 + h_6 + h_7 + h_8)\cos(\ell\varphi),$$

$$g_{4k}^i(x, y) = (h_1 - h_2 + h_3 - h_4 + h_5 - h_6 + h_7 - h_8)\sin(\ell\varphi),$$

$$g_{4k+1}^r(x, y) = (h_1 - h_4 - h_5 + h_8)\cos(\ell\varphi) + (h_2 - h_3 - h_6 + h_7)\sin(\ell\varphi),$$

$$g_{4k+1}^i(x, y) = (h_1 + h_4 - h_5 - h_8)\sin(\ell\varphi) + (h_2 + h_3 - h_6 - h_7)\cos(\ell\varphi),$$

$$g_{4k+2}^r(x, y) = (h_1 - h_2 - h_3 + h_4 + h_5 - h_6 - h_7 + h_8)\cos(\ell\varphi), \qquad (7.30)$$

$$g_{4k+2}^i(x, y) = (h_1 + h_2 - h_3 - h_4 + h_5 + h_6 - h_7 - h_8)\sin(\ell\varphi),$$

$$g_{4k+3}^r(x, y) = (h_1 - h_4 - h_5 + h_8)\cos(\ell\varphi) + (-h_2 + h_3 + h_6 - h_7)\sin(\ell\varphi),$$

$$g_{4k+3}^i(x, y) = (h_1 + h_4 - h_5 - h_8)\sin(\ell\varphi) + (-h_2 - h_3 + h_6 + h_7)\cos(\ell\varphi),$$

where $\ell = 4k + m$. Experimental results show that this algorithm actually saves almost eight times the computation time.

7.5.3   Boundary-based methods

We can use the boundary-based methods for the orthogonal moments in a similar way as for the geometric moments. An example of such a method was published in reference [32] for the Legendre moments of binary images. We use Green's theorem in the form

$$\iint\limits_{\Omega} \frac{\partial g(x,\, y)}{\partial x}\, \mathrm{d}x\, \mathrm{d}y = \oint\limits_{\partial\Omega} g(x,\, y)\, \mathrm{d}y, \tag{7.31}$$

where $g(x,\, y)$ is an arbitrary function having continuous partial derivatives and $\partial\Omega$ is a piecewise smooth, simple closed-curve boundary of object Ω. If we set

$$\frac{\partial g(x,\, y)}{\partial x} = P_m(x)P_n(y) \tag{7.32}$$

and substitute the recurrence relation (6.11) into the formula for the integral of the Legendre polynomial (7.27), we obtain

$$\int P_m(x)\, \mathrm{d}x = \frac{1}{m+1}(x P_m(x) - P_{m-1}(x)). \tag{7.33}$$

Consequently, we obtain for the Legendre moment λ_{mn}

$$\lambda_{mn} = \frac{(2m+1)(2n+1)}{4(m+1)} \oint\limits_{\partial\Omega} P_n(y)(x P_m(x) - P_{m-1}(x))\, \mathrm{d}y. \tag{7.34}$$

We can either track the boundary sequentially or pass through the image row by row and search for the boundary points. In the second version, the algorithm turns close to the delta method.

7.6   Generalization to n dimensions

In a continuous domain, geometric moments of n-D images are defined as

$$m_{p_1 \cdots p_n} = \int\limits_{-\infty}^{\infty} \int\limits_{-\infty}^{\infty} \cdots \int\limits_{-\infty}^{\infty} x_1^{p_1} x_2^{p_2} \cdots x_n^{p_n} f(x_1, x_2, \ldots, x_n)\, \mathrm{d}x_1\, \mathrm{d}x_2 \cdots \mathrm{d}x_n. \tag{7.35}$$

Straightforward calculation of the corresponding discrete moments

$$\tilde{m}_{p_1 \cdots p_n} = \sum_{i_1=1}^{N} \sum_{i_2=1}^{N} \cdots \sum_{i_n=1}^{N} i_1^{p_1} i_2^{p_2} \cdots i_n^{p_n} f_{i_1 \cdots i_n} \tag{7.36}$$

requires $\mathcal{O}(N^n)$ operations with N being the size of a hypercube containing the object. If $n \geq 3$, efficient algorithms become even more important than in a 2D case.

Fortunately, almost all methods mentioned in the previous sections can be readily extended to n dimensions. There is no need to develop special "high-dimensional" algorithms. Decomposition methods for binary images can be adapted easily; simply just replace the terms "square" and "rectangle" by "hypercube" and "hyperblock," respectively, when speaking about the elementary shapes of the decomposition. The moments of a hypercube can be calculated in $\mathcal{O}(1)$ time by symbolic integration as in 2D. All algorithms for the decomposition itself such as the delta method, the quadtree decomposition and the morphological decomposition, can be easily modified, too.

If $n > 2$, Green's theorem is called the *divergence theorem* or *Gauss–Ostrogradsky theorem* but its meaning is basically the same: it states that the outward flux of a vector field through a closed surface is equal to the volume integral of the divergence of the region inside the surface. Consequently, the methods based on Green's theorem also have n-D modifications which are easy to derive (see reference [33] for an n-D version of the Philips method and reference [34] for another example).

The methods that employ a polygonal approximation of the object boundary in 2D also have a 3D analogy, where the boundary surface is approximated by a polyhedron. Special algorithms exist for efficient computing polyhedra moments [35, 36].

We can find a few methods in the literature that were developed specifically for 3D moment computation without having a counterpart in 2D. For instance, Li and Ma [37] introduced special linear transforms allowing 3D binary moments to be calculated by means of 2D moments of an artificially created graylevel image.

7.7 Conclusion

In this chapter, we have reviewed existing efficient techniques for calculating image moments. We mentioned all the main categories of method – algorithms for fast computation of geometric moments of binary and graylevel images and algorithms for efficient evaluation of orthogonal moments. The reader may now ask the question whether or not the sophisticated "efficient" algorithms are actually better than a direct calculation from the definition formulas. The answer is not so simple and should be found for each category separately.

As for the decomposition methods of binary images, one should bear in mind that the decomposition itself takes a considerable time, depending neither on the number of the moments to be calculated nor on their orders. These methods may pay off only when a large number of moments of the image are calculated or if the blockwise representation is also used for another purpose (for instance, for compression). This is particularly true for complicated (but effective) decomposition methods.[2] The power of the decomposition methods depends also on the image itself. If the image does not contain large "natural" blocks, any decomposition method can perform even worse than the direct calculation (a chessboard is an illustrative example of an object where decomposition cannot help at all).

The boundary-based methods also require a certain constant-time overhead needed for detecting and representing the boundary but this time is usually much lower than the time of decomposition. The efficiency depends substantially on the boundary shape – for simple boundaries the speed-up could be really significant. Again, a chessboard is an example where these methods perform poorly. Some of these methods allow certain image-independent remedial variables to be pre-computed that can be further used for any image in the application.

---

[2]Many authors do not include the decomposition time in comparative experiments which incorrectly privileges time-consuming decomposition methods yielding a low number of blocks. Such an approach is, in practice, clearly misleading when used for a method selection.

The method of intensity slicing has limited usage for graylevel images with only few intensity levels and with large areas inside them. The approximation methods for graylevel images have even less practical importance. The overhead time for decomposing and piecewise approximating the image is high and the moment values are not calculated accurately, which prevents these methods from being used in practice.

In the case of orthogonal moments, the purpose is slightly different – first, we look for a numerically stable algorithm while its speed is a minor criterion. The use of proper recurrent relations (along with possible improvements due to the kernel symmetry) is highly recommended.

References

[1] Pratt, W. K. (2007) *Digital Image Processing*. New York: Wiley Interscience, 4th edn.

[2] Šonka, M., Hlaváč, V. and Boyle, R. (2007) *Image Processing, Analysis and Machine Vision*. Toronto: Thomson, 3rd edn.

[3] Lin, W. G. and Wang, S. (1994) "A note on the calculation of moments," *Pattern Recognition Letters*, vol. 15, no. 11, pp. 1065–70.

[4] Flusser, J. (2000) "Refined moment calculation using image block representation," *IEEE Transactions on Image Processing*, vol. 9, no. 11, pp. 1977–8.

[5] Zakaria, M. F., Vroomen, L. J., Zsombor-Murray, P. and van Kessel, J. M. (1987) "Fast algorithm for the computation of moment invariants," *Pattern Recognition*, vol. 20, no. 6, pp. 639–43.

[6] Jiang, X. Y. and Bunke, H. (1991) "Simple and fast computation of moments," *Pattern Recognition*, vol. 24, no. 8, pp. 801–6.

[7] Li, B.-C. and Shen, J. (1991) "Fast computation of moment invariants," *Pattern Recognition*, vol. 24, no. 8, pp. 807–13.

[8] Leu, J. G. (1991) "Computing a shape's moments from its boundary," *Pattern Recognition*, vol. 24, no. 10, pp. 949–57.

[9] Dai, M., Baylou, P. and Najim, M. (1992) "An efficient algorithm for computation of shape moments from run-length codes or chain codes," *Pattern Recognition*, vol. 25, no. 10, pp. 1119–28.

[10] Li, B. C. (1993) "A new computation of geometric moments," *Pattern Recognition*, vol. 26, no. 1, pp. 109–13.

[11] Singer, M. H. (1993) "A general approach to moment calculation for polygons and line segments," *Pattern Recognition*, vol. 26, no. 7, pp. 1019–28.

[12] Philips, W. (1993) "A new fast algorithm for moment computation," *Pattern Recognition*, vol. 26, no. 11, pp. 1619–21.

[13] Yang, L. and Albregtsen, F. (1996) "Fast and exact computation of Cartesian geometric moments using discrete Green's theorem," *Pattern Recognition*, vol. 29, no. 11, pp. 1061–73.

[14] Spiliotis, I. M. and Mertzios, B. G. (1998) "Real-time computation of two-dimensional moments on binary images using image block representation," *IEEE Transactions on Image Processing*, vol. 7, no. 11, pp. 1609–15.

[15] Sossa-Azuela, J. H., Mazaira, I. and Zannatha, J. I. (1999) "An extension to Philip's algorithm for moment calculation," *Computación y Sistemas*, vol. 3, no. 1, pp. 5–16.

[16] Wu, C.-H., Horng, S.-J. and Lee, P.-Z. (2001) "A new computation of shape moments via quadtree decomposition," *Pattern Recognition*, vol. 34, no. 7, pp. 1319–30.

[17] Sossa-Azuela, J. H., Yáñez-Márquez, C. and Díaz de León Santiago, J. L. (2001) "Computing geometric moments using morphological erosions," *Pattern Recognition*, vol. 34, no. 2, pp. 271–6.

[18] Tang, G. Y. (1982) "A discrete version of Green's theorem," *IEEE Transactions on Pattern Analysis and Machine Intelligence*, vol. 4, no. 3, pp. 242–9.

[19] Flusser, J. (1998) "Fast calculation of geometric moments of binary images," in *Proceedings of the 22nd Workshop on Pattern Recognition and Medical Computer Vision OAGM'98* (Illmitz, Austria) (M. Gengler, ed.), pp. 265–74, ÖCG.

[20] Chen, K. (1990) "Efficient parallel algorithms for the computation of two-dimensional image moments," *Pattern Recognition*, vol. 23, nos. 1–2, pp. 109–19.

[21] Chung, K. L. (1996) "Computing horizontal/vertical convex shape's moments on reconfigurable meshes," *Pattern Recognition*, vol. 29, no. 10, pp. 1713–17.

[22] Papakostas, G. A., Karakasis, E. G. and Koulouriotis, D. E. (2008) "Efficient and accurate computation of geometric moments on gray–scale images," *Pattern Recognition*, vol. 41, no. 6, pp. 1895–1904.

[23] Chung, K.-L. and Chen, P.-C. (2005) "An efficient algorithm for computing moments on a block representation of a grey-scale image," *Pattern Recognition*, vol. 38, no. 12, pp. 2578–86.

[24] Mukundan, R. (2004) "Some computational aspects of discrete orthonormal moments," *IEEE Transactions on Image Processing*, vol. 13, no. 8, pp. 1055–59.

[25] Zhu, H., Shu, H., Zhou, J., Luo, L. and Coatrieux, J.-L. (2007) "Image analysis by discrete orthogonal dual Hahn moments," *Pattern Recognition Letters*, vol. 28, no. 13, pp. 1688–1704.

[26] Zhu, H., Shu, H., Liang, J., Luo, L. and Coatrieux, J.-L. (2007) "Image analysis by discrete orthogonal Racah moments," *Signal Processing*, vol. 87, no. 4, pp. 687–708.

[27] Al-Rawi, M. (2008) "Fast Zernike moments," *Journal of Real-Time Image Processing*, vol. 3, nos. 1–2, pp. 89–96.

[28] Chong, C.-W., Paramesran, R. and Mukundan, R. (2003) "A comparative analysis of algorithms for fast computation of Zernike moments," *Pattern Recognition*, vol. 36, no. 3, pp. 731–42.

[29] Zhou, J. D., Shu, H. Z., Luo, L. M. and Yu, W. X. (2002) "Two new algorithms for efficient computation of Legendre moments," *Pattern Recognition*, vol. 35, no. 5, pp. 1143–52.

[30] Amayeh, G. R., Erol, A., Bebis, G. and Nicolescu, M. (2005) "Accurate and efficient computation of high order Zernike moments," in *Proceedings of the First International Symposium on Advances in Visual Computing ISVC'05* (Lake Tahoe, Nevada), LNCS vol. 3804, pp. 462–9, Springer.

[31] Hwang, S.-K. and Kim, W.-Y. (2006) "A novel approach to the fast computation of Zernike moments," *Pattern Recognition*, vol. 39, no. 11, pp. 2065–76.

[32] Mukundan, R. and Ramakrishnan, K. R. (1995) "Fast computation of Legendre and Zernike moments," *Pattern Recognition*, vol. 28, no. 9, pp. 1433–42.

[33] Flusser, J. (1998) "Effective boundary-based calculation of object moments," in *Proceedings of the International Conference on Image and Vision Computing '98 New Zealand IVCNZ'98* (Auckland, New Zealand) (R. Klette, G. Gimel'farb and R. Kakarala, eds), pp. 369–74, The University of Auckland.

[34] Yang, L., Albregtsen, F. and Taxt, T. (1997) "Fast computation of three-dimensional geometric moments using a discrete divergence theorem and a generalization to higher dimensions," *Graphical Models and Image Processing*, vol. 59, no. 2, pp. 97–108.

[35] Li, B. C. (1993) "The moment calculation of polyhedra," *Pattern Recognition*, vol. 26, no. 8, pp. 1229–33.

[36] Sheynin, A. V. and Tuzikov, S. A. (2001) "Explicit formulae for polyhedra moments," *Pattern Recognition Letters*, vol. 22, no. 10, pp. 1103–09.

[37] Li, B. C. and Ma, S. D. (1994) "Efficient computation of 3D moments," in *Proceedings of the 12th International Conference on Pattern Recognition ICPR'94* (Jerusalem, Israel), vol. I, pp. 22–6, IEEE Computer Society.

8

Applications

8.1 Introduction

In this chapter, the main categories of image processing problem, where moments have their uncontentious role, are introduced. The goal is to develop a notion of the practical applicability of moments and, last but not least, to provide a reference collection of the various approaches. There are illustrative experiments described in detail to give a notion of the main ideas that lie behind them. For each topic, several published papers are listed, together with a short overview of the proposed methods. They represent distinctive solutions for given problems. At the same time, advice for the successful application of moments is proffered. For any utilization, it is important to be aware of their limitations – high sensitivity to segmentation error, especially in the case of higher order moments, their global characteristics and tied negative implications for their applicability on potentially occluded objects and on objects with a nonuniform background. These facts should be considered when moments are chosen as a solution to the problem.

The classes of algorithm described include object representation and recognition, image registration, robot navigation and watermarking. Then, the medical and forensics applications are introduced and, finally, applications where moments do not constitute the typical solution are listed. Among these, problems such as image retrieval, optical flow or edge detection are described. The key issues of individual classes of algorithm are mentioned.

The aim of this book is not to give a comprehensive explanation of the methodology in its full complexity. All sections cover solutions based only on moments and their modified versions, even though approaches based on different ideas could be more feasible.

8.2 Object representation and recognition

An invariant description of an object and its application for the object classification is probably the most common utilization of moments and the corresponding invariants. Recognition of objects and patterns independent of their position, size, orientation and other variations in geometry and colors has been the aim of much recent research. Finding efficient invariant object descriptors is the key to solving this problem.

Moments and Moment Invariants in Pattern Recognition   Jan Flusser, Tomáš Suk and Barbara Zitová
© 2009 John Wiley & Sons, Ltd

Several groups of features have been used for this purpose, such as simple visual features (edges, contours, textures, etc.), Fourier and Hadamard coefficients, differential invariants and moment invariants, among others. In general, object descriptors should fulfill several conditions:

- *discriminability* – the descriptor reflects interclass variations (two objects from two different classes have different descriptions);

- *robustness* – the descriptor is not influenced by intraclass variations (two objects from one class have the same descriptions);

- *invariance* – the descriptor is invariant with respect to expected geometric or radiometric degradations (an object and its rotated and blurred version have the same descriptions);

- *independence* – any of the descriptors from the set cannot be formulated using only the other descriptors from the set.

The moment invariants and studies concerning their behavior for various experimental settings show that they can be used efficiently for object description and within reasonable scope fulfill given conditions. Of fundamental importance is the quality of object segmentation, due to the fact that higher order moments, especially, are sensitive to segmentation errors.

As stated previously, it is important to choose the proper class of invariants with respect to expected image degradations and, eventually, take into consideration a method for the moment computation acceleration depending on the type and amount of data to be analyzed (Chapter 7). The matching part of the method can then be realized using various approaches for the correspondence estimation, starting from standard minimum distance classification, clustering methods, support vector machines (SVMs), to neural networks, hidden Markov models (HMMs) or the Bayesian classifier. The moment description is used standalone or can be combined together with other descriptive features (for example, geometric, Fourier or differential invariants).

There is one broad application area of second-order moments for the object description and classification which is executed before the very description starts. The moments up to the second order are often used to bring selected parts of the scene (segmented objects, detected affine invariant regions) into the normalized position. Then, other types of feature descriptor can be applied that are not necessarily moment-based. It is often a question of the numerical properties of involved descriptors if the object normalization or the choice of higher order invariance is the best solution. Examples of methods for region detection that are subject to such object normalization can be found in references [1], [2] or [3].

No dedicated experiment on object recognition is described here, because many illustrative examples can be found in previous chapters of this book. There are applications of several classes of moment invariant, tested under simple as well as complex degradations. Moreover, even experiments in this chapter described in other sections can often be interpreted as object classification tasks. An overview of method variations follows.

Flusser and Suk [4] described an application of AMIs for character recognition. They work with a set of handwritten capital letters. The comparison of recognition ability of four Hu invariants and four AMIs are described and the results showed AMIs' better robustness with respect to the variations due to handwriting. Matching was realized using the minimum distance classifier.

A description system for texture recognition by means of the ZMs computed from the correlation functions of individual color channels was given in reference [5]. The method is independent of rotation, scale and illumination. The key to the method is the relation between illumination, rotation and scale changes in the scene and a specific transformation of correlation function of ZM matrices that can be estimated from a color image. Campisi et al. [6] proposed the use of a set of independent rotation moment invariants based on moments up to the fourth order computed from the texture model. The model is created as an output of a linear system driven by a binary image. The latter retains the morphological characteristics of the texture and it is specified by its spatial autocorrelation function (ACF). They show that moment invariants extracted from the ACF of the binary excitation are representative enough to capture the texture for classification purposes. Moreover, the proposed scheme has been shown to be robust with respect to noise. The geometric moments, Legendre moments and complex moments were utilized for texture description and unsupervised texture segmentation in reference [7]. All the moments were evaluated from the Gabor decompositions of texture patches.

Various systems for the classification of a certain category of things were proposed. The fish recognition system was designed by Zion et al. in reference [8]. They proposed to evaluate Hu moment invariants and their combinations to be able to distinguish between three fish species. The images were acquired while the fish were swimming in an aquarium with their sides to the camera. The moments were computed both on the whole body of the fish as well as on its parts (head, tail). This in-vivo identification of fish species regardless of size and 2D orientation produced satisfying results. The fish identification was also the aim of reference [9]. Here, however, White et al. use moments only for the estimation of the fish orientation. Gope et al. [10] introduced AMIs for the classification of marine mammals. Thiel et al. [11] proposed to apply the Hu invariants for the recognition of blue-green algae for measuring the water quality. Fischer and Bunke [12] classified unicellular algae found in water by means of AMIs.

The classification of plant leaves was described in reference [13]. First, a modified version of the watershed segmentation method combined with presegmentation and the morphological operation, which can handle partially occluded leaves, is run to segment leaf images. Then, seven Hu geometric moments and 16 ZMs are extracted as shape features from segmented binary images after leafstalk removal. In addition, a moving center hypersphere classifier is designed. The average rate of correct classification achieved was up to 92.6%.

Al-Batah et al. [14] employed the Hu, Zernike and AMIs evaluated on the binarized boundary and area to form the feature vector for the description of the aggregates in concrete, which are very important for the final quality of concrete. They proposed a new method for the automatic classification of the aggregate shapes, based on the described feature vector and the artificial neural network. Li and Huang [15] proposed a method for recognition of defect patterns in semiconductor manufacturing. The produced circuits have to meet the specifications and this ought to be tested as quickly as possible. A method based on a selforganizing map and SVM for wafer bin map classification is proposed. First, they distinguish between the systematic and random wafer bin map distribution. After smoothing descriptive features – co-occurrence matrix characteristics and Hu moment invariants – are extracted, the wafer bin maps are then clustered and manufacturing defects classified. The proposed method can transform a large number of wafer bin maps into a small group of specific failure patterns. The experimental results show over 90% classification accuracy and outperform the backpropagation neural network.

For remote sensing applications, Keyes and Winstanley [16] tested the abilities of the moment invariants to characterize topographical data. They computed Hu invariants but using the shape boundaries only. They intend to classify objects on large-scale maps. Their conclusion is that the moments are fairly reliable at distinguishing certain classes of topographic object; however, the authors would recommend fusing them with the results of other techniques.

Apatean *et al.* [17] were involved in the detection of an obstacle in a traffic scene situation. Detection of the obstacle (a pedestrian or a vehicle) is often very difficult due to the complex outdoor environment and the variety of appearances of the obstacle. They intended to achieve classification into five classes (standing person, unknown posture, motor bike, tourism car and utility car). Visible and infrared images were used as input data and for the recognition system, features of different type, among them seven geometric moments up to the fourth order, were evaluated and fed to the classification module – k-nearest neighbor algorithm with $k = 1$ and SVM with radial basis function kernel. Accuracy rates above 92% were achieved during the experiments.

In reference [18] the method for vehicle detection in high-resolution satellite images was described. First, the image segmentation of the input data was realized, working with the multispectral images, panchromatic images and a road network. Then, the two-stage classification was evaluated. After primary analysis, the following feature set was chosen: length, width, compactness, elongation, rectangularity, boundary gradient, spatial spread, contrast, smoothness, region mean, gradient mean, variance and Hu moments. The authors concluded that for the whole system the segmentation is the most critical part. Zhenghe *et al.* [19] derived the modified version of the standard Hu moment invariants and applied them for recognition of road traffic signs. The feature vectors were classified using the backpropagation neural network.

There are several papers on recognition of aircrafts and ships by means of moments. This undertaking commenced with the well-known paper by Dudani *et al.* [20] using rotational moment invariants. Reeves *et al.* [21] modified Dudani's proposed approach. They presented methods for the identification of a 3D object from a 2D image recorded at an arbitrary viewing angle and range. The 3D aircraft models from the database were represented by several 2D projections. They concerned themselves with the so-called standard moments (normalized moments). Several experiments are described with combinations of silhouette and boundary moments and different normalization techniques. Alves *et al.* [22] tried to classify ship types from an infrared silhouette. They computed Hu invariants both from the boundary silhouette and from the region, which were then fed into the neural-network classifier. The proposed method has a problem with the ship's occlusions owing to globally computed invariant features.

The Mokhtarian and Abbasi method [23] stems from reference [20]; moreover, they also use the database of aircraft images. They select the image subset of optimal views for multiview freeform object recognition. They address how to choose the best representative views for each object in a database and their optimum number that can enable the accurate recognition of that object from any single arbitrary view of the object. They propose representing each object effectively by its boundary and consecutively compute from it the curvature scale space, Hu moment invariants and Fourier descriptors. The selection algorithm is described, using the representative subset of the most typical views from the used features point of view. The number of retained views varies, depending on the complexity of the object and the measure of the expected accuracy. For recognition of 3D objects, Xu and Grebin [24] proposed extracting several features (texture and color characteristics, Hu moment invariants and AMIs) for 2D views of 3D objects. These feature vectors are then presented to a backpropagation neural network for training. The trained network can recognize 3D objects when provided with feature vectors of unseen views. A 100% correct rate of recognition was achieved when training views were presented for every $10°$.

The distinct application of the moments, in this case the ZMs, is the signature verification. Here [25], the moments are used to describe individual strokes (curves) of the signature. The signature contour is segmented into a fixed number of small curves and then the shape features can be separately computed for each curve. The magnitude of the ZM captures shape information in a rotationally invariant form, while the complex angle records a rotation angle with respect to the origin, which is a significant characteristic for the signature. Thus, both magnitude and angle are used as features. The ZMs up to the sixth order were evaluated. The proposed signature verification method demonstrates strong invariance among genuine signatures. Lin and Li [26] also made use of the ZMs, but in their normalized version [27]. They verify the Chinese signatures, taking the whole signature as a classified object. Wang *et al.* [28] employed Krawtchouk and Chebyshev moments as representatives of the discrete orthogonal moments for Chinese handwritten symbols recognition using a discrete-time HMM framework. The individual symbols are divided into smaller subparts from which the moment representation is computed. They conclude that the performance of discrete orthogonal moments is much better than that of continuous orthogonal moments (Zernike and Legendre). The recognition accuracy of Chebyshev moments is a little better than that of Krawtchouk moments.

Face recognition algorithms compose a specific subgroup of object description and recognition methods. The application of moments and moment invariants for face recognition is not a standard solution and it can even be disputed due to the conflict between the 2D nature of most invariants and 3D human faces and, moreover, due to often large variations between individual images of the same face (hairdress, glasses, to name the most important). However, there are many papers on this topic, covering many situations in the human face recognition framework. Foon *et al.* used ZMs together with wavelet transform for face recognition [29]. There is a comparative study on the moment application [30]. Various moment invariants have been used to extract features from human face images for recognition application: Hu, Zernike, Pseudo Zernike, normalized Zernike, normalized pseudo-Zernike, normalized geometric moment invariants and others. PZMs yield the best recognition accuracy of 95% on the AT&T face database. From these same authors, an application of PZMs for face recognition together with various classifiers can be found here [31]. The performance of a k-nearest neighbor algorithm where $k = 1$ and 3, SVM and HMM classifiers were studied. The best recognition rate of 91% was achieved with the HMM-based solution.

Zhu *et al.* applied Hu moment invariants for the recognition of facial expression [32]. They computed them locally, around eyes, nose and mouth to avoid the dependence on changes. Moreover, they used shift-variant, not centralized versions of moments thus they were able to capture eventual displacements due to the movements of facial muscles. The very recognition was done using HMM. They claim a very high recognition rate but, however, with strong limitations on head movements, glasses, etc. Zhi and Ruan [33] proposed comparing solutions for recognition of facial expression, which is a challenging task of intelligent human–computer interaction. The applied moments include Hu, Zernike, wavelet and Krawtchouk moments. Experimental results show that wavelet moments outperform other moment-based methods in facial expression recognition.

An interesting application of ZMs is described in reference [34]. As stated previously, ZMs are robust in the presence of a noise and they exhibit the rotational invariant property. They can also provide nonredundant shape representation because of their orthogonal basis and they reflect the overall shape of the regions. In the cited paper, they try to detect human ears. The final verification, where it has to be determined whether or not a detected ear is a true ear, is realized using ZMs. A small set of the lower-order ZMs captures the global general shape of an object, which is what is needed here. The magnitudes of the ZMs are used for the ear shape representation. If the similarity between the detected candidate and the ear template is high enough, then the claim is validated.

As is apparent from the approaches mentioned above, Hu invariants attract the greatest attention for the object description and consecutive classification, even though they are invariant to simple geometric deformations (TRSs), they are not mutually independent and, moreover, using just a finite set of geometric moments (which are the basic blocks of Hu invariants) the discriminability can be violated (two objects, whose moment representation differs only in higher order moments). In spite of all this, the recognition systems often give reasonable success rates. The more complex invariants (AMIs, the invariants based on the orthogonal moments) have already found their position in pattern recognition awareness and from the latest comparative studies [30], their greater efficiency is evident. The choice of object descriptor is often not directly connected to the choice of classifier, there are various combinations of many approaches, so to generalize about what is the best pipeline is beyond the scope of this book.

8.3   Image registration

Many variants of image registration [35] are based on the exploitation of moments and moments invariants. Image registration is a process of overlaying two or more images of the same scene taken at different times, from different viewpoints and/or by different sensors. It geometrically aligns the reference and the sensed images. The distortions between the images are introduced due to different imaging conditions. The image registration task appears, for example, in remote sensing, in medicine and in computer vision, to name just a few. The *feature-based registration* estimates the correspondence between distinctive objects (points, closed boundary areas, fixed-sized windows) in the reference and the sensed images. To characterize these features, the methods often employ some kind of invariants. These invariant descriptors override the influence of distortions, introduced by a geometric transform during the image acquisition process, and also possible discrepancies in graylevels. Moment invariants are an example of such invariant representation of the detected object, which are frequently used for feature matching.

The decision on which class the moment invariant has to be used depends on the type of present geometric and even radiometric degradations. In real situations a perspective deformation is present in acquired images that can often be successfully modeled by an affine transform, if the distance from the camera to the scene is much larger than the object's size. From the previous statement the usability of AMIs is substantiated. If, also, radiometric degradations are expected, it is well advised to consider the use of combined invariants. For instance, if we expect contrast changes between images, then invariants to contrast should be used. If blurring may be present in the images, we use convolution invariants, which are robust to main prospective kinds of blur, such as out-of-focus blur, motion blur and atmospheric turbulence blur and which at the same time provide robustness with respect to geometric degradation.

The second important decision which has to be made is about the area from which the moments will be calculated. We can use fixed-size windows localized on the regular mesh over the image or around detected salient points such as corners, road crossings, line endings, etc., or even the entire images can be used for moment calculation. The limitations of this approach originate in the relation between the window shape versus the kind of geometric deformation present. First, if the whole image is used for the computation, the method will function only if the reference and sensed images have a 100% overlap. Any scene occlusion or a difference in a field of view would harm the invariance, so this is not a recommended approach. The simple rectangular window suits the registration of images which differ locally only by translation. If images are deformed by more complex transformations, this type of window is not able to cover the same parts of the scene in the reference and the sensed images (the rectangle can be transformed to some other shape, depending on the deformation). Several authors proposed using circular windows for mutually rotated images [36]. However, the comparability of such simple-shaped windows is also violated if more complicated geometric deformations such as similarity or perspective transforms are present. Recent methods for invariant neighborhood selection can handle even affine deformations; see, for example, local affinely invariant regions [1]. Another possibility for selecting the area from which the moments are computed is not to use the gray values but first to segment the images and select several significant closed-boundary regions, such as lakes or fields on remote sensing images (see Figures 8.5, 8.6). The moments are then computed only from the binary regions and are insensitive to the particular colors. The robustness of this approach depends largely on the quality of the segmentation algorithm. In the case of reliable segmentation, we usually obtain better results than by calculating moments directly from the graylevels.

Two image registration applications, demonstrating different moment-based approaches, are described in detail below.

8.3.1 Registration of satellite images

The first example aims to register two satellite images of the same area (Czech Republic, northwest of Prague). However, the datasets were taken at different times and using different sensors – the first image was acquired by the Landsat 5 satellite while the second image was taken by the French satellite SPOT. They have different spatial as well as spectral resolutions – the Landsat image has a spatial resolution of 30 m and seven spectral bands, while the SPOT image has a resolution of 20 m and three spectral bands. In the preprocessing phase, the three most important components were computed from the Landsat dataset using principal component analysis (PCA) (Figure 8.1), the SPOT channel data were used directly (Figure 8.2). No correction to skew caused by the Earth's rotation was made on the SPOT dataset. As part of the preprocessing stage, the image denoising, based on minimizing the Mumford–Shah functional [37], was applied to both images (see Figure 8.3 for Landsat and Figure 8.4 for SPOT). The denoising makes the following segmentation easier.

In this experiment, the moments were not computed directly from original images but from segmented salient regions. From both images 15 regions were segmented by means of region growing. The very segmentation was followed by rejection of too small and too large regions, which are not favorable for region matching. Sufficient overlap of these two object sets was achieved, which is vital for the following correspondence estimation. One should

Figure 8.1 Registration of satellite images: Landsat image – synthesis of its three principal components.

Figure 8.2 Registration of satellite images: SPOT image – synthesis of its three spectral bands.

Figure 8.3   Registration of satellite images: Landsat image – denoising based on the minimization of the Mumford–Shah functional.

Figure 8.4   Registration of satellite images: SPOT image – denoising based on the minimization of the Mumford–Shah functional.

be aware of the time distance between the reference and the sensed images (the images were acquired in different years and even in different seasons), hence the variations in the region intensities and shapes; thus, some imperfections in the area detection and segmentation may appear. For these reasons the regions were binarized after the segmentation, so that contrast normalization was not necessary.

For each detected region from both images, the normalized moments τ_{pq} up to the eighth order were computed and its counterparts were searched by the minimum distance in the feature space. The three most forceable pairs of the regions from the reference and sensed images were found. Having the three matched regions, the coordinates of their centroids were used for the computation of the preliminary affine transform parameters. After this primal alignment, additional matching pairs were searched by minimum distance with thresholding in the spatial domain. Six other region pairs were matched. Finally, the centroids of matching regions were used as control points for the last image registration. The result can be seen in Figures 8.5 and 8.6, where the three most forceable regions have numbers in a circle, the other matching regions are simply numbered. The final result of the image registration can be seen in Figure 8.7, where the Landsat and the registered SPOT images are superimposed.

For test purposes, we repeated the region matching again by the AMIs. For the same detected regions the AMIs were computed. We used 15 independent invariants up to the fifth order. The four most forceable pairs of the regions were found. Three of them were the same as in the previous case, thus the recognition ability of both approaches are comparable.

In the described experiment the normalized moments proved to be robust enough with respect to the geometric deformation present on the satellite data. Even the different times

Figure 8.5 Registration of satellite images: Landsat image – segmented regions. The regions labeled by numbers in a circle were matched in the feature space, the other labeled regions were matched in the image space.

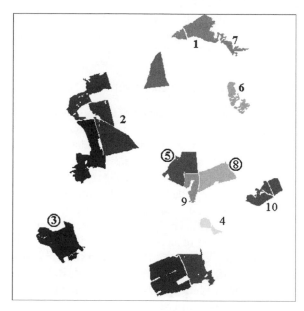

Figure 8.6 Registration of satellite images: SPOT image – segmented regions. The regions labeled by numbers in a circle were matched in the feature space, the other labeled regions were matched in the image space.

Figure 8.7 Registration of satellite images: the superimposed Landsat and the registered SPOT images.

of the data acquisitions do not disturb their invariance. The issue of the area for the moment computation was resolved using the closed boundary region segmentation, which ensures data correspondence.

8.3.2   Image registration for image fusion

In the second described experiment, the application of the moment invariants for image fusion is shown. The aim of the image fusion is to combine several images of the scene of low quality (blurred and noisy images of low resolution) and produce one output image with better resolution and with blur and noise removed or diminished (for an example of input images, see Figure 8.9 and for the corresponding result, see Figure 8.12). This can be achieved due to the fact that each input image is degraded in a slightly different way and the image fusion combines them in such a way that the qualities are emphasized and the degradations are removed. One condition has to be satisfied before the image fusion can start – the input images have to be registered so that the same objects in the scene are geometrically aligned across the images. Then, the fusion analyzes different versions of the same object, captured on the individual input images (see Figure 8.8 for the image fusion flowchart). As stated previously, a registration method able to handle blurred and noisy data is required here that is not fulfilled by majority of traditional registration techniques. The registration algorithms often expect the noise and blur removal ahead of their main process loop. However, the feature-based method using combined blur invariants can handle such situations owing to their simultaneous invariance with respect to blur and geometric transform.

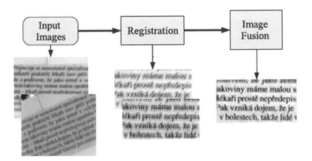

Figure 8.8 Image registration for image fusion: image fusion flowchart.

The proposed registration algorithm first detects feature points in all input images by means of the feature detector [38] (Figure 8.10), robust with respect to the geometric as well as to the radiometric degradations – blur. Then, the combined blur invariants are computed for the circular neighborhood of each detected distinctive point. The geometric parameters of necessary alignments among individual input images are found using the same correspondence estimation method as in the previous experiment. After the image registration (Figure 8.11) is processed, the image fusion via multichannel deconvolution (for more details, see reference [39]) can be computed and the resulting image of higher quality is created (Figure 8.12). None of the input images show comparable quality in terms of edge sharpness, noise level and detail visibility (Figure 8.13).

Figure 8.9 Image registration for image fusion: examples of several low-quality images of the same scene – they are blurred, noisy and of low resolution.

Figure 8.10 Image registration for image fusion: low quality images of the same scene with detected distinctive points.

Figure 8.11 Image registration for image fusion: low-quality images after image registration.

The registration methods can be distinguished by the choice of particular moment descriptor and region, from which they are computed. Goshtasby [36] published one of the

Figure 8.12 Image registration for image fusion: the output image of the image fusion. The resulting image has better resolution and the blur and noise are removed or diminished.

Figure 8.13   Image registration for image fusion: scaled version of an input image for comparison. The image does not show comparable quality in terms of edge sharpness, noise level and detail visibility.

first algorithms – he used Hu invariants, computed from the circular neighborhood. Morcover, he discussed the matters of the moment value normalization in his paper. When moment invariants of different scales are used in the similarity estimation by correlation, the feature with the largest scale would dominate the correlation value. He proposed a normalization related to the size of image and the order of the used moments. As a second improvement he introduced speedup of the very registration by splitting the similarity estimation into two stages. First, just the zeroth-order moment is used for localization of likely matches and then only on the selected positions is full feature representation computed. Li *et al.* [40] proposed using contours of the objects as salient features and their correspondence is found by matching their basic geometric features (the perimeter and the longest and shortest distances from boundary to the centroid), then by matching the first two Hu moment invariants and finally by means of the chain code correspondence. Dai and Khorram's [41] approach to image registration combines the moment-based descriptor with improved chain-code matching. The robustness of the minimum distance classifier is improved by matching in the image, using the rule of root-mean-square error. S. X. Hu *et al.* [42] discussed the Hu invariants-based registration algorithm, computed on either detected corners or distinctive edge points. They proposed another way of normalizing the computed invariant values – using the standard deviation of the absolute value of corresponding moment invariant difference. The improved similarity measure should exclude more outliers than the metric-Euclidean distance. Tuytelaars and Van Gool applied generalized color moments, computed from the local affinely invariant regions [1].

In Shi *et al.* [43], the moment descriptors with higher level invariance are exploited. They applied AMIs for the registration of satellite images. For the computation they use circular neighborhoods located at the position of the points detected by means of the redundant wavelet transform. They obtain good results; however, it is important to mention here that the possibility to register affinely deformed images is limited – if the choice of area for moment computation is a fixed-size circle, then in the case of more profound affine deformation it would not cover the same area, which is important for the moment invariance assurance. This fact was the motivation for the usage of segmented closed boundary regions in reference [44] for the registration of the satellite images. Here, the AMIs are computed on the segmented areas of reasonable size. Bentoutou *et al.* [45] amplified the invariance of the descriptors even to blurring, as in the second described experiment. Their registration of remote sensing data uses the feature points from the reference image detected by thresholding the gradient magnitude of the reference image and by applying the improved version of the Harris corner detector [46]. Then, the corresponding points in the sensed image are found using the minimum distance of the invariant description, computed from the reference point and tested candidates from the sensed image. They propose this version when even subpixel accuracy can be achieved. In reference [47], the registration of mutually rotated and blurred images was proposed. The matching is performed by means of combined blur invariants, invariant to rotation, translation and to image blurring by any symmetric point spread function. Consequently, the images can be registered blurred images directly without any de-blurring.

A group of methods are based on ZMs or PZMs. Yang and Guo [48] described a registration method based on the Harris corner feature points [46] and PZMs expressing the neighborhoods centered on feature points. The matching is then performed using RANSAC [49]. Badra *et al.* [50] utilized the rotational properties of the ZMs. Using the phase change of ZMs and the shift Fourier theorem, they estimate the rotation plus scaling parameters and translation parameters, respectively, in the iterative manner until the final registration is achieved. They present panoramic mosaics that were automatically generated using this approach. They claim that their algorithm can easily handle images with little overlap and scale differences. Yasein and Agathoklis [51] also applied ZMs, computed around the distinctive points found using the scale-interaction of Mexican-hat wavelet decomposition of the registered and sensed images. The very matching is done by estimating the correlation coefficients of the ZMs' vectors, so, contrary to the previous approach, the ZMs' invariance is used and not their ability to capture the parameters of the rotation and the scale.

Other variations in the registration methods use the ability of low-order moments to capture the orientation and shape of the objects. Sato and Cipolla [52] computed directly, without correspondence estimation, the parameters of the present geometric deformations (an affine transform was expected) using the circular moments of distribution of the line features orientation. They combined moments and the scale-space representation of the images. Brivio *et al.* [53] modeled shadow structures in mountain images by means of their reference ellipses. The ellipses are described here by their area, inclination of the main axis and ellipticity. All these attributes are functions of moments. The very matching is realized using a multivalue logical tree. There is a big group of registration methods based on prenormalization of the detected affine invariant distinctive regions that is realized by the matrix of second-order moments. Then, the matching follows based on rotationally invariant features. More on this topic can be found in section 8.2.

There exist applications of moments even for the registration of 3D data. Jaklič and Solina [54] applied moments for 3D registration, in this case the range images. They showed that objects that are compositions of super-ellipsoids can be computed as simple sums of moments of individual parts and that the 3D geometric moment of a globally deformed super-ellipsoid in a general position and orientation can be computed as a linear combination of 3D geometric moments of the corresponding nondeformed super-ellipsoid in the canonical coordinate system. They use a standard technique to find centers of gravity and principal axes in pairs of range images while third-order moments are used to resolve the four-way ambiguity. Shen and Davatzikos [55] solve the 3D registration of brain images, expecting the elastic model of present geometric difference. An attribute vector, formed by a set of geometric moment invariants defined on each voxel is used to describe the tissue maps to reflect the underlying anatomy at different scales. The proposed method (the hierarchical attribute matching mechanism for elastic registration – HAMMER) captures the image similarities using moment representation, contrary to other volumetric deformation methods, which are typically based on maximizing directly the image similarities.

As can be seen from the above discussion, there are many variations in the image registration scheme using moments. Either they use only low-order moments for estimation of the main object orientations, or they describe invariantly selected distinctive parts of the scene, to provide input to the matching algorithm and thus to help estimate the parameters of the present geometric deformation. In all cases, it is extremely important to compute the moments from the regions in the images that correspond to the same part of the scene.

8.4   Robot navigation

One of the key issues of robotics is to provide a robot with the information about the outer world and about its tasks, so that mobile robots can act autonomously without human help. A large part of this problem is connected to the pattern recognition/matching, where moment invariants represent one of the possible solutions.

There are several options as how to tackle the robot's notion about the surrounding world. If the robot's working space is complicated or is not static, then a *dynamic approach* can be the best solution. The information about the robot trajectory and the previous world representation is used for the estimation of the current robot position. The robot can construct its own world map based on the data acquired by its sensors. After an incorporation of the robot motion trajectory and the sensor data into the evolving world representation, the robot is able to make a decision about its current position and the next action to take.

However, in some tasks, the *landmark-based approach* is more appropriate. Here, the map of the robot's world can be assumed as a-priori known (either learned by the robot in the preprocessing phase, or directly given to the robot). The robot makes the decision about its current position and next action by matching the new collected sensor data with the stored reference world representation. So-called *landmarks* – objects in the scenes which are found to be distinctive – are often used for the robot position estimation. Using such an approach, no representation of the surrounding world is necessary, so complex and often memory-demanding descriptions can be avoided. Moreover, landmarks can be used to impart certain types of information to the robot. For example, recovery from failure during robot navigation can be based on finding landmarks.

There are two classes of landmarks used. The first consists of natural parts of the world – doors, straight lines, corners, etc., while in the other, artificial signs are placed in the robot's environment. In both cases the landmarks are detected by the robot's visual sensors during its movement, and compared against the stored database. If the artificial marks are used, we are not limited by the natural layout of objects.

Moments and moment invariants are used in the landmark-based algorithms, where they are involved in landmark classification. The question to be asked is which moment invariants should be used, in other words what kind of geometric deformation can we expect? The robot equipped with the common type of camera and moving in the usual scene introduces to the landmark acquisition process the projective deformation and an additive noise. A good choice for such situations is the affine moment invariants, which were proven to be robust under this deformation. Two examples of robot navigation systems together with proposed landmark sets will be presented. The robustness of the algorithms even in a situation of more complex geometric degradation is shown.

8.4.1   Indoor robot navigation based on circular landmarks

In this application, a view-invariant recognition of circular landmarks for mobile robot navigation is proposed, together with the recognition system based on the AMIs. The robustness of the system with respect to an additive random noise and to various viewing angles was tested and the discriminability and stability of the recognition model was proven by real situation experiments.

In general, a landmark set should be designed to be capable of bounding the mark shape with information, which should be passed to the robot (concerning the robot's position, crossing, obstacle, special task) and at the same time the landmark should effectively be detectable. The proposed mark set consists of patterns formed by two concentric circles with equal outer and different inner radii (Figure 8.14), allowing easy localization of the marks in the complex environment. The information, which has to be imparted to the robot, can be encoded into the ratio of the mark's inner and outer radii. This way of encoding ensures the independence of the acquired information on the robot's position with regard to the mark.

Figure 8.14 Indoor robot navigation based on circular landmarks: two examples of proposed landmarks, formed by two concentric circles with equal outer and different inner radii.

The discriminability of the system depends on both the marks themselves and also on the recognition model used. Since the conditions for approximation of the perspective projection by an affine transform are met (see Chapter 3 for an explanation), the AMIs can provide robust enough recognition. An important limitation of the AMIs' use should be considered. Many AMIs have theoretically zero values in the case of radially symmetric objects. Thus, it is important to be aware of this fact and to use only those invariants which do not suffer from this property. The proposed recognition model computes for a given mark image three nonzero AMIs and finds the most similar mark from the database using the minimum distance similarity method. Before the very moment evaluation, the images of the landmarks are binarized in order to eliminate different lighting conditions.

The ability of this recognition model to make correct decisions in situations of a nonzero viewing angle or in the presence of an additive random zero-mean noise is investigated experimentally. The experimental setting is shown in Figure 8.15.

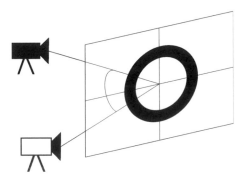

Figure 8.15 Indoor robot navigation based on circular landmarks: experimental setting. The landmark was placed in the indoor environment. The camera is scanning the landmark with different viewing angles (0°–80°).

The first experiment deals with noise robustness. A mark image was corrupted by an additive random zero-mean noise with a uniform distribution. The camera viewing angle was fixed, the values of the added noise standard deviation (STD) were chosen from the interval 0 ± 250. The minimum distances between the acquired mark images and corresponding reference mark from the database were computed in the feature space of the three AMIs. The first misclassification occurred with STD$= 232$, so the noise robustness proved to be sufficient. Even with high values of STD the distance to the corresponding reference mark was shown to be much smaller than the distance to the other reference marks.

To verify the stability of the AMIs-based recognition system with respect to the viewing angle change in real situations, the following experiment was carried out. The landmark was placed in the indoor environment (Figure 8.16). The camera viewing angle varies from 0° to 80° with steps of 5°, so, owing to the introduced projective deformation the landmark shape changed from the circular to an ellipse-like shape (Figure 8.16). In all cases the radii ratio of the test landmark was always correctly recognized, so the viewing angle robustness proved to be sufficient, too.

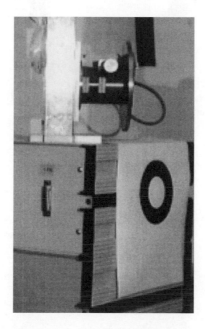

Figure 8.16 Indoor robot navigation based on circular landmarks: the real indoor scene used in the second experiment. Note the projective deformation of the landmark owing to a nonzero viewing angle.

Although the AMIs are invariant only under the affine transform, which is an approximation of the perspective transform occurring in the robot vision system, their recognition ability was proven to be high enough for the described experiment. The stability of AMIs under a random zero-mean noise present in the acquisition process of mark images also leads to good results. Even high values of STD had a rather small influence on the recognition results of AMIs.

The next application shows the robustness of the AMIs to even more complex geometric deformation than the perspective transform, which is introduced by the fish-eye lens camera.

8.4.2 Recognition of landmarks using fish-eye lens camera

Here, the recognition of artificial landmarks using the AMIs is investigated, but in this case the robot is equipped with a fish-eye lens camera. Such a camera provides more information than that with a conventional lens. In particular, the robot navigation in indoor environments can be improved using the fish-eye lens because of its wide angle of view (approximately 180° in the diagonal direction) and the ability to obtain information even from a very close robot's surrounding [56]. However, such a camera introduces complex geometric deformation of the acquired images (Figure 8.18).

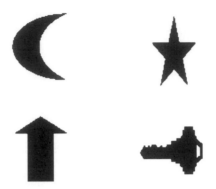

Figure 8.17   Recognition of landmarks using fish-eye lens camera: the database of used landmark shapes (MOON, STAR, ARROW, KEY).

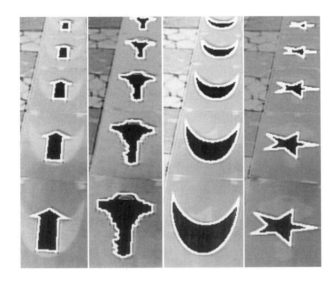

Figure 8.18   Recognition of landmarks using fish-eye lens camera: the examples of landmark images with detected boundaries. The introduced nonlinear geometric deformation is apparent.

The landmark set consists of four shapes – MOON, STAR, ARROW and KEY – which have different properties of curvature, number of sides, holes, etc., making them easy to

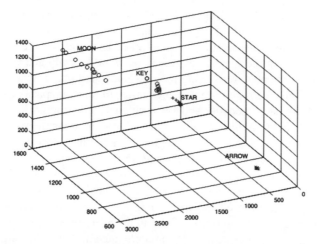

Figure 8.19   Recognition of landmarks using fish-eye lens camera: the AMIs' feature subspace corresponding to the three chosen invariants. The feature vectors, corresponding to the segmented landmarks are plotted. Every cluster is labeled with the reference shape type.

differentiate (Figure 8.17), even in the case of complex geometric deformation due to the fish-eye lens. It is known (and shown in the previous application) that the AMIs are robust under the perspective transform if the distance between the camera and the object is significantly larger than the size of the object. We assume that under similar conditions the AMIs-based recognition model can be robust enough under the fish-eye lens deformation, too.

The discriminability of the AMIs was tested on images, an example of which is shown in Figure 8.18 (the deformation introduced by the fish-eye lens is apparent). For every shape several images were acquired with the landmark located at specified positions. The landmarks were segmented and corresponding AMIs were computed for all possible settings. Three AMIs confirmed the expectation that AMIs can be robust enough even under the perspective and fish-eye lens degradations, while the other three seem to be too sensitive to the present type of deformation.

In Figure 8.19 the AMIs' feature subspace corresponding to the three chosen invariants is plotted together with the feature vectors, corresponding to the segmented landmarks. The MOON and ARROW clusters are well separated in the feature space; on the other hand the KEY and STAR clusters are close to each other in the feature space. A misclassification can appear in the case of wrong segmentation. However, in most cases the distance to the corresponding representatives is much smaller than to other representatives. It can be seen that the landmarks are classified correctly using the minimum distance classifier even when the introduced degradation is perceptible.

The proposed recognition model showed sufficient stability and robustness even under the perspective transform together with the deformation introduced by the fish-eye lens camera. Although the AMIs are invariant theoretically only under the affine transform and we worked with images transformed by much more complex geometric deformations, the AMIs' recognition ability proved to be high enough even in such cases.

One of the first papers [57] to apply moment invariants for mobile robot localization proposed to use natural landmarks. Integrated sonar, vision and infrared input data were fused for sensor-based spatial representation of the scene. The approximate robot position was determined by comparing the features extracted from the recently acquired images against the world representation, prepared during the pre-processing phase, when the robot creates grid maps of the distinctive regions that it encounters in its workspace. The area and seven Hu invariants [58] for each modality and for each analyzed structure were computed. The experiment was held in ten different rooms and ten different doorways and it proved a 94% recognition rate of the rooms and a 98% recognition rate of the doorways.

Similarly to described applications, Xua *et al.* [59] proposed using AMIs for unmanned aerial vehicles (UAVs) navigation – automatic landing on the ship's deck. They use the infrared radiation images of the scene with inserted targets (the T-shaped cooperative objects put on the runway). The infrared radiation images feature a high temperature difference between the target and background, which help to precisely segment the cooperative object. After segmentation the four simplest AMIs are computed and used for target identification. The achieved recognition rate was 97.2% even when the present geometric deformation was perspective and not just an affine one.

Another example of the application where the natural landmarks are used is the work of Lee and Ko [60]. They do not insert any new signs into the robot's surrounding; they work directly with distinctive parts of the scene, where the robot should be navigated. They call the method gradient-based local affine invariant feature extraction (G-LAIFE). The places where the invariants should be computed were identified using the Harris corner detector [46] and then invariants were evaluated in the close neighborhood based on the regular grid of squared windows. Beforehand the acquired images were transformed from the direct intensity domain into the normalized gradient magnitudes of image intensity to decrease the dependability on the image information change. They tested the lengths of the feature vector from 4 to 10 and proved the assumption that invariants based on higher-order moments are more sensitive. The feature vector length 6 outperformed others.

The variant solution of the recognition system for robot navigation, based on ZMs, is described in reference [61]. Lin *et al.* make use of local distinctive structures for robot navigation. They extract highly robust and repeatable features representing unique structures with the strongest response in both spatial and scale domains. Instead of the geometric moment invariants, they proposed to compute rotation invariant weighted ZMs on the normalized image patches. They use parts of the scene as the landmarks. The experiment showed good system performance on object recognition as well as on indoor topological navigation, even in the case of severe occlusions and scale changes.

There are other variants suitable to integrate the moments into the robot navigation. Wells and Torras [62] proposed to use simple functions of geometric moments up to the third order to characterize statistical variations in the object's projection in the acquired image when the camera position is changed. They chose such functions which can express individual geometric deformations but are not invariant with respect to them. The very features were computed on binary images corresponding to the thresholded versions of the acquired data (three different thresholds were applied). Feedforward neural networks learned the mapping between the feature variations and pose displacements, when the robot moved. In this experiment, the geometric moments outperformed the eigenfeatures (Karhunen–Loeve features) and pose–image covariance vectors, even though they use the 2D geometric moments (not the moment invariants) for the description of the projections of 3D movements.

They improved the performance of the method by feature subset selection using mutual information.

Variantly, Celaya *et al.* [63] used the geometric moments to normalize detected regions to a patch of given size to make its description invariant to changes in scale and to moderate perspective deformations so the recognition algorithm need not take these deformations into account. The geometric central moments up to the second order are computed and the equivalent ellipse axes are found providing a rotation angle to align the region and scale factors for the x and y dimensions.

Application of moments and moment invariants for the robot navigation makes use of their recognition abilities even under perspective transform and their stability with respect to an additive random noise. The moments are used for both the image invariant representation and the image normalization. In all cases the landmark-based approach for the mobile robot navigation is assumed, using either natural or artificial landmarks.

8.5 Image retrieval

Nowadays, falling prices of all kinds of image sensor and widespread use of the end-user handheld cameras produce a growing stream of images resulting in data organization difficulties, namely with respect to efficient data-searching. One approach is to tag the acquired images and then ask text-based queries. However, this methodology is time-consuming and can easily be biased by the human factor and operator's understanding of the image content. The variant approach is based directly on the image data. The search engine looks for the images which are similar to the query image, given as the input; or the query image can be expressed as a collection of basic colored shapes at the desired position in the scene, representing the rough sketch of the goal image. This solution for an image-based lookup in the databases has a big advantage over the first one; there is no loss of information due to the pipeline *image – text tag – image*. Moreover, it is known that one image can tell more than a thousand words. The weak point of the image-based search is how algorithmically to express the human notion of the similarity of two images. This is the key issue of the algorithms, belonging to the category of content-based image retrieval (CBIR) applications [64], when using various image descriptors, similarity measures and specialized data structures on which the image-based search is realized. The moments and moment invariants are used for object description, as described previously. In the same way, they can be employed in the context of CBIR.

In 1995, Stricker and Orengo [65] proposed to compute color moments for image retrieval–today, one of the most often used applications of moments in CBIR [66]. These moments are computed from the color histogram of the image and can express the overall color characteristics of the image. They are computed either directly from the whole image, or locally, on the regular grid or even on the segmented objects.

There are various kinds of moment utilized for the CBIR systems, depending on the type of data to be searched, the expected image degradations or the authors preferences. The well-known QBIC (IBM) [67] image-retrieval system employs geometric moments up to the fifth order for the description of the shape variances of objects. The normalized Hu moment invariants scaled to the same range were applied in reference [68] together with a histogram of edge directions. In the refinement phase the edge deformable models were employed to solve ambiguities. The proposed system was tested on the trademark image database.

Wei *et al.* [69] decided to use ZMs. They proposed a system for trademark image retrieval in the trademark registration system. They detect edges and realize shape normalization. Then global and local features are extracted. In this way they capture the main shape variations and details, respectively. Curvature and distance to the object centroid are used as local features, whereas first fourth-order ZMs are employed to capture the global differences. ZMs can be found in reference [70], too. They investigate texture and shape features for CBIR by combining Gabor filters for texture extraction and ZMs for shape description. The combination of the proposed descriptors outperformed their individual application.

Shao and Brady [71] applied the salient regions detection followed by their invariant description for CBIR. The detected regions have very high repeatability under various viewpoints and illumination changes. They are detected according to local entropy and scale selection. Each region is then described by the Hu moment invariants, five higher-order invariants from reference [72], and generalized color moment invariants [73]. The minimum distance classifier is used for the similarity estimation. The method proved to be effective in retrieving a variety of cluttered images with partial occlusion. Banerjee *et al.* [74] designed a CBIR method based on the normalized geometric moments up to the second order evaluated around visually significant point features. The distinctive point clusters are extracted using a fuzzy set theoretic approach. The moments are computed from the detected regions of interest as well as from the whole image in a normalized RGB model. The authors tried to select an optimal set of features suitable for retrieving a set of images perceptually relevant to the query image. It is shown that a single set of features may not be suitable for all types of query. Tuytelaars and van Gool [75] made use of locally defined AMIs [73], computed on the affinely invariant regions to be able to deal with large changes in viewpoint. The system is robust to occlusions and changes in background as well as to illumination changes.

The CBIR system for texture retrieval is described in reference [76]. The proposed texture descriptors are modified ZMs, invariant to translation, scaling and rotation. First, the power spectrum of an original texture image is calculated for translation invariance and then the power spectrum image is normalized for scale invariance. Finally, modified ZMs are calculated for rotation invariance.

In reference [77] the authors pay attention to the issue of 3D retrieval. For feature characteristics, a method combining a distance histogram and geometric moment invariants is proposed to improve retrieval performance. The mutual information and Euclidean metrics estimate the similarity between features and, finally, the SVM classification is applied. They demonstrate the applicability on the shred database search. The 3D object retrieval is also the subject of a paper by Mademlis *et al.* [78]. They introduce weighted 3D Krawtchouk moments for efficient 3D analysis that are suitable for content-based search and retrieval applications. Due to the relatively high spatial frequency components of Krawtchouk low-order moments, they can capture sharp shape changes of the object. Thus, proposed weighted 3D Krawtchouk moments can form a very compact and highly discriminative descriptor vector.

Di Ruberto and Morgera [79] compared the applicability of various moment-based features for the object retrieval. Their set of descriptors consists of Hu, Flusser and Taubin invariants, Legendre and ZMs, and generic Fourier descriptors. The experiments were run on the database of binary objects. Similarly, Yadav *et al.* [80] analyzed the applicability of Fourier descriptor, Legendre moment descriptor and wavelet ZM descriptor as descriptors for CBIR. The Euclidean distance was evaluated as a similarity measure. The experiments proved that the Zernike-based descriptors outperform other techniques.

The practical application of the CBIR approach for the image data retrieval is presented in reference [81]. The authors design the CBIR for the retrieval of the tree leaves images (Figure 8.20). The selected feature set consists of the basic geometric descriptors such as eccentricity, aspect ratio or elongation. The more informative features are Hu invariants, which proved their efficiency during the experiments. The Euclidean distance is used as the similarity measure in a *k-NN* classifier with $k = 10$. The website providing the listed method is active.

Figure 8.20 Image retrieval: Pauwels *et al.* [81] proposed the CBIR system for the retrieval of tree leaves images (a), based on simple geometric features together with Hu invariants; binarized versions of the leaves (b).

As is apparent, many variants of CBIR algorithms use the recognition ability of moments. The integral part of the system design is the choice of classifier, which is often the standard minimum distance classifier. Often, the conclusion that ZMs outperform the others is achieved.

8.6 Watermarking

Digital watermarking is a class of methods for copyright protection, which could rank it among the image forensic applications. However, it is mentioned here separately, because of its importance and its spread. Historically, a watermark was an image printed by water instead of ink, which causes thickness variations in the paper. The watermark then appears as various shades of lightness/darkness when viewed by transmitted light. The used symbol for the watermark referred to the creator/owner of the document. Today's aim of digital watermarking remains the same – we want to encode new information (watermark) into the original signal (host image). The original image is stamped in this way; if the image data are copied, the digital stamp is carried as well. The watermark can be visible or the brightness change at the place of the embedded watermark can be reduced so it becomes invisible. Watermarks can be used for copyright protection and for active authentication of the images. For an example of a watermarked image, see Figure 8.21.

Figure 8.21   Watermarking: the watermarked image. The watermark "Lednice castle" is apparent. This is an example of the visible watermark.

An important issue for copyright protection is the robustness of watermarks against various attacks including geometric transforms (at least the similarity transform), multiple JPEG compressions, cut outs, etc., to name a few examples. As in the previous section, only the solutions based on the moments will be considered. The OG moments are usually the key components of proposed algorithms. ZMs are the most frequent, but other families of moments are mentioned too, such as Krawtchouk (for their better reconstruction properties) or Chebyshev moments. Some approaches are even based on Hu moment invariants. The moments can be used either for the very watermark insertion itself, when the values of moment invariants (often ZMs) are appropriately changed or they are applied for the deformation inhibition, when the watermarked image is transformed into the normalized position and thus the influence of present geometric deformations is decreased.

8.6.1   Watermarking based on the geometric moments

Alghoniemy and Tewfik [82] applied the geometric moments directly in the creation of the invariant watermark. They considered the image I as watermarked if a predefined function of the invariants has a certain value N within a given tolerance; otherwise the image is declared unwatermarked. They propose modifying the host image in such a way that the previous claim is fulfilled. Various moment invariants can be used, depending on the expected geometric deformations. The predefined function can be any function of the invariants, linear or nonlinear (for example, a weighted sum of the invariants can be used). Also, for security purposes, a secret key should be introduced in the function design.

For image modification, a proper perturbation image $\triangle I$, where $\triangle I = \beta \Pi(I)$ should be constructed and added to the host image I (Figures 8.22 and 8.23). Π is the mapping function which serves as a noise added to the host image with weight β, which is controlled by feedback to ensure N. The authors propose $log(I)$ as an appropriate choice for Π. For the decoding, the equality of the invariants' value to N is simply checked. The decoder does not need the host image for the decoding which can be advantageous (only the N and the error tolerance have to be known). However, this watermarking approach returns a YES – NO answer; thus, the method offers 1-bit capacity only, which can be insufficient. Other disadvantages of the method are its inability to preserve fidelity, i.e. the watermarked image will create contrast and brightness variations compared to the host image (Figure 8.23) and the method vulnerability to the change of the aspect ratio, cropping and to attacks that would change the geometric moments, such as a histogram equalization. To improve the robustness, the image can be a priori normalized. A similar approach was applied in reference [83], where the Hu moment invariants are computed on video object detected by snakes.

Figure 8.22 Watermarking based on the geometric moments: the original host image to be watermarked by the method [82].

An example of a watermark insertion scheme can be found in reference [84]. They proposed using invariance properties of circularly orthogonal moments (ZMs and PZMs). The proposed method is robust against geometrical deformations: image rotation, scaling,

Figure 8.23 Watermarking based on the geometric moments: the corresponding watermarked image. There is no visible pattern inserted into the image; only intensity variations are apparent. These changes are a known disadvantage of the method [82].

flipping, as well as other image manipulations such as lossy compression, additive noise and lowpass filtering. Here, the watermark is encoded as a sequence of binary digits and is inserted into the host image using the quantization of the ZM's/PZM's magnitudes through quantization index modulation. Part of the included binary code contains the information about the unit disk used for the watermark encoding. The set of suitable low-order ZMs and PZMs with an appropriate computational accuracy is selected and modified using the watermark binary representation and dither modulation. The quantization step influences the final quality of a watermarked image – it determines the tradeoff between the visibility and robustness of a watermark. A larger step gives better watermark robustness but makes the watermark more visible, and vice versa. The desired step can be estimated using the required peek signal-to-noise ratio (PSNR) of the image. The watermarked image is then the sum of the original host image and the watermark signal reconstructed from the difference between the modulated and original ZMs and PZMs.

For data extraction, the test image, which can be a distorted version of the original host image, is analyzed with the watermark after some possible manipulations, such as rotation and scaling. First the unit disk region where the watermark was encoded is located, extracting the first part of the included binary code. Then the binary code is retrieved using the invariant magnitudes of the same subset of ZMs and PZMs, the same method of quantization and a minimum distance decoder. The comparison of the extracted code with the watermark can indicate whether or not the test image is the distorted version of the host image.

In reference [85], a similar approach is applied, based on a combination of PZMs, Krawtchouk moments and support vector regression (SVR). The PZMs of the host image represented in polar coordinates are computed. Magnitude values of selected low-order PZMs are quantized in order to encode the digital watermark. Then, the Krawtchouk moments of test image form the input to the trained SVR to obtain an estimate of the geometric parameters. They are used for the geometric rectification of the test image, followed by extraction of the watermark. In this way the robustness against filtering, sharpening, noise adding and JPEG compression is achieved and even the geometric attacks such as TRS transform, cropping and a combination of various attacks can be handled successfully.

Zhang *et al.* [86] proposed using Chebyshev moments for watermark embedding, and for the decoding they use the independent component analysis (ICA) method for blind source separation. The watermark is generated randomly, independent of the original image and embedded by modifying perceptual invariant Chebyshev moments of the host image. The watermarked image is created using the inverse Chebyshev transform of the weighted sum of Chebyshev moments of the host and the watermark image. The weight can influence the intensity of the watermark so it should be set according to the requirements of the human visual system. The approach proposed in this paper withstands rotation, scaling and translation attacks as well as filtering, additive noise and JPEG compression. Similarly, they proposed a given scheme but with the Krawtchouk moments, once again using ICA [87].

Several authors proposed divergent application of moments for image watermarking [82, 88–90]. They used central geometric moments for the image normalization with respect to the flipping, translation, scale and orientation of the image preceding the very watermark encoding and testing. This approach should provide robustness with respect to various geometric attacks; however, it cannot resist, in general, copy and crop attacks. Parameswaran and Anbumani [88] combined such moment normalization with the wavelet-based watermark encoding. Lu [89] used the discrete cosine transform (DCT) decomposition for watermark insertion, and in reference [90] they combined the image normalization with the multibit watermarking system based on the spread spectrum direct-sequence code division multiple access approach [91].

Two main possibilities of moment applications for image watermarking have been mentioned. The first covers all algorithms, where the moments and the moment invariants (both orthogonal and geometric) computed from the host and watermark images are modified in order to embed the watermark into the host image. In the second class of algorithms, the image is first transformed into the normalized position using moments to diminish the influence of the geometric transformations. Afterwards, various watermarking methods can be applied.

8.7   Medical imaging

There is wide application of moments and their invariants in medical imaging. In fact, there can be found similar tasks as in the previous sections (object recognition and image registration, to name a few); however, this domain is often perceived as a specific application category and that is why it is mentioned separately. The main divergences inherent to this kind of method are caused by the characteristics of the input data. They are often 3D, sometimes with quite low resolution, the noise level can be very high and the contrast and the data information content are often too low. Due to the previous statements, the general image processing methods have to be adjusted to fit given limitations.

The experiment illustrating medical image applications covers the topic of evaluation of the scoliosis progress (a medical condition in which a spine displays the tendency to an S-like shape and may also be rotated).

8.7.1 Landmark recognition in the scoliosis study

The rotation moment invariants from Chapter 2 were applied here to analyze the medical images shown in Figure 8.24. The goal of the medical project was to measure and evaluate changes to the woman's back and spine arising from pregnancy (see reference [92] for details of this study). All measurements were noninvasive using Moire contour graphs and specific landmarks attached to the body. This technique allows 3D measurements from a sequence of 2D images.

One of the subtasks was to measure the progress of scoliosis. In order to do so, circular black landmarks were glued on the woman's back and a template-matching algorithm was used for automatic localization of the landmarks and their centers. The values of four moment invariants c_{11}, $c_{20}c_{02}$, c_{22} and c_{33} were evaluated for a sliding window across the whole image. The results obtained were correlated with the moment invariants of the templates and the positions of individual landmarks were located using the correlation peaks.

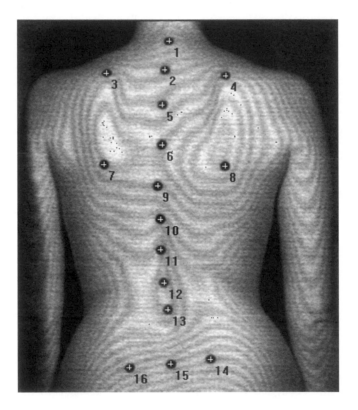

Figure 8.24 Landmark recognition in the scoliosis study: an example of the human body with attached landmarks and Moire contour graphs.

As illustrated in Figure 8.24, the method yielded accurate localization of all landmarks and produced no false matches (the crosses denote identified landmark centers). Having detected landmarks, the extent of the *S*-like deformation can be evaluated. Note that the application of the Hu invariants failed. They were unable to distinguish between the landmarks and some parts of the Moire strips for all cases.

As stated earlier, medical applications do not have any unifying task definition, so the following overview is diverse. Ruggeri and Pajaro [93] applied various moment invariants in order to decide which layer of the eye cornea is seen in the confocal microscope. The proposed layer recognition is based on the cell shapes, which are distinctive for individual layers. They use Hu invariants computed on binarized data and classify them by an artificial neural network. To avoid binarization, which can bias the result, an alternative cell-shape description was investigated, based on ZMs. The results achieved with ZMs outperformed those with Hu moment invariants and, moreover, no binarization was necessary. The moment-based object description is also used in reference [94]. They combined the Hu moment invariants together with the Bayes classifier [95] to classify the human embryos according to their quality for transfer. Bentoutou *et al.* [96] applied moment invariants for the image preprocessing for digital subtraction angiography, when blood vessels in a sequence of X-ray images are visualized. The registration of image sequences is based on local similarity detection according to combined moment invariants, followed by thin-plate spline image warping. The proposed technique can handle both slow and sudden motions.

Sommer *et al.* [97] proposed a method for comparing protein-binding sites. They use 3D geometric moment invariants as feature vectors for the binding description. For many molecules, coordinates of atoms are available from X-ray crystallography or nuclear magnetic resonance (NMR) experiments. They fitted a 3D Gaussian onto the center of each atom and summed over all the Gaussians. In this way the density of a molecule is approximated and used further as the molecule shape, from which a feature vector is computed. They found the moment invariants as a robust technique for comparing densities on a global level, useful also for other applications in structural bioinformatics. Similarly, Hung *et al.* [98] constructed a method for protein function identification, such as of the SARS coronavirus. They applied Hu moment invariants as feature descriptors of protein fragments. Hu invariants are exploited in reference [99], too. Here, Avci and Varol applied them on segmented and binarized images, for classification of human parasite eggs. The recognition system is based on the multiclass SVM.

Banada *et al.* [100] were looking for a solution to detection and classification of bacterial pathogens, the topic of which is crucial for securing food supply. They use a light-scattering sensor together with the recognition system based on the magnitudes of ZMs and Haralick texture descriptors. The very correspondence was estimated using SVM. The detection and identification of the bacterial colonies consisting of the progeny of a single parent cell is achievable in real-time. The success rate for all bacteria categories was over 90%.

Medical image processing often has to deal with 3D data owing to the characteristics of the investigated objects. In reference [101], the authors described the morphometry of cortical sulci (depressions or fissures in the surface of an organ) using 3D geometric moments and their invariants with respect to translation, rotation and uniform change of scale. They demonstrate the potential of such an approach for the characterization of the brain activities on the PCA of the 12 first invariants computed for 12 different deep brain structures manually segmented for seven different brains. Finally, this invariant description was used to find correlates of handedness and sex among the shapes of 116 different cortical sulci automatically identified in the database of 142 brains. They plan to apply 3D moment invariants to study the influence of cognitive, genetic and pathologic features on the shapes of various other cerebral objects. Ng *et al.* [102] intended to characterize spatial distribution changes in functional magnetic resonance imaging (fMRI) activation statistics under different experimental conditions. They computed 3D geometric moment invariants with respect to the translation, rotation and change of scale from the dataset recorded from eight healthy subjects performing internally- or externally-cued finger-tapping sequences. Voxel-based activation statistics were characterized in several regions of interest. The 3D moment invariants analysis revealed the difference between externally- and internally-cued tasks.

Zacharaki *et al.* [103] proposed to use the aforementioned HAMMER method [55] to deformably register 3D brain data acquired from magnetic resonance. They limit themselves to the application of low-order geometric moments as the feature descriptors. The temporal tracking of coronaries using a multislice computed tomography dynamic sequence is solved in reference [104] using 3D moment descriptors. Here, Laguitton *et al.* proposed to apply the geometric moments and local cylindrical approximation of the vessel. In this way they plan to capture local volume variations and to estimate its displacement along the image sequence. The main idea is to find given points in the successive volumes. The locally defined vessel model, a cylinder, is constructed with parameters, computed from the moments up to the second order. The resulting positions inside the vessel are then adjusted considering neighboring points from the vessel central axis.

An example of a content-based image retrieval system (CBIR) for medical purposes is that proposed by Jones *et al.* [105]. They concentrate on the database of thermal images, using the idea that a query on images showing symptoms of any given disease will provide visually similar output images which will usually also represent similarities in medical terms. Their CBIR system is based on the set of Hu moment invariants computed from graylevel thermal images.

An important subgroup of medical imaging algorithms is the class of image reconstruction utilizing data from different modalities, such as X-ray computed tomography or emission tomography. Milanfar *et al.* [106] proposed a variational framework for the tomographic reconstruction of an image from the maximum likelihood (ML) estimates of its orthogonal Legendre moments. They show that in a linear manner they can compute these estimated moments and their error statistics directly from given noisy and possibly sparse projection data. Moreover, they show that this iterative refinement results in superior reconstructions of images from very noisy data as compared with the well-known filtered back-projection (FBP) algorithm. Basu and Bresler [107] offered solutions for situations when not all viewing angles are known (for example, involuntary motion of the patient can result in this uncertainty). They address the problem of determining the view angles directly from the projection data itself in the 2D parallel beam case. They showed that when the projections are shifted by some random amount which must be jointly estimated with the view angles, unique recovery of both the shifts and view angles is possible. They exploit the Helgasson–Ludwig consistency conditions, which relate the geometric moments of the object to the moments of the projections.

Moments and moment invariants are used in medical applications in two main roles. They can be employed as the descriptors capable of capturing the shape variations or other structure changes, leading to data classification or data alignment. Alternatively, they are used as tools for data reconstructions from different medical imaging modalities such as magnetic resonance, computed tomography or emission tomography. Often, the 3D moments are applied because of the volumetric nature of the analyzed medical data.

8.8   Forensic applications

Moment theory and namely the moment invariants play an important role in detailed analysis of image-based information utilized in forensic science applications. Image analysis helps to provide forensic evidential examinations of photographs, documents, handwriting, shoeprints, fingerprints, etc. Nowadays the dominant approach for image capturing is digital imaging; thus, images and videos are widespread. Their use can be either helpful – for person identifications, archiving purposes – or they can be the very subject of a crime act – image forgeries and manipulations. Image-processing software, easily available today, can provide almost anybody with effective tools for image misuse and the necessity for detection methods of such image tampering grows. Another specific area where image-processing methods can be very useful is biometric identification (fingerprints, iris, etc.) for security applications or for probate proceedings.

The application of moments for forensics makes use of their invariance (an image verification) and their recognition ability (a person identification by means of various biometrics such as fingerprints, palmprints, etc.). The following detailed example introduces the first case, when the moment blur invariants help to detect near-duplicated image regions.

8.8.1   Detection of near-duplicated image regions

Often, the naked eye is not able to detect artificially manipulated image data (Figures 8.25 and 8.26). Mahdian and Saic [108] proposed a method for the automatic detection and localization of near-duplicated regions in digital images. Copy-move forgery detection methods have to take into account that copied parts can be retouched or processed by other localized image-processing tools. During this image-tampering, the near-duplicated regions are created in an image. Some regions of the image are used for covering unwanted parts of the original image. They are partially blurred, especially their borders, to conceal traces of pasting copied versions of other parts of the image. Mainly due to this blurring, it is difficult to apply standard similarity measures such as cross-correlation for the tamper localization. The proposed method is based on blur moment invariants and on detection of duplicated regions in the image, since the duplicated regions may be indicators of this type of forgery. The moment blur invariants allow successful detection of the copy-move forgery even when blur degradation and arbitrary contrast changes are present in the duplicated regions. The lossy format such as JPEG can be treated. Moreover, these invariants work well with the presence of an additive noise, which is important because the additive noise is often introduced to the image to make the detection even more difficult. The proposed method belongs to the category of *passive* (blind) approaches. In contrast to an *active* approach (such as the watermarking) the passive methods do not need any a-priori information about the tested image, which can be a big advantage in many cases.

Thus, a given image I can contain an arbitrary number of duplicated regions of unknown location and shape and the task is to determine the presence of such regions in the image and to localize them. The authors suppose that the blurring present can be modeled by convolution with a shift-invariant symmetric energy-preserving point spread function so the blur moment invariants can be applied. The proposed copy-move forgery detection method works with the image tiled into the overlapping blocks. For each block the blur moment invariants are evaluated and then the principal component transformation of all acquired block representations is run to decrease the dimensionality of the feature space. They use 24 normalized blur invariants up to the seventh order. Then, the blocks' similarity is analyzed, using the hierarchical k-D tree structure for an efficient identification of all blocks which are in a desired similarity relation with the analyzed block. The k-D tree is a commonly used structure for searching for nearest neighbors. The main idea of the block similarity step is that a duplicated region consists of many neighboring duplicated blocks. If two similar blocks in the analyzed space are found and their neighborhoods are also similar to each other, there is a high probability that they are duplicated.

Figure 8.27 shows an example of a feature space, where black dots represent overlapping blocks. All similar blocks are found for each tested block (similar to the nearest neighbors search) and their neighborhood is verified. In this step, similar blocks with different neighbors are eliminated. Finally, a duplicated regions map is created (Figure 8.28) showing the image parts, which are probably copied and smoothly pasted in other positions of the image.

In comparison to other methods based mainly on DCT and PCA, the description of duplicated regions by moment invariants has better discriminating properties in cases where the regions were modified by blurring or by contrast changes. The results are not affected by an additive zero mean noise. However, the proposed approach, like other existing methods, has a problem with uniform areas since it looks for identical or similar areas in the image from its definition.

Figure 8.25 Detection of near-duplicated image regions: an example of the tampered image with near-duplicated regions.

Fingerprint, palmprint and iris recognition belong among the group of moment applications using their recognition ability for person identification. Yang *et al.* in reference [109] described the Hu invariants-based fingerprint recognition. They compute invariants from four

Figure 8.26 Detection of near-duplicated image regions: an example of the tampered image with near-duplicated regions – the original image.

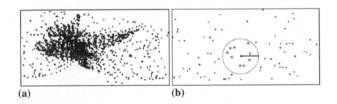

(a)   (b)

Figure 8.27 Detection of near-duplicated image regions: a 2D feature space. (a) Black dots represent overlapping blocks, which have to be analyzed. (b) The method finds all similar blocks to each block and analyzes their neighborhood. In other words, all blocks inside a circle, which has the analyzed block as centroid, are found. The radius r is determined by the similarity threshold.

subimages, invariantly selected from the enhanced fingerprints and matched by means of the absolute distance matching and the backpropagation neural network. The modified version of the approach can be found in reference [110], where the Hu moment invariants are computed for surrounding of each detected minutiae point to decrease the dependence on the subimage localization (fingerprint minutiae points are local ridge characteristics that occur at either a ridge bifurcation or a ridge ending).

Both previous methods apply moment descriptors on already enhanced fingerprint images. The preprocessing of the fingerprints is important due to the low contrast and low quality these images often have. Almansa and Lindeberg [111] capitalized on the properties of

Figure 8.28 Detection of near-duplicated image regions: the estimated map of duplicated regions.

the second-order moments to indicate the local image contrast, to measure the degree of anisotropy and to reflect the dominant orientation (the orientation most orthogonal to the gradients in a neighborhood of the point, which will be used for estimating the ridge orientation). They proposed the mechanism for shape-adapted smoothing based on the second-order moment descriptors and, moreover, an automatic scale selection based on normalized derivatives. The shape adaptation procedure estimates the strength of the fingerprint smoothing according to its local ridge structure, which allows the interrupted ridges to be joined without destroying singularities such as branching points. They produce detailed estimates of the ridge orientation field and the ridge width, as well as a smoothed graylevel version of the input image. The minutiae detection is presented on the post-processed fingerprint which shows good performance.

The gait is another biometrics feature used for person identification. Other biometrics technologies, such as face, hand and fingerprint recognition, are useless when the tested person is distant. The main idea behind the application of moments is to divide the silhouette at each frame of the sequence in a predefined way, to compute a set of features based on the moments for each subpart of the image, and then, finally, somehow to reduce the acquired feature vector to efficiently represent the gait sequence. Lee and Grimson [112] represented each orthogonal view video silhouette of the human walking motion by a fixed number of ellipses and they computed the feature vector based on the second-order moments. They aggregated these feature vectors over time and finally processed them using the Fourier transform to obtain frequency based information. Shi *et al.* [113] proposed the silhouette subdivision in a multiscale way, where the moments are computed for each resolution. They use first-order moments, which limits the characterization ability of their method.

The dimensionality of their feature space is reduced using PCA and the recognition is carried out by SVM.

AlGarni and Hamiane [114] proposed using Hu moment invariants for shoeprint classification against the database in order to link suspects to crime scenes. They work with binarized shoeprint images, variously rotated and with various resolution. Their experiments showed a good recognition rate; however, it is important to note that the method would fail in the case of a noncomplete shoeprint, which is often the case.

Iris patterns are random, which makes them one of the most reliable biometric features. On the other hand, due to the complex iris image structure and its intra-class variations, it is difficult to represent the iris efficiently. Nabti and Bouridane [115] described a multiscale edge detection approach based on a preprocessing localization step followed by filtering with special Gabor filters run on the wavelet-decomposed image. Finally, a feature vector representation using Hu moment invariants is created. The experimental results have shown that the proposed system is comparable to the best iris recognition systems. Their method is distinct in its computation of moments on the results of Gabor and wavelet decomposition, which should emphasize the main iris features.

Besides fingerprints, palmprints are typical biometric features for image forensic applications. Once again the moment invariants act as invariant features used for recognition of the palmprint papillary lines structures. Besides the Hu moment invariants [116], applied on the Gabor filtered images, the application of the ZMs is proposed [117]. Pang *et al.* [118] compared the effectiveness of utilizing three various OG moments, namely ZMs, PZMs and Legendre moments, in the application of palmprint verification. The orthogonal property of these moments is capable of characterizing independent features of the palmprint image and thus has minimum information redundancy in a moment set, opposite to Hu moment invariants, which consist of geometric moments. Among all the moments, PZMs perform best. They compute moments directly on the palmprint images. High-order ZMs are used to describe hand shape [119]. They compute ZMs for binarized and segmented fingers and palm separately better to capture the geometry of individual parts of the hand. In the latter approach the papillary lines are not taken into account; only the shape variation of the palm and the fingers are utilized for person identification.

The moment ability to invariantly represent the object is used for person identification by means of various biometrics. It is important to realize that moment invariance is preserved only if the original image and deformed image depict the same scene, so in the case of a partial shoeprint or partial fingerprint, the algorithms could fail. The near-duplicated region detection uses the invariance with respect to the radiometrical deformations which are rare among descriptors and due to this, the proposed algorithm can achieve very good results even in situations where other methods break down.

8.9 Miscellaneous applications

Besides the large categories of moment application there are various other image-processing domains where moment theory comes to the fore and offers new viewpoints. These are not so distinguished as the previous topics covered in this section; however, they deserve to be mentioned. Several examples of such idea upgrading are covered in the following.

8.9.1   Noise-resistant optical flow estimation

Optical flow estimation is an important tool to track motion in a sequence of images with many applications; for example, recovering surface structures, change detection and various medical problems. The objective of the optical flow computation is to obtain measurements describing the 2D pixel motion in images, known as the *motion field*. However, such motion is only the projection of the relative 3D motion between the camera and the scene surfaces in 3D space on the image plane. The accuracy of such a model is greatly limited due to the approximating nature of the optical flow, which is only the approximation of the motion field depending on the observed motion of points in the image. Most optical flow methods are very sensitive to the quality of input images. They prefer images with constant brightness and texture of the image, close sampling (small motion) and noise-free scenes. Unfortunately, in many applications, for example in medical imaging, the images are very noisy, thus preventing use of standard methods (for example, reference [120]) relying on local computation of derivatives in each pixel. Moments were utilized to achieve higher robustness with respect to noise and small perturbations. In reference [121], the authors took advantage of the stability of rotation-invariant ZMs with respect to noise. They modified the method [120] to work with the values of invariants based on ZMs, instead of the pixel values. The linear equations system, which forms the solution for optical flow, is then formed over each small local region based on the principle of conservation of those invariant features. They obtain a more robust estimation at the expense of a slight decrease in spatial resolution.

Kharbat *et al.* [122] proposed applying geometric moments for the optical flow stabilization with respect to the intensity changes. They describe the pixel values by the content of their neighborhood using geometric moments. Then, the description of each pixel is normalized and made insensitive to fluctuations of intensity. In this way, the resulting optical flow is much less sensitive to visual phenomena such as varying illumination, specular reflections and shadows.

Figure 8.29 shows an application of this method to a fast sequence of heart images taken by magnetic resonance (MR) scanner Philips Intera 1.5 T (Philips Medical Systems). The images are marked by the harmonic phase imaging (HARP) technique resulting in a typical gridlike pattern, which helps to improve the precision of the applied optical flow algorithm. It is apparent that the method using ZMs exhibits for fewer artifacts and irregularities of the optical flow field, especially in the areas lacking distinctive texture, than the standard method [120] (Figure 8.30).

8.9.2   Focus measure

An interesting application of moments is their use for focus/defocus measuring. In the real world, the image acquisition process faces blurring degradation from various sources such as atmospheric blurring, out-of-focus blurring and motion blurring, to name the most common. The presence of some level of blur in images is inevitable; however, its amount and its strength can be different. Having a relative focus measure (answering the question, which image is the least blurred) is important for ordering an image sequence according to the amount of blur. This is a common task, for example, in astronomy, where long sequences of images (typically, several hundreds or even thousands) of the same object are taken by ground telescopes. The images are blurred by atmospheric turbulence; the amount of the blur may

Figure 8.29 Noise-resistant optical flow estimation: an example of using ZMs to stabilize optical flow computation in the case of noisy magnetic resonance images of the heart.

Figure 8.30 Noise-resistant optical flow estimation: the result of the standard method [120] is shown, for the comparison with the moment-based approach.

quickly vary as the atmospheric conditions change. Our task is to select several best images that are further used (Figure 8.31).[1]

Moreover, the performance of image-processing algorithms can be influenced by proper setting of parameters, describing the amount of present blurring or certain similar characteristics. All these are reasons why it is important to know how to evaluate the degree of blur degradation in the acquired images, in both absolute (answering the question how much is the image blurred) and relative ways.

Various focus measures have been reported in the literature. Most of them try to emphasize high frequencies of the image and measure their quantity. The main idea behind that is that the blurring suppresses high frequencies regardless of the particular point spread function. The most popular approaches are based on variance of image graylevels or on the image derivatives (L_1 norm and L_2 norm of image gradient, L_1 norm of second derivatives, and energy of image Laplacian) [123].

Moments were used for the focus measure design in several ways. Zhang *et al.* [124] proposed a measure based on the second-order or on the fourth-order geometric central moments of an image sequence and on their linear combinations, such as $\mu_{20} + \mu_{02}$ or $\mu_{40} + \mu_{22} + \mu_{04}$. These quantities are often used inversely as a focus measure. The focus measure increases as the image blur level decreases and reaches its maximum when the image is exactly focused (see Figure 8.32 for the measure behavior on the image sequence, whose example is in Figure 8.31). They are even able to express the blur property of the imaging system owing to the independence of the method on the imaged objects. The algorithm also returns an explicit expression of the parameter of the point spread function, modeling the present blurring. However, the functionality of the proposed method is violated when the actual blurring is not exactly the convolution. One of the common deviations from the model is the cropping on the boundaries. Yap and Raveendran [125] based their measure on Chebyshev moments. They define the measure as the ratio of the norm of the high-order moments to that of the low order. In this way, they characterize the low and high spatial-frequency components of an image, respectively. The focus measure is monotonic and unimodal with respect to image blurring, which is important. Additionally, it is invariant to contrast changes due to the differences in the intensities of illumination and robust under Gaussian and salt-and-pepper noise. Wee *et al.* [126] defined their full-reference measure (defined with respect to the reference image) using the symmetric geometric moments (SGMs)

$$
s_{pq} = \int_{-1}^{1} \int_{-1}^{1} x^p y^q f(x, y) \, dx \, dy,
$$

where $f(x, y)$ is an image or an image block mapped into $\langle -1, 1 \rangle \times \langle -1, 1 \rangle$. The authors divided the tested image into several blocks and for each of them computed all SGMs up to the fourth order. Then they evaluated the score of the image as the average of the quality measures of each block, computed using SGMs of the tested and the reference images. However, we have to admit that even though the mentioned publications show various ways in which moments can help to estimate the blur level, the idea of using moments for focus measure design is quite minor.

---

[1] In astronomy, such selecting of a few good images from a large set of observations is called "lucky imaging".

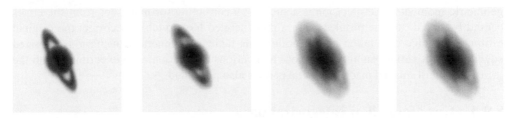

Figure 8.31 Focus measure: example frames from image sequence (Saturn images), where each image is differently blurred and corrupted.

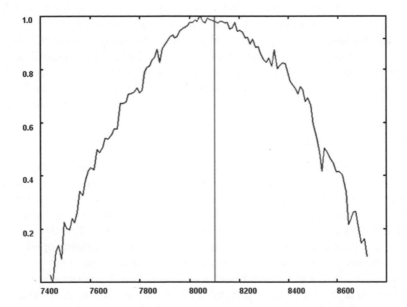

Figure 8.32 Focus measure: the graph depicting the focus measure [124] based on the second-order moments for individual image frames (represented along the x-axis). The value of the focus measure is represented on the y-axis. The ideal position of the image plane (the most focused image) is represented by the vertical line.

8.9.3 Edge detection

Edge detection is a base element of many image-processing tasks. Moments found their application even here. In reference [127], using orthogonal Fourier–Mellin moments, edge detection with subpixel accuracy was proposed, based on moments with lower radial orders and on the interrelationships among their different orders and degrees. Experimental results show that high edge location accuracy can be achieved (the authors claim 0.16 pixel for straight lines with noise and 0.23 pixel for curves with noise). The authors of reference [128] detected step edges with subpixel accuracy, too. Their approach is based on a set of Zernike orthogonal complex moments. Lyvers *et al.* [129] derived an edge operator based on 2D geometric moments. The precision is achieved by correcting for many deterministic errors caused by nonideal edge profiles using a pregenerated lookup table to correct the original estimates of edge orientation and location. This table was generated using a synthesized edge located at various subpixel locations and various orientations. An experiment with the measurement of imaged machined metal parts is also presented.

8.9.4 Gas–liquid flow categorization

In reference [130] the authors presented a novel approach for analyzing 2D flow field data. They employed geometric invariant moments for the purpose of interactively exploring flow field data. They modified moment invariants for vector fields in order to extract and visualize 2D flow patterns, invariant under translation, scaling and rotation. They used moments up to the third order and related the invariant values to specific structures in the flow fields. The moment evaluation is carried out for all positions and for all possible scales in the so-called moment pyramid model, to capture all different structures. Similarly, Zhou *et al.* [131] applied Hu invariants and co-occurrence matrices for the gas–liquid flow regime classification. They used multifeature fusion and a SVM recognition kernel. A database of the various kinds of flow was created, feature vectors evaluated and finally, SVM was trained. The feature sets, reduced using the rough sets theory, capture variations between seven typical flow regimes.

8.9.5 3D objects visualization

An interesting application from the computer graphics area is the simplification of meshes [132] for 3D surface visualization. Mesh simplification is important for accurate as well as plausible virtual reality scene generation and, specifically, the simplification metric is a vital part of the whole topic. The authors proposed using surface geometric moments as well as volume geometric moments for mesh complexity measuring. The objective function is defined as the difference between moments computed on the original mesh and those evaluated on the simplified mesh. These metrics are used in an edge collapse scheme. For a given maximum order and the number of triangles required, the optimal mesh with a minimum moment difference from the original mesh can be determined. As an example from the material analysis application area, O'Connor *et al.* [133] used second-order moments for 3D modeling of the inner structure of the solid materials based on X-ray scans of individual objects. They first detect regions of interest and then approximate them by best-fit ellipsoids, which also provide the main orientation of the inner parts.

8.10 Conclusion

The main application areas of moments and moment invariants have been named, together with illustrative examples and many variations. The main properties of moments used during these applications are their invariance under geometric and radiometric degradations, their ability to capture the shape characteristics of individual objects and their competence in image normalization. The Hu group of geometric moment invariants are often the first choice for many authors, even though other classes of moment invariant can offer better properties in particular tasks. Vital issues of moment applications are proper selection of regions for moment computation, their high sensitivity to segmentation errors, their breakdown in the case of object occlusion, and discrepancy between the features defined in the 2D and 3D natures of many objects described. In practice, not many applications described in scientific journals are using moment computation acceleration. In spite of the issues mentioned, moments are still the best solution to many problems, as is apparent from the publication dates of many of the papers listed.

References

[1] Tuytelaars, T. and van Gool, L. (2000) "Wide baseline stereo based on local, affinely invariant regions," in *Proceedings of the British Machine Vision Conference BMVC'00*, pp. 412–22, BMVA.

[2] Matas, J., Chum, O., Urban, M. and Pajdla, T. (2002) "Robust wide baseline stereo from maximally stable extremal regions," in *Proceedings of the British Machine Vision Conference BMVC'02*, pp. 384–96, BMVA.

[3] Mikolajczyk, K. and Schmid, C. (2004) "Scale and affine invariant interest point detectors," *International Journal of Computer Vision*, vol. 60, no. 1, pp. 63–86.

[4] Flusser, J. and Suk, T. (1994) "Affine moment invariants: a new tool for character recognition," *Pattern Recognition Letters*, vol. 15, no. 4, pp. 433–6.

[5] Wang, L. and Healey, G. (1998) "Using Zernike moments for the illumination and geometry invariant classification of multispectral texture," *IEEE Transactions on Image Processing*, vol. 7, no. 2, pp. 196–203.

[6] Campisi, P., Neri, A., Panci, G. and Scarano, G. (2004) "Robust rotation-invariant texture classification using a model based approach," *IEEE Transactions on Image Processing*, vol. 13, no. 6, pp. 782–91.

[7] Bigiin, J. and Hans du Buf, J. M. (1994) "N-folded symmetries by complex moments in Gabor space and their application to unsupervised texture segmentation," *IEEE Transactions on Pattern Analysis and Machine Intelligence*, vol. 16, no. 1, pp. 80–7.

[8] Zion, B., Shklyar, A. and Karplus, I. (2000) "In-vivo fish sorting by computer vision," *Aquacultural Engineering*, vol. 22, no. 3, pp. 165–79.

[9] White, D., Svellingen, C. and Strachan, N. (2006) "Automated measurement of species and length of fish by computer vision," *Fisheries Research*, vol. 80, no. 2–3, pp. 203–10.

[10] Gope, C., Kehtarnavaz, N., Hillman, G. and Würsig, B. (2005) "An affine invariant curve matching method for photo-identification of marine mammals," *Pattern Recognition*, vol. 38, no. 1, pp. 125–32.

[11] Thiel, S. U., Wiltshire, R. J. and Davies, L. J. (1995) "Automated object recognition of blue-green algae for measuring water quality – a preliminary study," *Water Research*, vol. 29, no. 10, pp. 2398–404.

[12] Fischer, S. and Bunke, H. (2001) "Automatic identification of diatoms using decision forests," in *Proceedings of the Second International Workshop on Machine Learning and Data Mining in Pattern Recognition MLDM'01*, LNAI vol. 4571, pp. 173–83, Springer.

[13] Wang, X.-F., Huang, D.-S., Dua, J.-X., Xu, H. and Heutte, L. (2008) "Classification of plant leaf images with complicated background," *Applied Mathematics and Computation*, vol. 205, no. 2, pp. 916–26.

[14] Al-Batah, M. S., Isa, N. A. M., Zamli, K. Z., Sani, Z. M. and Azizli, K. A. (2009) "A novel aggregate classification technique using moment invariants and cascaded multilayered perceptron network," *International Journal of Mineral Processing*, vol. 92, nos 1–2, pp. 92–102.

[15] Li, T.-S. and Huang, C.-L. (2009) "Defect spatial pattern recognition using a hybrid SOM–SVM approach in semiconductor manufacturing," *Expert Systems with Applications*, vol. 36, no. 1, pp. 374–85.

[16] Keyes, L. and Winstanley, A. (2001) "Using moment invariants for classifying shapes on large-scale maps," *Computers, Environment and Urban Systems*, vol. 25, no. 1, pp. 119–30.

[17] Apatean, A., Rogozan, A. and Bensrhair, A. (2008) "Objects recognition in visible and infrared images from the road scene," in *Proceedings of the International Conference on Automation, Quality and Testing, Robotics AQTR'08* (Cluj-Napoca, Romania), vol. 3, pp. 327–32, IEEE.

[18] Eikvil, L., Aurdal, L. and Koren, H. (2009) "Classification-based vehicle detection in high-resolution satellite images," *ISPRS Journal of Photogrammetry and Remote Sensing*, vol. 64, no. 1, pp. 65–72.

[19] Zhenghe, S., Bo, Z., Zhongxiang, Z., Meng, W. and Enrong, M. (2008) "Research on recognition method for traffic signs," in *Proceedings of the Second International Conference on Future Generation Communication and Networking FGCN'08*, pp. 387–90, IEEE Computer Society.

[20] Dudani, S. A., Breeding, K. J. and McGhee, R. B. (1977) "Aircraft identification by moment invariants," *IEEE Transactions on Computers*, vol. 26, no. 1, pp. 39–45.

[21] Reeves, A., Prokop, R., Andrews, S. and Kuhl, F. (1988). "Three-dimensional shape analysis using moments and Fourier descriptors," *IEEE Transactions on Pattern Analysis and Machine Intelligence*, vol. 10, no. 6, pp. 937–43.

[22] Alves, J., Herman, J. and Rowe, N. C. (2004) "Robust recognition of ship types from an infrared silhouette," in *Proceedings of the Command and Control Research and Technology Symposium CCRTS'04* (San Diego, California), p. 18, CCRP Press.

[23] Mokhtarian, F. and Abbasi, S. (2005) "Robust automatic selection of optimal views in multi-view free-form object recognition," *Pattern Recognition*, vol. 38, no. 7, pp. 1021–31.

[24] Xu, Y. and Qi-cong, P. (2008) "3D object recognition using multiple features and neural network," *Conference on Cybernetics and Intelligent Systems CIS'08*, pp. 434–9, IEEE.

[25] Chen, S. and Srihari, S. (2005) "Use of exterior contours and shape features in off-line signature verification," in *Proceedings of the Eighth International Conference on Document Analysis and Recognition ICDAR'05*, p. 5, IEEE Computer Society.

[26] Lin, H. and Li, H.-Z. (1996) "Chinese signature verification with moment invariants," in *International Conference on Systems, Man, and Cybernetics SMC'96*, vol. 4, pp. 2963–8, IEEE Computer Society.

[27] Belkasim, S. O., Shridhar, M. and Ahmadi, M. (1991) "Pattern recognition with moment invariants: A comparative study and new results," *Pattern Recognition*, vol. 24, no. 12, pp. 1117–38.

[28] Wang, X., Yang, Y. and Huang, K. (2006) "Combining discrete orthogonal moments and DHMMs for off-line handwritten Chinese character recognition," in *Proceedings of the Fifth International Conference on Cognitive Informatics ICCI'06*, vol. 2, pp. 788–93, IEEE Computer Society.

[29] Foon, N. H., Pang, Y.-H., Jin, A. T. B. and Ling, D. N. C. (2004) "An efficient method for human face recognition using wavelet transform and Zernike moments," in *Proceedings of the International Conference on Computer Graphics, Imaging and Visualization CGIV'04* (Beijing, China), pp. 65–9, IEEE Computer Society.

[30] Nabatchian, A., Abdel-Raheem, E. and Ahmadi, M. (2008) "Human face recognition using different moment invariants: A comparative study," in *Proceedings of the Congress on Image and Signal Processing CISP'08*, pp. 661–6, IEEE Computer Society.

[31] Nabatchian, A., Makaremi, I., Abdel-Raheem, E. and Ahmadi, M. (2008) "Human face recognition using different moment invariants: A comparative study," in *Proceedings of the Third International Conference on Convergence and Hybrid Information Technology ICCIT'08*, pp. 933–6, IEEE Computer Society.

[32] Zhu, Y., de Silva, L. C. and Ko, C.C. (2002) "Using moment invariants and HMM in facial expression recognition," *Pattern Recognition Letters*, vol. 23, nos 1–3, pp. 83–91.

[33] Zhi, R. and Ruan, Q. (2008) "A comparative study on region-based moments for facial expression recognition," in *Proceedings of the Congress on Image and Signal Processing CISP'08* (Sanya, Hainan, China), pp. 600–4, IEEE Computer Society.

[34] Prakash, S., Jayaraman, U. and Gupta, P. (2009) "A skin-color and template based technique for automatic ear detection," in *Proceedings of the Seventh International Conference on Advances in Pattern Recognition ICAPR'09* (Kolkata, India), pp. 213–16, IEEE Computer Society.

[35] Zitová, B. and Flusser, J. (2003) "Image registration methods: a survey," *Image Vision Computing*, vol. 21, no. 11, pp. 977–1000.

[36] Goshtasby, A. (1985) "Template matching in rotated images," *IEEE Transactions on Pattern Analysis and Machine Intelligence*, vol. 7, no. 3, pp. 338–44.

[37] Mumford, D. and Shah, J. (1989) "Optimal approximation by piecewise smooth functions and associated variational problems," *Communications on Pure and Applied Mathematics*, vol. 42, no. 5, pp. 577–685.

[38] Zitová, B., Kautsky, J., Peters, G. and Flusser, J. (1999) "Robust detection of significant points in multiframe images," *Pattern Recognition Letters*, vol. 20, no. 2, pp. 199–206.

[39] Šroubek, F., Cristóbal, G. and Flusser, J. (2007) "A unified approach to superresolution and multichannel blind deconvolution," *IEEE Transactions on Image Processing*, vol. 16, no. 9, pp. 2322–32.

[40] Li, H., Manjunath, B. S. and Mitra, S. K. (1992) "Contour-based multisensor image registration," in *26th Asilomar Conference on Signals, Systems and Computers ACSSC'92*, vol. 1, pp. 182–6, IEEE Computer Society.

[41] Dai, X. and Khorram, S. (1999) "A feature-based image registration algorithm using improved chain-code representation combined with invariant moments," *IEEE Transactions on Geoscience and Remote Sensing*, vol. 37, no. 5, pp. 2351–62.

[42] Hu, S. X., Xiong, Y.-M., Liao, M. Z. W. and Chen, W. F. (2007) "Accurate point matching based on combined moment invariants and their new statistical metric," in *Proceedings of the International Conference on Wavelet Analysis and Pattern Recognition ICWAPR'07* (Beijing, China), pp. 376–81, IEEE Computer Society.

[43] Shi, A., Tang, M., Huang, F., Xu, L. and Fan, T. (2005) "Remotely sensed images registration based on wavelet transform using affine invariant moment," in *Proceedings of the Second International Symposium on Intelligent Information Technology Application IITA'08* (Shanghai, China), vol. 1, pp. 384–9, IEEE Computer Society.

[44] Flusser, J. and Suk, T. (1994) "A moment-based approach to registration of images with affine geometric distortion," *IEEE Transactions on Geoscience and Remote Sensing*, vol. 32, no. 2, pp. 382–7.

[45] Bentoutou, Y., Taleb, N., Kpalma, K. and Ronsin, J. (2005) "An automatic image registration for applications in remote sensing," *IEEE Transactions on Geoscience and Remote Sensing*, vol. 43, no. 9, pp. 2127–37.

[46] Harris, C. and Stephens, M. (1988) "A combined corner and edge detector," in *Proceedings of the Fourth Alvey Vision Conference AVC'88*, pp. 147–51, University of Manchester.

[47] Flusser, J., Zitová, B. and Suk, T. (1999) "Invariant-based registration of rotated and blurred images," in *Proceedings of the International Geoscience and Remote Sensing Symposium IGARSS'99* (Hamburg, Germany) (I. S. Tammy, ed.), vol. 2, pp. 1262–4, IEEE.

[48] Yang, Z.-L. and Guo, B.-L. (2008) "Image registration using feature points extraction and pseudo-zernike moments," in *International Conference on Intelligent Information Hiding and Multimedia Signal Processing IIH-MSP'08*, pp. 752–5, IEEE Computer Society.

[49] Fischler, M. A. and Bolles, R. C. (1981) "Random sample consensus: a paradigm for model fitting with applications to image analysis and automated cartography," *Graphics and Image Processing*, vol. 24, no. 6, pp. 381–95.

[50] Badra, F., Qumsieh, A. and Dudek, G. (1998) "Rotation and zooming in image mosaicing," in *Proceedings of the Fourth Workshop on Applications of Computer Vision WACV'98*, pp. 50–5, IEEE Computer Society.

[51] Yasein, M. S. and Agathoklis, P. (2005) "Automatic and robust image registration using feature points extraction and Zernike moments invariants," in *Proceedings of the Fifth International Symposium on Signal Processing and Information Technology ISSPIT'05* (Athens, Greece), pp. 566–71, IEEE Computer Society.

[52] Sato, J. and Cipolla, R. (1995) "Image registration using multi-scale texture moments," *Image and Vision Computing*, vol. 13, no. 5, pp. 341–53.

[53] Brivio, P., Ventura, A., Rampini, A. and Schettini, R. (1992) "Automatic selection of control points from shadow structures," *International Journal of Remote Sensing*, vol. 13, no. 10, pp. 1853–60.

[54] Jaklič, A. and Solina, F. (2003) "Moments of superellipsoids and their application to range image registration," *IEEE Transactions on Systems, Man, and Cybernetics – Part B: Cybernetics*, vol. 33, no. 4, pp. 648–57.

[55] Shen, D. and Davatzikos, C. (2002) "Hammer: hierarchical attribute matching mechanism for elastic registration," *IEEE Transactions on Medical Imaging*, vol. 21, no. 11, pp. 1421–39.

[56] Zitová, B., Ríos, H., Gutierrez, J. M. and Marin, A. (1999) "Recognition of landmarks distorted by fish-eye lens," in *Proceedings Simposio Iberoamericano de Reconocimiento de Patrones*, pp. 611–22.

[57] Courtney, J. and Jain, A. (1994) "Mobile robot localization via classification of multisensor maps," in *Proceedings IEEE International Conference on Robotics and Automation*, vol. 2, pp. 1672–78.

[58] Hu, M.-K. (1962) "Visual pattern recognition by moment invariants," *IRE Transactions on Information Theory*, vol. 8, no. 2, pp. 179–87.

[59] Xua, G., Zhang, Y., Jia, S., Chenga, Y. and Tiana, Y. (2009) "Research on computer vision-based for UAV autonomous landing on a ship," *Pattern Recognition Letters*, vol. 30, no. 6, pp. 600–5.

[60] Lee, J. and Ko, H. (2008) "Gradient-based local affine invariant feature extraction for mobile robot localization in indoor environments," *Pattern Recognition Letters*, vol. 29, no. 14, pp. 1934–40.

[61] Lin, Z., Kim, S. and Kweon, I. S. (2005) "Robust invariant features for object recognition and mobile robot navigation," in *Proceedings of the IAPR Conference on Machine Vision Applications MVA'05* (Tsukuba Science City, Japan), pp. 611–22, Oxford University Press.

[62] Wells, G. and Torras, C. (1998) "Selection of image features for robot positioning using mutual information," in *Proceedings of the International Conference on Robotics and Automation ICRA'98*, vol. 4, pp. 2819–26, IEEE Robotics and Automation Society.

[63] Celaya, E., Albarral, J.-L., Jiménez, P. and Torras, C. (2007) "Natural landmark detection for visually-guided robot navigation," in *Proceedings of the Tenth Congress of the Italian Association for Artificial Intelligence AI*IA'07*, LNCS vol. 4733, pp. 555–66, Springer.

[64] Rui, Y., Huang, T. S. and Chang, S.-F. (1999) "Image retrieval: current techniques, promising directions, and open issues," *Journal of Visual Communication and Image Representation*, vol. 10, no. 1, pp. 39–62.

[65] Stricker, M. and Orengo, M. (1995) "Similarity of color images," in *Proceedings of the Storage and Retrieval for Image and Video Databases III*, vol. 2420, pp. 381–92, SPIE.

[66] Mostafa, T., Abbas, H. M. and Wahdan, A. A. (2002) "On the use of hierarchical color moments for image indexing and retrieval," *IEEE International Conference on Systems, Man and Cybernetics*, vol. 7, p. 6.

[67] Flickner, M. *et al.* (1995) "Query by image and video content: the QBIC system," *Computer*, vol. 28, no. 9, pp. 23–32.

[68] Jain, A. K. and Vailaya, A. (1998) "Shape-based retrieval: a case study with trademark image databases," *Pattern Recognition*, vol. 31, no. 9, pp. 1369–90.

[69] Wei, C.-H., Li, Y., Chau, W.-Y. and Li, C.-T. (2009) "Trademark image retrieval using synthetic features for describing global shape and interior structure," *Pattern Recognition*, vol. 42, no. 3, pp. 386–94.

[70] Fu, X., Li, Y., Harrison, R. and Belkasim, S. (2006) "Content-based image retricval using Gabor-Zernike features," in *Proceedings of the 18th International Conference on Pattern Recognition ICPR'06* (Hong Kong), p. 4, IEEE Computer Society.

[71] Shao, L. and Brady, M. (2006) "Invariant salient regions based image retrieval under viewpoint and illumination variations," *Journal of Visual Communication and Image Representation*, vol. 17, no. 6, pp. 1256–72.

[72] Li, Y. (1992) "Reforming the theory of invariant moments for pattern recognition," *Pattern Recognition*, vol. 25, no. 7, pp. 723–30.

[73] Mindru, F., Tuytelaars, T. and van Gool, L. (2004) "Moment invariants for recognition under changing viewpoint and illumination," *Computer Vision and Image Understanding*, vol. 94, no. 1–3, pp. 3–27.

[74] Banerjee, M., Kundu, M. K. and Maji, P. (2009) "Content-based image retrieval using visually significant point features," *Fuzzy Sets and Systems*, p. 19, doi:10.1016/j.fss.2009.02.024 (in press).

[75] Tuytelaars, T. and Van Gool, L. (1999) "Content-based image retrieval based on local affinely invariant regions," in *Third International Conference on Visual Information Systems VISUAL'99* (Amsterdam, the Netherlands), LNCS vol. 1614, pp. 493–500, Springer.

[76] Sim, D.-G., Kim, H.-K. and Park, R.-H. (2004) "Invariant texture retrieval using modified Zernike moments," *Image and Vision Computing*, vol. 22, no. 4, pp. 331–42.

[77] Lu, K., He, N. and Xue, J. (2009) "Content-based similarity for 3D model retrieval and classification," *Progress in Natural Science*, vol. 19, no. 4, pp. 495–9.

[78] Mademlis, A., Axenopoulos, A., Daras, P., Tzovaras, D. and Strintzis, M. G. (2006) "3D content-based search based on 3D Krawtchouk moments," in *Proceedings of the Third International Symposium on 3D Data Processing, Visualization, and Transmission DPVT'06*, pp. 743–9, IEEE Computer Society.

[79] Di Ruberto, C. and Morgera, A. (2008) "Moment-based techniques for image retrieval," in *19th International Conference on Database and Expert Systems Application DEXA'08* (Turin, Italy), pp. 155–9, IEEE Computer Society.

[80] Yadav, R. B., Nishchal, N. K., Gupta, A. K. and Rastogi, V. K. (2008) "Retrieval and classification of objects using generic Fourier, Legendre moment, and wavelet Zernike moment descriptors and recognition using joint transform correlator," *Optics & Laser Technology*, vol. 40, no. 3, pp. 517–27.

[81] Pauwels, E. J., de Zeeuw, P. M. and Ranguelova, E. B. (2009) "Computer-assisted tree taxonomy by automated image recognition," *Engineering Applications of Artificial Intelligence*, vol. 22, no. 1, pp. 26–31.

[82] Alghoniemy, M. and Tewfik, A. H. (2004) "Geometric invariance in image watermarking," *IEEE Transactions on Image Processing*, vol. 13, no. 4, pp. 145–53.

[83] Tzouveli, P., Ntalianis, K. and Kollias, S. (2006) "Video object watermarking using Hu moments," in *13th International Conference on Systems, Signals and Image Processing IWSSIP'06*, p. 6, IEEE.

[84] Xin, Y., Liao, S. and Pawlak, M. (2007) "Circularly orthogonal moments for geometrically robust image watermarking," *Pattern Recognition*, vol. 40, no. 12, pp. 3740–52.

[85] Wang, X.-Y., Xu, Z.-H. and Yang, H.-Y. (2009) "A robust image watermarking algorithm using SVR detection," *Expert Systems with Applications*, vol. 36, no. 5, pp. 9056–64.

[86] Zhang, L., Qian, G., Xiao, W. and Ji, Z. (2007) "Geometric invariant blind image watermarking by invariant Tchebichef moments," *Optics Express*, vol. 15, no. 5, pp. 2251–61.

[87] Zhang, L., Xiao, W., Qian, G. and Ji, Z. (2007) "Rotation, scaling, and translation invariant local watermarking technique with Krawtchouk moments," *Chinese Optics Letters*, vol. 5, no. 1, pp. 21–4.

[88] Parameswaran, L. and Anbumani, K. (2006) "A robust image watermarking scheme using image moment normalization," *Transactions on Engineering, Computing and Technology*, vol. 13, no. 5, pp. 239–43.

[89] Lu, C.-S. (2005) "Towards robust image watermarking: combining content-dependent key, moment normalization, and side-informed embedding," *Signal Processing: Image Communication*, vol. 20, no. 2, pp. 129–50.

[90] Dong, P., Brankov, J. G., Galatsanos, N. P., Yang, Y. and Davoine, F. (2005) "Digital watermarking robust to geometric distortions," *IEEE Transactions on Image Processing*, vol. 14, no. 12, pp. 2140–50.

[91] Cox, I. J., Kilian, J., Leighton, F. T. and Shamoon, T. (1997) "Secure spread spectrum watermarking for multimedia," *IEEE Transactions on Image Processing*, vol. 6, no. 12, pp. 1673–87.

[92] Jelen, K. and Kusová, S. (2004) "Pregnant women: Moiré contourgraph and its semiautomatic and automatic evaluation," *Neuroendocrinology Letters*, vol. 25, nos 1–2, pp. 52–6.

[93] Ruggeri, A. and Pajaro, S. (2002) "Automatic recognition of cell layers in corneal confocal microscopy images," *Computer Methods and Programs in Biomedicine*, vol. 68, no. 1, pp. 25–35.

[94] Morales, D. A., Bengoetxea, E. and Larranaga, P. (2008) "Selection of human embryos for transfer by Bayesian classifiers," *Computers in Biology and Medicine*, vol. 38, nos 11–12, pp. 1177–86.

[95] Duda, R. O., Hart, P. E. and Stork, D. G. (2001) *Pattern Classification*. New York: Wiley Interscience, 2nd edn.

[96] Bentoutou, Y., Taleb, N., Chikr El Mezouar, M., Taleb, M. and Jetto, J. (2002) "An invariant approach for image registration in digital subtraction angiography," *Pattern Recognition*, vol. 35, no. 12, pp. 2853–65.

[97] Sommer, I., Müller, O., Domingues, F. S., Sander, O., Weickert, J. and Lengauer, T. (2007) "Moment invariants as shape recognition technique for comparing protein binding sites," *Bioinformatics*, vol. 23, no. 23, pp. 3139–46.

[98] Hung, C.-M., Huang, Y.-M. and Chang, M.-S. (2006) "Alignment using genetic programming with causal trees for identification of protein functions," *Nonlinear Analysis*, vol. 65, no. 5, pp. 1070–93.

[99] Avci, D. and Varol, A. (2009). "An expert diagnosis system for classification of human parasite eggs based on multi-class SVM," *Expert Systems with Applications*, vol. 36, no. 1, pp. 43–8.

[100] Banada, P. P., Huff, K., Bae, E., Rajwa, B., Aroonnual, A., Bayraktar, B., Adil, A., Robinson, J. P., Hirleman, E. D. and Bhunia, A. K. (2009) "Label-free detection of multiple bacterial pathogens using light-scattering sensor," *Biosensors and Bioelectronics*, vol. 24, no. 6, pp. 1685–92.

[101] Mangin, J.-F., Poupon, F., Duchesnay, E., Riviere, D., Cachia, A., Collins, D. L., Evans, A., and Regis, J. (2004) "Brain morphometry using 3D moment invariants," *Medical Image Analysis*, vol. 8, no. 3, pp. 187–96.

[102] Ng, B., Abugharbieh, R., Huang, X. and McKeown, M. J. (2006) "Characterizing fMRI activations within regions of interest (ROIs) using 3D moment invariants," in *Proceedings of the Computer Vision and Pattern Recognition Workshop CVPRW'06*, p. 8, IEEE Computer Society.

[103] Zacharaki, E. I., Hogea, C. S., Shen, D., Biros, G. and Davatzikos, C. (2009) "Non-diffeomorphic registration of brain tumor images by simulating tissue loss and tumor growth," *NeuroImage*, vol. 46, no. 3, pp. 762–74.

[104] Laguitton, S., Boldak, C. and Toumoulin, C. (2007) "Temporal tracking of coronaries in multi-slice computed tomography," in *29th Annual International Conference on IEEE Engineering in Medicine and Biology Society*, pp. 4512–15.

[105] Jones, B., Schaefer, G. and Zhu, S. (2004) "Content-based image retrieval for medical infrared images," in *26th Annual International Conference on IEEE Engineering in Medicine and Biology Society*, pp. 1186–87.

[106] Milanfar, P., Karl, W. C. and Willsky, A. S. (1996) "A moment-based variational approach to tomographic reconstruction," *IEEE Transactions on Image Processing*, vol. 5, no. 3, pp. 459–70.

[107] Basu, S. and Bresler, Y. (2000) "Uniqueness of tomography with unknown view angles," *IEEE Transactions on Image Processing*, vol. 9, no. 6, pp. 1094–1106.

[108] Mahdian, B. and Saic, S. (2007) "Detection of copy-move forgery using a method based on blur moment invariants," *Forensic Science International*, vol. 171, nos 2–3, pp. 180–9.

[109] Yang, J., Min, B. and Park, D. (2007) "Fingerprint verification based on absolute distance and intelligent BPNN," in *Proceedings of Frontiers in the Convergence of Bioscience and Information Technologies*, pp. 676–81.

[110] Yang, J., Shin, J., Min, B., Lee, J., Park, D. and Yoon, S. (2008) "Fingerprint matching using global minutiae and invariant moments," in *Proceedings of the Congress on Image and Signal Processing CISP'08*, pp. 599–602, IEEE Computer Society.

[111] Almansa, A. and Lindeberg, T. (2000) "Fingerprint enhancement by shape adaptation of scale-space operators with automatic scale selection," *IEEE Transactions on Image Processing*, vol. 9, no. 12, pp. 2027–42.

[112] Lee, L. and Grimson, W. E. L. (2002) "Gait analysis for recognition and classification," in *Proceedings of the Fifth International Conference on Automatic Face and Gesture Recognition FGR'02* (Washington, D.C.), pp. 155–62, IEEE Computer Society.

[113] Shi, C.-P., Li, H.-G., Lian, X. and Li, X.-G. (2006) "Multi-resolution local moment feature for gait recognition," in *Proceedings of the International Conference on Machine Learning and Cybernetics*, pp. 3709–14.

[114] AlGarni, G. and Hamiane, M. (2008) "A novel technique for automatic shoeprint image retrieval," *Forensic Science International*, vol. 181, no. 1, pp. 10–14.

[115] Nabti, M. and Bouridane, A. (2008) "An effective and fast iris recognition system based on a combined multiscale feature extraction technique," *Pattern Recognition*, vol. 41, no. 3, pp. 868–79.

[116] Wang, S. and Xu, Y. (2008) "A new palmprint identification algorithm based on Gabor filter and moment invariant," in *IEEE Conference on Cybernetics and Intelligent Systems*, pp. 491–6.

[117] Kong, J., Li, H., Lu, Y., Qi, M. and Wang, S. (2007) "Hand-based personal identification using k-means clustering and modified Zernike moments," in *Proceedings of the Third International Conference on Natural Computation*, pp. 651–6.

[118] Pang, Y.-H., Andrew, T. B. J., David, N. C. L. and San, H. F. (2003) "Palmprint verification with moments," in *Journal of WSCG (Winter School of Computer Graphics)* (Plzeň, Czech Republic), vol. 12, nos 1–3, pp. 325–32, University of West Bohemia, Plzeň, Czech Republic.

[119] Amayeh, G., Bebis, G., Erol, A. and Nicolescu, M. (2009) "Hand-based verification and identification using palm–finger segmentation and fusion," *Computer Vision and Image Understanding*, vol. 113, no. 4, pp. 477–501.

[120] Lucas, B. D. and Kanade, T. (1981) "An iterative image registration technique with an application to stereo vision," in *Proceedings of the International Joint Conference on Artificial Intelligence IJCAI'81*, vol. 2, pp. 674–9, University of British Columbia, Vancouver, Canada.

[121] Ghosal, S. and Mehrotra, R. (1994) "Robust optical flow estimation," in *International Conference on Image Processing ICIP'94*, pp. 780–4, IEEE Computer Society.

[122] Kharbat, M., Aouf, N., Tsourdos, A. and White, B. (2008) "Robust brightness description for computing optical flow," in *Proceedings of the British Machine Vision Conference BMVC'08*, p. 10, BMVA.

[123] Kautsky, J., Flusser, J., Zitová, B. and Šimberová, S. (2002) "A new wavelet-based measure of image focus," *Pattern Recognition Letters*, vol. 23, no. 14, pp. 1785–94.

[124] Zhang, Y., Zhang, Y. and Wen, C. (2000) "A new focus measure method using moments," *Image and Vision Computing*, vol. 18, no. 12, pp. 959–65.

[125] Yap, P. and Raveendran, P. (2004) "Image focus measure based on Chebyshev moments," *IEE Proceedings on Vision, Image and Signal Processing*, vol. 151, no. 2, pp. 128–36.

[126] Wee, C.-Y., Paramesran, R. and Mukundan, R. (2007) "Quality assessment of Gaussian blurred images using symmetric geometric moments," in *Proceedings of the International Conference on Image Analysis and Processing ICIAP'07*, pp. 807–12, IEEE Computer Society.

[127] Bin, T. J., Lei, A., Jiwen, C. and Dandan, K. W. L. (2008) "Subpixel edge location based on orthogonal Fourier–Mellin moments," *Image and Vision Computing*, vol. 26, no. 4, pp. 563–9.

[128] Ghosal, S. and Mehrotra, R. (1993) "Orthogonal moment operators for subpixel edge detection," *Pattern Recognition*, vol. 26, no. 2, pp. 295–306.

[129] Lyvers, E., Mitchell, O., Akey, M. and Reeves, A. (1989) "Subpixel measurements using a moment-based edge operator," *IEEE Transactions on Pattern Analysis and Machine Intelligence*, vol. 11, no. 12, pp. 1293–1309.

[130] Schlemmer, M., Heringer, M., Morr, F., Hotz, I., Hering-Bertram, M., Garth, C., Kollmann, W., Hamann, B. and Hagen, H. (2007) "Moment invariants for the analysis of 2D flow fields," *IEEE Transactions on Visualization and Computer Graphics*, vol. 13, no. 6, pp. 1743–50.

[131] Zhou, Y., Chen, F. and Sun, B. (2008) "Identification method of gas-liquid two-phase flow regime based on image multi-feature fusion and support vector machine," *Chinese Journal of Chemical Engineering*, vol. 16, no. 6, pp. 832–40.

[132] Tang, H., Shu, H. Z., Dillenseger, J. L., Bao, X. D. and Luo, L. M. (2007) "Moment-based metrics for mesh simplification," *Computers & Graphics*, vol. 31, no. 5, pp. 710–18.

[133] O'Connor, A., Mulchrone, K. F. and Meere, P. A. (2009) "WinDICOM: A program for determining inclusion shape and orientation," *Computers and Geosciences*, vol. 35, no. 6, pp. 1358–68.

9

Conclusion

The theory of moments and moment invariants has been covered step by step in eight chapters. Our aim was to address the main categories of moments, their computational issues and their applicability in various domains. In this way we aspired to provide both material for theoretical study and references for algorithm design. The text is based on our profound experience with moments gained from fundamental research in the area of moment invariants, from teaching graduate and postgraduate university courses, and from tutorials given at major international conferences. In this book, we wanted to share our knowledge and experience with the readers.

The individual classes of moments were introduced, comprising of geometric moments, complex moments and a wide variety of orthogonal moments such as Zernike, Chebyshev and Legendre moments, to name but a few. They were presented together with comments on their advantages and relevant weak points. Illustrative examples accompany the theoretical parts. The core section of the book is dedicated to moment invariants with respect to various geometric and radiometric degradations. We consider these properties most interesting from the image-processing point of view. In all cases, we have presented a consistent theory allowing an automatic derivation of invariants of any orders; in some cases we even demonstrated several alternative approaches. While the invariants to translation, rotation and scaling have been widely known for many years, the affine invariants, the convolution invariants and namely the invariants to elastic transformations are recently developed features. Some of their properties have never been published prior to this book.

Moments are valued for their representative object description, as we have seen in Chapter 8, dedicated to applications. The application areas mentioned make use of the moments' ability to express shape variations and at the same time to stay invariant under common image deformations. For time-optimal applications of moments, methods and algorithms used for their computations were reviewed to ensure the time efficiency of the solutions based on moments.

Here, finally, we would like to summarize the positive as well as negative issues of moment applicability. As follows from previous chapters, moments are by definition global features, computed from the whole image or from the whole chosen region of interest (ROI). The global nature of moments results in their low robustness with respect to unexpected disturbances in the image or in the individual object shapes, respectively. Object occlusion is a typical example. Even a small local change in the image could affect all image moments. Moreover, the choice of the ROIs and their proper localization is crucial for successful moment utilization. Moments are sensitive to segmentation errors and to data correspondence of the analyzed parts of the scene, thus the choice of the segmentation algorithm has to be accomplished carefully. Data correspondence is not handled properly even in many published papers, where the authors are using ROIs of the wrong type with respect to the expected geometric degradations. An example of a bad choice is a fixed-size circular neighborhood in the case of expected scale change. After that, the ROIs content is no longer valid. Lastly, we would like to mention the vulnerability of certain moment invariants with respect to the violence of the degradation model. Convolution invariants are a typical example – if image blurring is not described by a space-invariant convolution or if the boundary effect is significant, they may become inapplicable. Fortunately, many of the shortcomings mentioned can be diminished by proper experimental setting.

The negative properties are often compensated by moment *plausible features*. They are accompanied by a well-developed mathematical theory providing us with an insight into the behavior of moments under various conditions. As we have aimed to show, moments can be made invariant to many common degradations, both geometric and radiometric, even simultaneously. They offer good discriminability and relatively high robustness to noise owing to their definition based on an integration over the whole ROI. Moreover, moments and their corresponding invariants form complete and independent sets of features for object description, providing in the end higher discriminability and computational time saving than dependent and/or incomplete systems. Among the positive properties of moments, we should recall many efficient algorithms allowing stable and fast evaluation.

We hope that this book provides readers with an insight into moment theory, giving them an idea of how to construct moment invariants, and helping them if they wish to use moments in object recognition tasks. The presented survey of various reference papers where moments have been successfully applied should be used as a motivation as to where and how moments can be used and which image-processing problems can be handled using these techniques.

Index